Springer Series on Naval Architecture, Marine Engineering, Shipbuilding and Shipping

Volume 27

Series Editor

Nikolas I. Xiros, University of New Orleans, New Orleans, USA

The Naval Architecture, Marine Engineering, Shipbuilding and Shipping (NAMESS) series publishes state-of-art research and applications in the fields of design, construction, maintenance and operation of marine vessels and structures. The series publishes monographs, edited books, as well as selected PhD theses and conference proceedings focusing on all theoretical and technical aspects of naval architecture (including naval hydrodynamics, ship design, shipbuilding, shipyards, traditional and non-motorized vessels), marine engineering (including ship propulsion, electric power shipboard, ancillary machinery, marine engines and gas turbines, control systems, unmanned surface and underwater marine vehicles) and shipping (including transport logistics, route-planning as well as legislative and economical aspects).

The books of the series are submitted for indexing to Web of Science.

All books published in the series are submitted for consideration in Web of Science.

Ievgen Bilousov • Roman Varbanets

Marine Internal Combustion Engines and Environmental Protection Issues

Springer

Ievgen Bilousov
Odesa National Maritime University
Odesa, Ukraine

Roman Varbanets
Odesa National Maritime University
Odesa, Ukraine

ISSN 2194-8445 ISSN 2194-8453 (electronic)
Springer Series on Naval Architecture, Marine Engineering, Shipbuilding and Shipping
ISBN 978-3-032-20939-9 ISBN 978-3-032-20940-5 (eBook)
https://doi.org/10.1007/978-3-032-20940-5

This Springer imprint is published by the registered company Springer Nature Switzerland AG
The registered company address is: Gewerbestrasse 11, 6330 Cham, Switzerland

Dedicated to the 95th anniversary of the Odessa National Maritime University

Preface

This publication is a logical continuation of a series of books devoted to the problems of modern marine power engineering. The first book in the series, *Modern Marine Internal Combustion Engines,* by Bilousov et al. (2020), published by Springer, provides a general overview of the current state of marine piston engine engineering. This new book, which is a logical continuation of our textbook *Internal Combustion Engines of Modern Marine Vessels* (Bilousov and Bulgalov, Astroprint, 2024), focuses on environmental protection issues in modern shipping. It partially expands on the information presented in previous books in the series, but primarily contains new information presented in the context of the issues raised. This approach allows the book to be viewed as a continuation of the main course, as well as a teaching aid for a separate course devoted to current environmental issues in the merchant navy. Today, issues of reducing the harmful impact on the environment during the operation of ships of the world merchant fleet are becoming especially relevant. Global processes that lead to climate change on the planet require an immediate response at all levels, from people making regulatory decisions at the legislative level to executors of these decisions directly on ships of the merchant fleet. First of all, this concerns ship mechanics who are engaged in the operation of ship power equipment, which is the main source of harmful emissions. When starting work on the book, the authors proceeded from the firm conviction that solving environmental problems is practically impossible without a deep understanding of their essence by ship personnel, without special knowledge about the impact of ship power equipment on the environment and possible ways to reduce this impact. This basic knowledge should form at the stage of training future ship engineers in maritime educational institutions, and this book can help with this.

Of course, the book does not pretend to be a complete presentation of all aspects related to environmental protection issues, but is intended only to form basic competencies in future ship engineers, which should form the basis of their environmental worldview. This worldview requires an understanding of the processes of formation of harmful substances in internal combustion engines, methods for reducing their emissions, ways of switching to existing and currently under development environmental technologies.

The authors express their deep gratitude to the esteemed reviewers: Doctor of Technical Sciences, Professor **Andriy Marchenko** (National Technical University "Kharkiv Polytechnic Institute"); Doctor of Technical Sciences, Professor **Sergey Sagin** (National University "Odesa Maritime Academy"); Doctor of Technical Sciences, Professor **Sergey Serbin** (National University of Shipbuilding) who spent their time to review the manuscript and expressed a number of valuable comments and suggestions, which were accepted with gratitude by the authors.

The authors also express their sincere gratitude to the management and entire staff of the Odessa National Maritime University for their support of this work and other projects by our team.

Our teams other projects can view by following this link: https://sites.google.com/view/modern-marine-engines/

Sincerely,

Odesa, Ukraine Ievgen Bilousov
Odesa, Ukraine Roman Varbanets

Competing Interests The authors have no competing interests to declare that are relevant to the content of this manuscript.

Contents

About the Authors

Ievgen Bilousov Doctor of Technical Sciences, is a Professor at the Department of Marine Power Systems and Complexes at Odesa National Maritime University. With more than 20 years' experience in maritime educational institutions, during which he held the positions of associate professor, head of department, dean of the faculty, and professor, he has also been an active author with more than 190 scientific and methodological works, including 5 monographs. His research interests cover the working processes of marine engines and their operation with traditional and alternative fuels.

Roman Varbanets Doctor of Technical Sciences, is a Professor and the Head of the Department of Marine Power Systems and Complexes at Odesa National Maritime University. With more than 20 years' experience in maritime educational institutions, he is also author of more than 100 scientific publications. Founder and developer of the DEPAS (Diesel Engine Performance Analysis System) diagnostic systems, his research interests include modeling, monitoring, and diagnostics of internal combustion engine processes and turbocharging systems.

Abbreviations[1]

ABS	Ammonium bisulfate
ACU	Actuation control unit
AICWS	Ammonia injector cooling water system
AIG	Ammonia injection grid
ASTM	American Society for Testing and Materials
AVPS	Ammonia vapor processing system
BDC	Bottom dead center
BSN	Bosch smoke number
CAN	Controller area network
CASS	Combustion air saturation system
CCU	Cylinder control unit
DCC	Dynamic combustion control
DCU	Decomposition unit
DCU	Distributed control unit
DF	Dual-fuel
DME	Dimethyl ether
ECA	Emission control area
ECS	Electronic control system
EEDI	Energy efficiency design index
EEVA	Electronic exhaust valve actuation
EGR	Exhaust gas recirculation
EIAPP	Engine International Air Pollution Prevention certificate
ELGI	Electronic gas injection
ELGI	Electronic liquid gas injection
ELGIB	Electronic gas injection block

[1] Different manufacturers may use different abbreviations to designate the same functional elements, and in some cases, the same abbreviation may be used for different elements. To avoid ambiguous interpretation, the text retains the original manufacturer designations while also providing the expanded names of the components within the context of each section.

ELWC	Electronic window control valve
ELWI	Electronic window valve
FAME	Fatty acid methyl esters
FBIV	Fuel booster injection valve
FBIV-A	Fuel booster injection valve for ammonia
FBIV-M	Fuel booster injection valve for methanol
FBIV-P	Fuel booster injection valve for propane
FGIV	Fuel gas injection valves
FGSS	Fuel gas supply system
FIS	Fuel injection system
FIVA	Fuel injection valve actuation
FSN	Filter smoke number
FVU	Fuel valve unit
GA	Gas admission
GAV	Gas admission valve
GI	Gas injector
GRU	Gas regulating unit
GVT	Gas valve train
GWP	Global warming potential
HAM	Humid air motor
HFO	Heavy fuel oil
HP	High-pressure systems
HP-DI	High-pressure direct injection
HVO	Hydrated vegetable oil
HVO	Hydrotreated vegetable oil
IAPP	International Air Pollution Prevention certificate
iCER	intelligent control by exhaust recycling
IMO	International Maritime Organization
IRENA	International Renewable Energy Agency
ISO	International Organization for Standardization
KDU	Knock detection unit
LGI	Liquid gas injection
LNG	Liquefied natural gas
LP	Low-pressure systems
LPG	Liquefied petroleum gases
MARPOL	Marine pollution (International Convention for the Prevention of Pollution from Ships)
MBIV	Micro-boosted injection valve
MBU	Methanol booster units
MCR	Maximum continuous rating
MCU	Main control unit
MDO	Marine diesel oil
MFPU	Methanol fuel pump unit
MGO	Marine gas oil
NBR	Nitrile butadiene rubber

OCV	Oil control valves
PAH	Polycyclic aromatic hydrocarbons
PFI	Port fuel injection
PM	Particulate matter
PVU	Pump vaporizer unit
Ro-Ro	Roll-on/roll-off
SCR	Selective catalytic reduction
SECA	Sulfur emission control areas
SFOC	Specific fuel oil consumption
SFWI	Stratified fuel-water injection
SGAV	Safe gas admission valve
SMU	Sensor multiplexer unit
SOB	Sealing oil block
SOF	Soluble organic fraction
SOGAV	Solenoid operated gas admission valve
SOLAS	Safety of Life at Sea (International Convention for the Safety of Life at Sea)
TCEV	Top-controlled exhaust valve
TDC	Top dead center
ULSFO	Ultra low sulfur fuel oils
VCR	Variable compression ratio
VLSFO	Very low sulfur fuel oils
VOC	Volatile organic compounds
WaCoReG	Water-cooled residual gas
WECS	Wärtsilä engine control system
WFE	Water-fuel emulsions
WV	Window valve
WVFC	Window valve force close

Chapter 1
Marine Internal Combustion Engines, Environmental Issues and Global Climate Change

Shipping plays a key role in the global economy, transporting more than 80% of the world's total trade. There are currently more than 100,000 shipping vessels in operation, of which about 6000 are large container ships, transporting approximately 10 billion tons of containers annually. Waterways literally encircle our entire planet (Fig. 1.1). Between 2016 and 2019 alone, the intensity of maritime transport increased by 18%. Water transport provides both regional and intercontinental movement of large volumes of various goods, being the most economical and efficient mode of transport in terms of fuel consumption per unit of transport work performed.

The high competitiveness of the world merchant fleet for many years was largely ensured by the use of heavy fuel, which was available and cheap. As of 2023, the annual consumption of these fuels was about 330 million tons (five–six million barrels per day), and if the current dynamics of growth in maritime transport continues, then by 2040 fuel consumption will increase by another 50%. This trend has led to the fact that in recent decades there has been increasing concern about the current practice of using traditional fossil fuels for shipping. First, this is due to obvious climate change, which most researchers explain by the anthropogenic impact of human economic activity on the environment. Climate change is associated with emissions of so-called greenhouse gases into the atmosphere, the main (but far from the only) of which is carbon dioxide—CO_2. Its significant emission into the Earth's atmosphere is due to the burning of fossil fuels, of which carbon makes up a significant portion. According to the International Maritime Organization (IMO), shipping, including international, domestic and fishing, accounted for about 3% of total anthropogenic greenhouse gas emissions between 2012 and 2018. In addition, marine internal combustion engines are a source of atmospheric pollution with harmful compounds such as nitrogen oxides (NO_x), sulfur oxides (SO_x), carbon monoxide CO, hydrocarbons (CH) and particulate matter (PM) (Fig. 1.2). Maritime

I. Bilousov, R. Varbanets, *Marine Internal Combustion Engines and Environmental Protection Issues*, Springer Series on Naval Architecture, Marine Engineering, Shipbuilding and Shipping 27,
https://doi.org/10.1007/978-3-032-20940-5_1

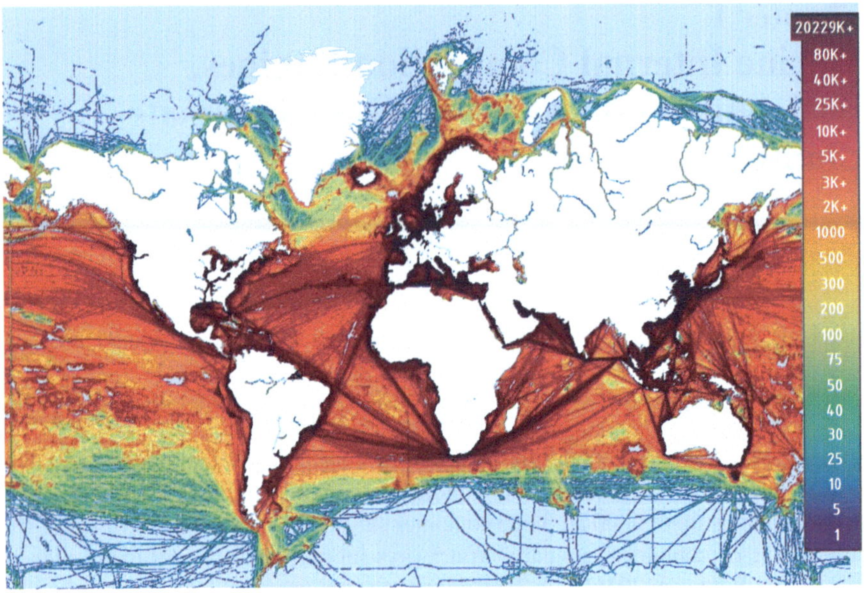

Fig. 1.1 Intensity of modern maritime transport (number of ships passing through a location with an area of 39.14 km^2 during the year) [1]

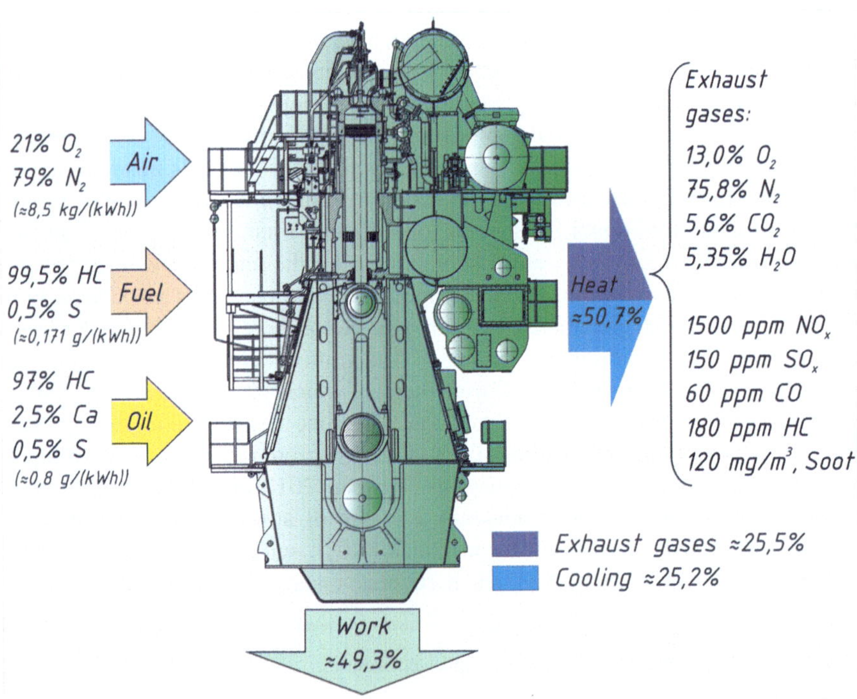

Fig. 1.2 Mass and energy balance of a low-speed marine engine

transport accounts for approximately 2.6% of CO_2, 5% of NO_x and 13% of SO_x of the total anthropogenic emissions of these components into the atmosphere. If nothing done, by 2050 the growth of world trade could lead to an increase in emissions from shipping by 50–250%. Over the same period, about 15% of global carbon dioxide emissions will be associated with shipping. Even today, the problem of air pollution is relevant in areas of intensive shipping, where air pollution can reach values significantly exceeding the statistical average. This explained primarily by the large aggregate capacities of ship power plants, which in some cases already exceed 100 MW [2].

Conventionally, the impact of internal combustion engines on the environment can divided into short-term and long-term:

- short-term, associated with emissions of harmful substances, which, being carcinogenic and chemically active, affect living organisms, including humans, vegetation, material objects of natural and artificial origin;
- long-term, associated with emissions of greenhouse gases and particulate matter, which lead to global climate change, melting of glaciers, shifting of climate zones, etc.

1.1 Energy Transformations and Chemical Processes in Internal Combustion Engines and Their Impact on the Environment

To ensure the working process, atmospheric air supplied to the engine (Fig. 1.2), in which the volumetric content of nitrogen (N_2) is approximately 79% and 21% is oxygen (O_2). Air also contains other gases, but their content can neglected when assessing the mass balance of the engine gas charge due to their low content.

In addition, to air, hydrocarbon fuel (CH), obtained mainly from oil, and supplied to the engine. In addition to hydrocarbon compounds, of which the fuel consists of 97.0–99.9%, it may contain sulfur (S), as part of organosulfur compounds entering the fuel from crude oil during its processing.

Together with the fuel, part of the lubricating oil burns in the engine cylinders, which like fuel, consists mainly of hydrocarbons and organosulfur compounds. In addition, the oils contain a small portion of alkaline compounds necessary to neutralize the combustion products of sulfur, which have a destructive effect on the structural elements of the engine. During the working process in the engine cylinders, the fuel components enter into a series of sequential-parallel chemical reactions with atmospheric oxygen, which have a total exothermic character. The thermal energy that released partially converted into mechanical work on the engine shaft, and partially discharged into the environment with exhaust gases and through the cooling system.

Products of complete combustion During the combustion of fuel, part of the oxygen in the air used to oxidize the hydrocarbons in the fuel, and part passes into the exhaust gases of the engine. The main products of fuel combustion are carbon dioxide (CO_2), formed because of the interaction of fuel carbon with atmospheric oxygen, and water vapor (H_2O), and formed because of the interaction of fuel hydrogen with atmospheric oxygen. At the engine outlet, the mass of gaseous combustion products increases by the amount of the mass of the burned fuel, so the relative concentration of nitrogen (N_2) in the exhaust gases decreases.

An approximate ratio of mass and energy flows for a modern marine two-stroke low-speed engine shown in Fig. 1.2.

Thus, carbon dioxide and water vapor are the main combustion products formed in the engine. They are also greenhouse gases responsible for global warming. And if the water vapor content in the atmosphere is effectively regulated by the processes of the water cycle, then the accumulation of CO_2 is a serious problem with delayed consequences.

Products of incomplete combustion In addition to the main combustion products, engines produce products of incomplete combustion that are quite dangerous from an ecological point of view, including carbon monoxide (CO), hydrocarbons (CH) and soot (C). The amount of these components is small, so they are measured in parts per million (ppm)—a unit of measurement of any relative values equal to 1×10^{-6} of the base indicator.

Carbon monoxide Incomplete combustion products, primarily CO, formed because of the limited time for the combustion process, when the complete oxidation reactions do not complete. Due to the large excess of air in marine diesel engines, the amount of CO molecules formed is small compared to other types of piston engines.

Hydrocarbon compounds Unreacted fuel molecules (CH) can enter the exhaust gases from areas adjacent to the combustion chamber walls, where the reaction rate significantly slowed down due to the lower temperature. A significant amount of CH can enter the working cylinder from the sub-needle volume of the injector nozzle after the combustion process is complete. Under high temperature conditions, hydrocarbons can change, forming new groups of compounds, including very toxic ones, such as aldehydes.

Soot (atomic carbon) The soot (C) formed in diesel engines is also a product of incomplete combustion of motor fuel. After the start of combustion, there is an intensive heating of the fuel-air environment in the internal space of the spray plume, which, under conditions of insufficient air, leads to the formation of atomic carbon. This is especially relevant for transient modes of engine operation, when the optimal fuel-air ratio violated in the combustion chamber. In these modes, soot is formed, which accompanied by the release of black smoke from the engine exhaust system.

Sulfur compounds When organosulfur compounds burn, the sulfur they contain interacts with oxygen to form several types of compounds that have the general designation SO_x. When these compounds interact with water, they form sulfuric (H_2SO_4) and sulfurous acids (H_2SO_3).

The destructive effect of sulfur compounds already observed in the engine. When interacting with air moisture (especially if the incoming air temperature incorrectly adjusted relative to the dew point), acids formed on the cylinder walls, which intensively interact with their surfaces. The result of this interaction is chemical corrosion, called low-temperature corrosion. This type of corrosion is especially activated when the fuel is waterlogged and the cylinder liner walls are at a low temperature, which can lead to condensation of water vapor contained in the purge air and exhaust gases. When using high-sulfur fuels, this type of corrosion is the main one—in some cases, up to 90% of the loss of liner mass can be associated with chemical corrosion processes. To reduce the harmful effects of sulfur compounds, alkaline additives added to lubricating oils, allowing the neutralization of most of the acids formed.

When released into the atmosphere and interacting with atmospheric moisture, sulfur compounds contribute to the formation of acid rain, which has a negative effect on living and non-living nature. The most effective way to reduce SO_x emissions with exhaust gases is to reduce the content of organosulfur compounds in the fuel.

In some cases, sulfur combustion products neutralized already at the engine outlet using alkaline reagents in special devices called scrubbers.

Nitrogen oxides The most toxic products that form during combustion in engines are nitrogen oxides—a group of nitrogen-oxygen compounds that collectively referred to as NO_x. Under high-temperature conditions in the combustion chamber, a stable nitrogen molecule N_2 can break down into two atoms, each of which is capable of attaching several oxygen atoms to it. NO_x emissions considered one of the main causes of photochemical smog. When combined with water vapor in the atmosphere, they form nitrous HNO_2 and nitric HNO_3 acids, and, together with sulfur oxides, are the cause of acid rain.

1.2 International Regulation of Greenhouse Gas and Harmful Substance Emissions from Ship Engines

Due to the growing attention to the problem of air pollution by ship engines, since 2005, the IMO has been consistently introducing increasingly strict restrictions on greenhouse gas and harmful substance emissions by ship internal combustion engines (Fig. 1.3). The tightening of restrictions concerns, on the one hand, a reduction in the standards for harmful substance emissions into the atmosphere, and on the other hand, the expansion of special emission controls areas.

The fundamental document regulating environmental protection issues in the field of shipping is the MARPOL 73/78 Convention. Its Appendix VI "Regulations for the Prevention of Air Pollution from Ships" contains the main provisions concerning the limitation of the concentration of harmful emissions in the exhaust gases of main and auxiliary marine diesel engines. At the same time, the Appendix itself includes the "Technical Code for the Control of Nitrogen Oxide Emissions from Ships". The Appendix entered into force in May 2005.

In accordance with Regulation 13 of Appendix VI, all ships of 400 tons gross tonnage and above, built after January 1, 2000, must undergo inspection and receive an International Air Pollution Prevention Certificate (IAPP) for a period of five years. Additionally, each marine diesel engine with a capacity of 130 kW or more must undergo inspection in accordance with the Technical Code and receive an Engine International Air Pollution Prevention Certificate (EIAPP), which relates to nitrogen oxide emissions and issued for a period of five years. Upon expiration of the aforementioned documents, an inspection on board and an extension of the relevant documents for the next five years provided.

In addition to the general international standards that apply everywhere, international maritime legislation also establishes special control zones (ECA—Emission Control Area). In these zones, which today include the 200-mile zones of the USA and Canada, the waters of the North and Baltic Seas, the English Channel, the Caribbean Sea and other regions, even more stringent requirements imposed on emissions (Fig. 1.4).

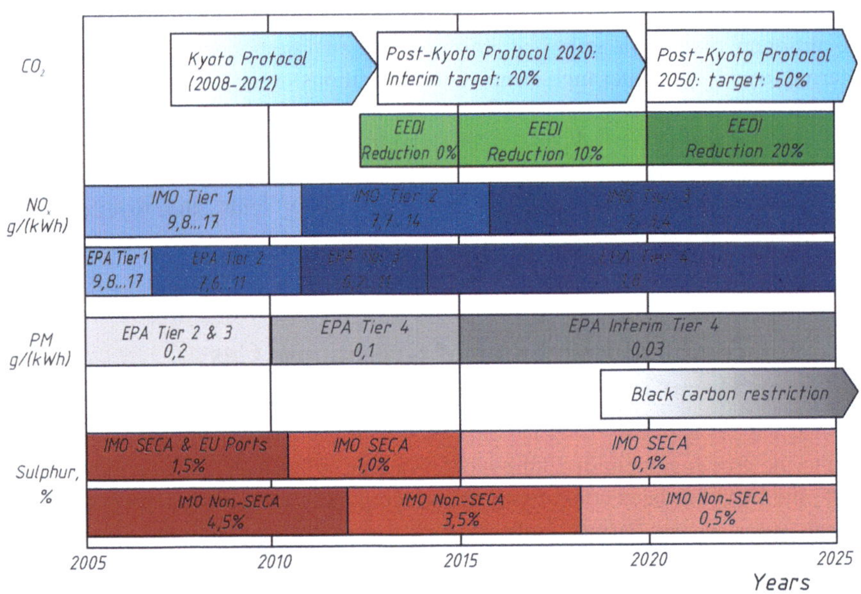

Fig. 1.3 Progressive measures to tighten emission standards for greenhouse gases and harmful substances from marine internal combustion engines (NO_x emissions given as a range for engines with different speeds) [3]

In order to stimulate manufacturers and companies operating ships to reduce the harmful impact on the environment, a system of economic factors of influence developed. These include fines for emissions, as well as complex systems of discounts and compensation for the implementation of measures aimed at reducing emissions of harmful substances.

1.2.1 Regulation of Greenhouse Gas Emissions

The first global agreement on environmental protection was the Kyoto Protocol. The protocol came into force on January 1, 2008, and lasted for five years until December 31, 2012 (Fig. 1.3). Countries that ratified this protocol committed to limiting or reducing average emissions of six types of gases (carbon dioxide, methane, hydrofluorocarbons, perfluorocarbons, nitrous oxide, sulfur hexafluoride) by 5.2% compared to 1990 levels. In the periods following the Kyoto Protocol, attempts also made to regulate issues related to greenhouse gas emissions, including in the shipping sector. Thus, in accordance with the Paris Agreement of the UN Climate Change Conference of 2015, the IMO adopted a strategy for the gradual reduction of greenhouse gas emissions by the shipping sector. The strategy includes two stages—a short-term goal of reducing the intensity of CO_2 emissions from transport operations by 40% by 2030 and a long-term goal of reducing annual greenhouse gas emissions by 50% and carbon emission intensity by 70% by 2050 compared to 2008 levels.

As of the beginning of 2024, Annex VI of the MARPOL 73/78 Convention regulates emissions of nitrogen oxides, sulfur and carbon dioxide [6]. The latter chemical compound is a greenhouse gas that leads to global warming. The introduction of the IMO limit on CO_2 emissions is a logical continuation of the global

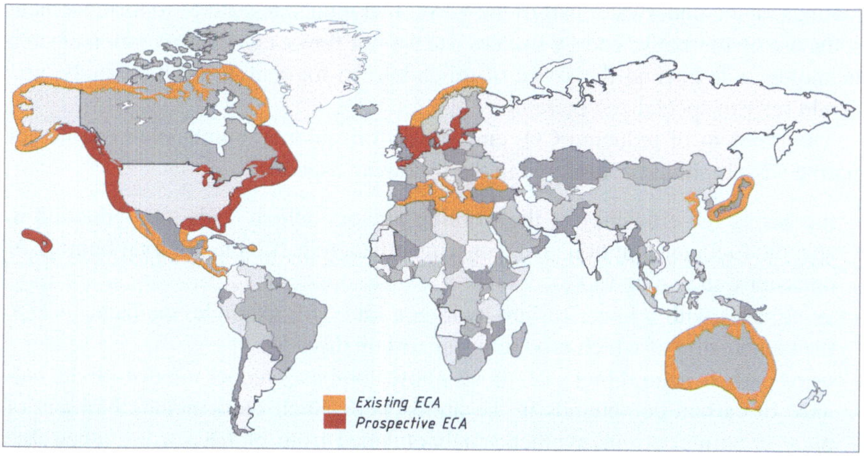

Fig. 1.4 Emission control area [4, 5]

environmental policy to reduce greenhouse gas emissions, which were first regulated by the Kyoto Protocol and then by the Paris Agreement. When considering the structure of the world fleet, the largest contribution to the total volume of CO_2 emissions made by three types of ships: container ships, bulk carriers and oil tankers, accounting for approximately 55% or about 1 billion tons [7].

The most stringent aggregate requirements for limiting carbon dioxide emissions from merchant ships came into force in 2009 with the introduction of the Energy Efficiency Design Index (EEDI) of ships.

The main regulatory document for calculating EEDI is IMO Resolution MEPC.245(66) [8]. This document provides a detailed methodology for determining the index, accompanied by calculation examples for specific cases. EEDI calculated using a formula based on the technical design parameters for a given ship.

Introduction EEDI is a technical measure aimed at encouraging the use of more energy-efficient equipment and engines in the design of new ships in order to make them more environmentally friendly. According to MEPC.245(66), from 1 January 2013, new ships must comply with the established EEDI level for a specific ship type. The level gradually tightened every five years, which stimulates continuous innovation and technical development of all components affecting the fuel efficiency of a ship, starting from the design stage. At the same time, the mechanisms for achieving a given EEDI level may be different, leaving the choice of technologies for use in a specific ship design to ship designers and builders. For a ship, EEDI provides a specific figure expressed in grams of carbon dioxide (CO_2) per mile travelled by the ship. Thus, the lower the EEDI, the more energy efficient the ship design is.

The practical result of EEDI calculation and verification, subject to compliance with the necessary conditions, is that the ship receives an International Energy Efficiency Certificate (IEE Certificate).

As part of the activities aimed at achieving the specified EEDI level, various technologies considered that allow reducing CO_2 emissions. For example, the use of coatings on the underwater part of the vessel that reduce resistance to its movement or the use of alternative energy sources, such as air flows, etc. Consideration of such technologies is beyond the scope of this book, so for a more detailed study, you should refer to specialized literature.

The problem of reducing CO_2 emissions from internal combustion engines of marine vessels should considered in the following aspects:

- increasing the efficiency of the working process, which leads to a reduction in specific fuel consumption, and, consequently, a reduction in the total number of emissions, including CO_2;
- using fuels with a lower carbon-hydrogen ratio (for example, methane—CH_4), the combustion of which produces less carbon dioxide;
- using carbon-neutral fuels, i.e. those whose combustion does not disturb the balance of carbon compounds in the atmosphere. Such fuels include biofuels of plant or animal origin, synthetic fuels obtained from biomass using renewable energy sources (for example: methanol, ethanol). When such fuels burned, car-

bon previously extracted from the atmosphere during the formation of biomass released into the atmosphere;
- use of carbon-free fuels. Currently, ammonia (NH_3) and hydrogen (H_2) considered as such fuels. These fuels can obtained or synthesized using renewable energy, which eliminates the release of carbon dioxide into the atmosphere when they are used.

In some cases, the extraction of carbon dioxide from engine exhaust gases for its subsequent disposal and burial considered as a separate approach to solving the problem. However, such methods still exist more in theoretical terms, and, today, are far from practical implementation.

It should note that the above methods for reducing CO_2 emissions most often used in combination, combining several approaches in one solution. Below, we will consider these methods and technical solutions based on them in more detail.

1.2.2 Regulation of Nitrogen Oxide Emissions

As already noted, among the harmful components that formed during the operation of engines, special attention is paid to nitrogen oxides—NO_x.

Getting into the atmosphere, nitrogen oxides participate in photochemical processes with the formation of high concentrations of ozone O_3 and contribute to the appearance of the so-called photochemical smog [9].

Despite the fact that shipping accounts for only 5% of the world's total anthropogenic emissions of nitrogen oxides, the requirements for environmental performance of ship engines are becoming more stringent from year to year. In the emissions of ship engines, the concentration of NO_x is usually high (500–1200 ppm), which is due to the peculiarities of the organization of the work process and the level of forcing. This is especially true for low-speed engines, in which nitrogen compounds formed most intensively. In areas of intensive shipping, air pollution with nitrogen oxides (NO_x) can reach critical values.

In the human body, NO_x forms compounds of nitric and nitrous acids, leading to irritation of the mucous membrane of the eyes and respiratory tract. At a volume concentration of 0.004–0.008% nitrogen oxides in the atmosphere, pulmonary edema occurs in humans.

The amount of NO_x emissions from marine engines regulated in accordance with Annex VI of the MARPOL 73/78 Convention, Regulation 13 [6]. This regulation limits the mass of NO_x emissions (equivalent mass of NO_2) per engine output energy for a specified speed. With respect to this annex (Fig. 1.5), the NO_x emission limits for Tier I, II and III described below.

1. Tier I was introduced in 2000:

   - 17.0 g/(kW × h) at engine speed $0 < n < 130$ min^{-1};
   - 45.0 $n^{-0.2}$ g/(kW × h) at engine speed $130 < n < 2000$ min^{-1};
   - 9.8 g/(kW × h) at engine speed $n > 2000$ min^{-1}.

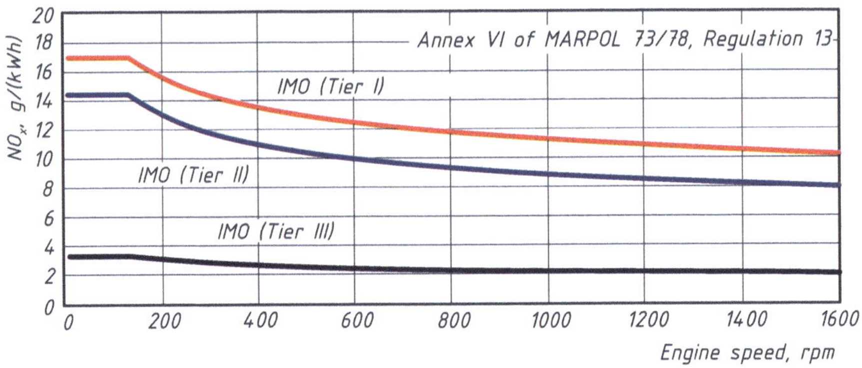

Fig. 1.5 Curves for the gradual tightening of NO_x emission standards for marine internal combustion engines [6]

2. Tier II was introduced in 2011:

   – 14.4 g/(kW × h) at engine speed $0 < n < 130\ \text{min}^{-1}$;
   – 44.0 $n^{-0.23}$ g/(kW × h) at engine speed $130 < n < 2000\ \text{min}^{-1}$;
   – 7.7 g/(kW × h) at engine speed $n > 2000\ \text{min}^{-1}$.

3. Tier III was introduced in 2016:

   – 3.4 g/(kW × h) at engine speed $0 < n < 130\ \text{min}^{-1}$;
   – 9.0 $n^{-0.2}$ g/(kW × h) at engine speed $130 < n < 2000\ \text{min}^{-1}$;
   – 2.0 g/(kW × h) at engine speed $n > 2000\ \text{min}^{-1}$.

Vessels built after 1 January 2016 must not exceed the Ter III limits when navigating within the ECA (Fig. 1.4), unless the zone designated after the date of construction of the vessel. When navigating outside the ECA, the Tier II limits must meet.

There are two main approaches to regulating NO_x emissions, which can be divided on into in-cylinder control and removal of nitrogen oxides from exhaust gases at the engine outlet [10].

The first approach includes methods aimed at influencing the engine operating process in order to reduce the formation of nitrogen oxides in the working cylinder.

The second approach includes methods aimed at decomposing nitrogen oxides into molecular nitrogen and other non-toxic compounds after the exhaust gases have left the engine.

To understand the essence of in-cylinder control methods, in the following chapters we will consider in more detail the mechanisms of nitrogen oxide formation in the engine working space. In-cylinder NO_x control methods are an important area for achieving IMO Tier III compliance, which provides for a 76% reduction in NO_x emissions compared to IMO Tier II standards.

1.2.3 Regulation of Sulfur Oxide Emissions

Sulfur oxides, denoted as SO_x, include sulfur monoxide (SO), sulfur dioxide (SO_2) and sulfur trioxide (SO_3). When released into the atmosphere, sulfur monoxide SO immediately oxidized to SO_2, which is a gas that does not decompose under atmospheric conditions. Consequently, when fuels containing sulfur compounds burned, SO_2 released into the atmosphere as a gas and SO_3 as Particulate Matter (PM). SO_x formation is commonly observed in cases of fossil fuels such as oil and coal and is an undesirable by-product since it is an air pollutant along with NO_x. In addition to NO_x, SO_x is also one of the causes of acid rain. In the atmosphere, SO_2 reacts with atmospheric oxygen O_2 to form SO_3. When dissolved in atmospheric moisture (H_2O), SO_3 and SO_2 form solutions of sulfuric (H_2SO_4) and sulfurous (H_2SO_3) acids, which are the cause of acid rain. Acid rain caused by SO_x is an indirect source of human diseases. In addition, SO_3 is a form of PM, the presence of which in the atmosphere can cause respiratory diseases such as bronchitis and asthma.

Over the past decades, MARPOL restrictions have become more stringent with regard to SO_x emissions. In addition to the general ECAs, IMO establishes emission control areas, which impose stricter controls to minimise emissions of sulphur compounds into the atmosphere from ships (Sulfur Emission Control Areas—SECA). The amount of SO_x emissions from marine engines regulated by MARPOL Annex VI, namely Regulation 14, paragraph A.2. Instead of directly regulating the amount of SO_x in emissions, Regulation 14 [11] sets a limit on the permissible sulphur content of fuel. The requirements become more stringent according to the stage of implementation and depending on the area, i.e. outside and inside ECAs and SECAs.

Currently (since 1 January 2020), Regulation 14 sets the permissible sulphur content of up to 0.50% by weight in fuel intended for use outside the ECA and SECA and up to 0.1% within these areas.

The use of fuels with sulphur content above 0.5% is permitted only for ships equipped with exhaust gas scrubber systems. Owners and operators of all other ships are required to use only low-sulphur marine fuels or their alternatives. Violation of these rules is fraught with high fines, detention of ships, and in some jurisdictions even imprisonment.

1.2.4 Regulation of Emissions of Hydrocarbon Compounds and Products of Incomplete Combustion

At present, neither the IMO nor other international or national regulatory bodies have set standards for emissions of hydrocarbons (CH) and carbon monoxide (CO) in the combustion products of marine engines. The content of these substances in exhaust gases largely depends on the completeness of fuel combustion. Leading manufacturers of marine engines pay considerable attention to reducing these emissions, based more on economic than on environmental considerations. However, these compounds have a significant impact on the human body, and their significant accumulation in the air can have serious negative consequences.

References

1. https://www.marinetraffic.com/en/ais/home/centerx:-58.3/centery:-53.4/zoom:2
2. Modern Marine Internal Combustion Engines. A Technical and Historical Overview. Bilousov, I, Bulgakov, M., Savchuk, V. Springer Series on Naval Architecture, Marine Engineering, Shipbuilding and Shipping. 2020. 385 p. ISBN 978–3–030-49748-4. https://doi.org/10.1007/978-3-030-49749-1
3. Lagarde, D: Alternative Fuels. STC Group, STC T&C, Rotterdam. 34 p
4. Oil spill risk assessment methodology for extreme conditions, including Arctic. Innovation and Networks Executive Agency. Horizon 2020 BG-2014-2015/BG2015–2. Project coordinator: Finnish Environment Institute (SYKE). 2019. 71 p
5. https://www.sia-partners.com/en/insights/publications/towards-cleaner-and-more-sustainable -shipping-required-actions
6. IMO MARPOL Annex VI Regulation 13. – Nitrogen Oxides (NO_x)
7. Olmer, N, et al.: Greenhouse gas emissions from global shipping, 2013–2015. Washington: ICCT, 2017. 38 p
8. MEPC.245 (66) 2014 Guidelines on the Method of Calculation of the Attained Energy Efficiency Design Index (EEDI) for New Ships
9. Hansen, J.P: Reduction of SO_2, NO_x and Particulate Matter from Ships with Diesel Engines. Hansen, J.P., Kaltoft, J., Bak F., Gørtz, J., Pedersen, M., Underwood, C.: Miljoministeriet Milj Styrelsen. København, 2013–114 p
10. Wilson, K: The Engine Users and the Implications of IMO Tier III Coming into Force. Wilson K., Associates: CIMAC Congress 2016, Helsinki Paper № 62. 6 p
11. IMO MARPOL Annex VI. Regulation 14. – Sulphur oxides (SO_x) and Particulate Matter (PM)

Chapter 2
Reducing Nitrogen Oxide Emissions from Marine Engines

2.1 Mechanisms of Nitrogen Oxide Formation in Marine Engines

Diatomic molecular nitrogen (N_2), which makes up the bulk of atmospheric air (79% by mass), is a relatively inert gas, but atomic nitrogen (N) can have ionization levels (called valence states) from +1 to +5. Thus, nitrogen can form several different oxides. Valence state refers to the number of electrons that are either deficient (positive valence) or excessive (negative valence) in the ion compared to a neutral molecule. The family of NO_x compounds and their properties are listed in Table 2.1. With the exception of NO_x from soils, lightning, and wildfires, NO_x is largely anthropogenic (i.e. generated by human activity). Biogenic sources account for less than 10% of the total NO_x emissions into the Earth's atmosphere.

Nitrogen monoxide (NO) and nitrogen dioxide (NO_2) make up the majority of nitrogen oxides NO_x (90–95%), other compounds are present in the exhaust gas in much smaller quantities. In the working cylinders of diesel engines, NO is mainly formed during the working process. When released into the atmosphere, it oxidized to NO_2 and in much smaller quantities to N_2O_5. Therefore, the current standards in accordance with the requirements of Annex VI of the MARPOL 73/78 Convention regulate only nitrogen dioxide (NO_2) as a general compound of the entire family, as the most common form of NO_x in the atmosphere generated by anthropogenic (human) activity. The remaining nitrogen oxides listed in Table 2.1 formed in very small quantities, so there are currently no standards for their emissions. There are two main mechanisms of NO_x formation in internal combustion engines, in which the specified compounds formed as result of various reactions:

– "thermal", in which NO_x is formed as a result of the interaction of atmospheric nitrogen with oxygen under high temperature conditions.

I. Bilousov, R. Varbanets, *Marine Internal Combustion Engines and Environmental Protection Issues*, Springer Series on Naval Architecture, Marine Engineering, Shipbuilding and Shipping 27,
https://doi.org/10.1007/978-3-032-20940-5_2

Table 2.1 Nitrogen oxides (NO_x)

Formula	Name	Valence of nitrogen	Melting point, °C	Boiling point, °C	Properties
N_2O	Nitrous oxide (nitrous oxide)	I	– 90.8	– 88.49	Colorless gas, soluble in water
NO N_2O_2	Nitrogen oxide Nitrogen dioxide	II	– 163.6	– 151.74	Colorless gas, slightly soluble in water
N_2O_3	Nitrogen trioxide (nitrous anhydride)	III	– 101.1	≈ 3	Black solid, soluble in water, decomposes in water
NO_2 N_2O_4	Nitrogen dioxide Nitrogen dioxide (nitrous anhydride)	IV	– 9.3	21.15	Reddish-brown gas, very soluble in water, decomposes in water
N_2O_5	Nitrogen pentoxide (nitrous anhydride)	VI		33 (subl.)	White solid crystals, well soluble in water, decomposes in water

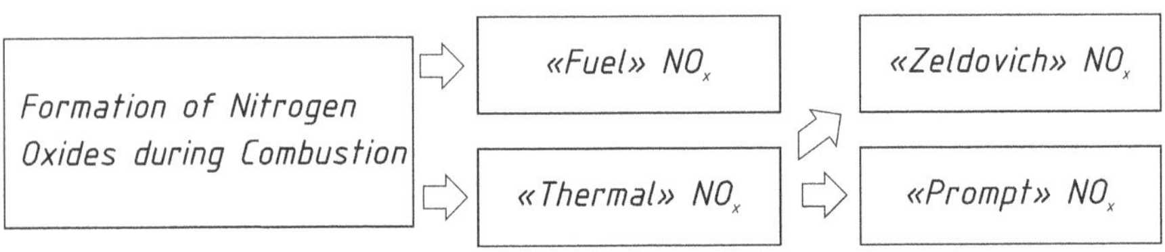

Fig. 2.1 Scheme of formation of nitrogen compounds during fuel combustion

– "fuel"—in which oxidation of nitrogen compounds contained in the fuel occurs (Fig. 2.1).

In turn, the "thermal" mechanism of NO_x formation divided into the "high-temperature Zeldovich mechanism" and the so-called "fast" mechanism, which also called "chemical" because NO_x formed because of the interaction of nitrogen with fuel radicals that arise during the combustion process.

The thermal mechanism of NO_x formation usually considered the main one. About 90–95% of all nitrogen oxide emissions from an engine are formed by this mechanism. The contribution of different formation mechanisms to the total NO_x content in engine exhaust gases shown in Fig. 2.2.

The mechanism of formation of "thermal" NO_x is extremely complex, includes hundreds of different chemical reactions, and depends on the stochastic parameters of air movement, the nature of the formation of the air-fuel mixture and

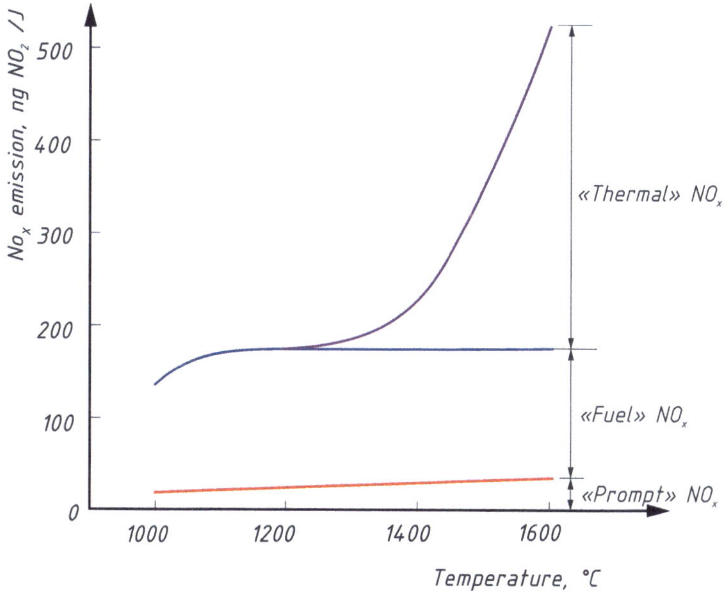

Fig. 2.2 Contribution of different NO_x formation mechanisms to the total emissions of these compounds with exhaust gases from marine engines (adapted from [1])

the distribution of temperatures in the combustion chamber. However, the basic mechanisms of formation of nitrogen compounds are well studied.

The high-temperature mechanism (thermal) of nitrogen oxide formation was investigated and described by Ya. B. Zeldovich in the mid-1940s. It was later supplemented by C.P. Fenimore and G.W. Jones.

During the studies, it was established that the rate of NO_x formation is primarily a function of temperature and the time that nitrogen is exposed to this temperature. At high temperatures, molecular nitrogen (N_2) and oxygen (O_2) the air dissociate to an atomic state and participate in reactions described by the well-known extended Zeldovich-Lavoie mechanism, which includes reactions for mixtures close to stoichiometric composition. This mechanism is considered the main mechanism of NO_x formation during combustion, and includes the following elementary stages [1]:

– formation of an oxygen radical (O-radical) at high temperatures

$$O_2 \leftrightarrow O + O;$$

– oxygen radical destroys the N_2 molecule to form NO and N-radicals

$$N_2 + O \leftrightarrow NO + N;$$

– the N radical destroys the O_2 molecule to form NO and the O radical

$$N + O_2 \leftrightarrow NO + O;$$

– radical N forms with radical OH and exchanges oxygen atom O

$$N + OH \leftrightarrow NO + H.$$

The reactions occur in both directions—that is, formation and decomposition. The first three chain reactions are called the Zeldovich mechanism. The fourth reaction was studied by C.P. Fenimore and G.W. Jones, so the process of NO formation, including this reaction, is called the extended Zeldovich mechanism, which was fully developed by G.A. Lavoie in 1970. In some sources, the extended mechanism of NO formation called the Zeldovich-Lavoie mechanism. The rate of NO_x formation is determined mainly by the oxygen concentration and the rate constant of the direct reactions. Figure 2.3 shows the dependences of the NO_x formation rate on temperature [1].

From the dependence shown in Fig. 2.3, it can seen that due to the high activation energy, which limits the reaction rates, the mechanism of formation of "thermal" NO_x strongly depends on the temperature, which has an exponential effect on the process. To a lesser extent, the rate of formation of NO_x increases with increasing pressure in the working cylinder. In the temperature range of 1400–2000 K, an increase in pressure in the range most typical for the combustion process in modern marine engines from 11.5 to 18.5 MPa leads to an increase in the formation of NO_x by 26.5% regardless of temperature.

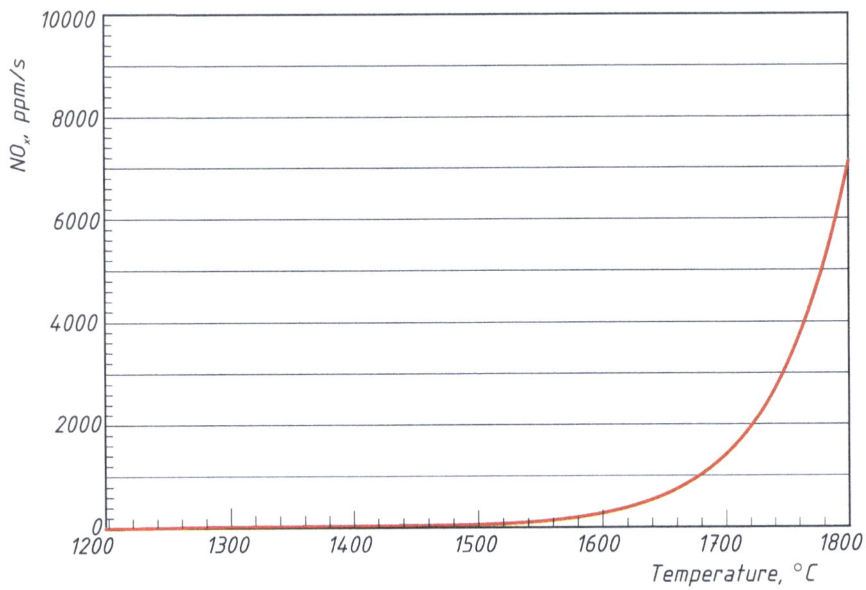

Fig. 2.3 Dependence of the rate of NO_x formation on temperature (adapted from [1])

In general, NO_x emissions for highly accelerated marine diesel engines are relatively high due to high local combustion temperatures.

Due to the obvious dependence of combustion and temperature in individual zones of the combustion chamber, the formation of NO_x emissions can controlled by monitoring the heat supply process. Experimental studies have shown that a decrease in temperature by 100 K reduces the rate of formation of "thermal" NO_x by 10 times [2]. Analysis of the kinetics of the formation of "thermal" NO_x indicates that the overall emission affected not only by temperature, but also by the time the reactants spend in the high-temperature zone.

Thus, the formation of NO_x in the flame can significantly reduce either by reducing the time and temperature, or by reducing the oxygen content in the combustion zone. Low-speed engines with a long combustion process and a high excess air coefficient have the highest NO_x emissions because oxygen has the ability to react with nitrogen for quite a long time.

Figure 2.4 shows the dependence of the time (*t*) required for the formation of "thermal" NO_x compounds by the Zeldovich mechanism depending on the temperature in the working cylinder of a low-speed marine diesel engine 6S60MC-C. It is evident that the onset of kinetic conditions for the reactions of "thermal" NO_x formation occur at the angular section of the cycle from 204 to 226° p.c. Considering that for the passage of 1° p.c. 0.00159 s is required, the formation of "thermal" NO_x occurs within 0.035 s, which is an order of magnitude greater than that of medium-speed ($\approx$ 0.0073 s, at a rotation speed of 500 min^{-1}) and high-speed ($\approx$ 0.0024 s, at a rotation speed of 1500 min^{-1}) marine engines.

The mechanism of formation of "fast" NO_x is not fully understood and cannot be explained by the Zeldovich mechanism. The contribution of "fast" NO_x to the total emission of nitrogen oxides is insignificant, but not so much that it can be

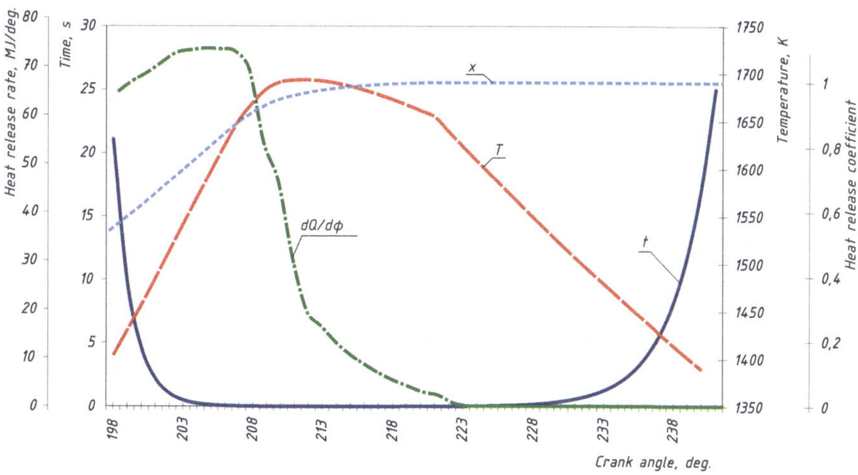

Fig. 2.4 Time (*t*) of formation of "thermal" NO_x compounds according to the Zeldovich mechanism in the working cylinder of a low-speed marine diesel engine 6S60MC-C [3]

neglected. This mechanism was first described by C.P. Fenimore, who studied the combustion of a stoichiometric methane-air mixture.

Reactions by this mechanism occur in fuel-rich zones at relatively low temperatures. They take considerably less time than is required for NO_x formation by the "thermal" mechanism. It has been established that "fast" nitrogen oxides are formed in a time interval of 10^{-4} s directly in the laminar flame front at temperatures of about 900–1000 K. The formation of "fast" NO_x is attributed to reactions of N with C, CH and CH_2 radicals, which are formed during the decomposition of fuel molecules.

At the first stage of combustion, carbon radicals from the fuel react with nitrogen according to the following equation:

$$N_2 + CH \leftrightarrow HCN + N$$

and several similar reactions involving CH and C radicals

$$N + O_2 \leftrightarrow NO + O,$$

after which HCN partially reacts with O_2 to form NO, according to the following equations

$$HCN + OH \leftrightarrow CN + H_2O,$$

$$NH + OH \leftrightarrow N + H_2O,$$

$$CN + O_2 \leftrightarrow NO + !O.$$

In this mechanism, HCN and N radicals formed as intermediates in combustion reactions. "Fast" NO_x ultimately formed by the oxidation of N and CN within the combustion flame.

The mechanism of formation of "fuel" NO_x, in which oxidation of nitrogen compounds contained in the fuel occurs, is determined by the content of these compounds in the fuel. In heavy fuel oils, N is present in the form of pyridine (C_5H_5N), quinoline (C_9H_7N), ammonia (NH_3), hydrogen cyanide (HCN), pyrrole (C_4H_5N), methylamine (CH_3NH_2), etc. These substances decompose during pyrolysis in the pre-flame zone to nitrogen-containing compounds with a low molecular weight: NH_3; HCN; CN, etc.

During combustion, nitrogen released from these compounds as a free radical and reacts with oxygen-containing substances to form NO and N_2, but the proportion of molecular nitrogen formed during these reactions is insignificant. Oxidation occurs quite quickly, over a time comparable to the combustion time.

Most of the "fuel" NO_x formed during combustion of fuels with a high content of nitrogen compounds, such as heavy residual fuels. Table 2.2 shows the nitrogen content in different types of fossil fuels.

Table 2.2 Content of nitrogen compounds in different types of fuel [3]

Fuel type	N content,% by weight
Diesel fuel	0.007–0.01
Motor fuel	0.12–0.13
Fuel oil	0.3–0.5
Liquefied petroleum gas, natural gas	0

The formation of "fuel" nitrogen oxides from nitrogen-containing compounds of the fuel begins already upon their contact with hot air at a temperature of 700–800 °C and has a weakly expressed linear dependence on temperature (Fig. 2.2). The formation of nitrogen oxides at a relatively low temperature explained by the lower energy expenditure on the conversion of nitrogen from compounds included in the fuel than on the splitting of the N_2 molecule and the formation of NO_x by the "thermal" mechanism.

2.2 Methods for Reducing Nitrogen Oxides in Exhaust Gases

Understanding the mechanisms of nitrogen oxide formation in the working space of internal combustion engines has allowed us to develop quite effective methods aimed at reducing them. As already noted, these methods can divide into intra-cylinder and extra-cylinder. Since the bulk of nitrogen oxides is formed by the "thermal" mechanism, most intra-cylinder methods are aimed at reducing the likelihood of conditions for the occurrence of NO_x formation reactions by this mechanism. However, for marine engines running on heavy fuels, it is important to reduce "fuel" NO_x by reducing the amount of nitrogen-containing compounds in the fuel.

Intra-cylinder methods usually aimed at reducing temperatures both in the engine working space as a whole and in local areas of the combustion chamber.

Extra-cylinder methods aimed at chemically decomposing nitrogen oxides formed in the working cylinder into molecular nitrogen and water. As a rule, different methods are used in various combinations, complementing each other.

2.2.1 Improving Mixture Formation and Combustion in the Engine

The most universal method that allows improving not only the environmental but also the engine's efficiency. The process of mixture formation in a diesel engine is understood as the entire complex of complex physical and chemical phenomena and processes that occur from the moment the first portions of fuel are injected into the combustion chamber until the last portion is ignited [4]. The quality of mixture formation usually means the uniformity of the fuel distribution throughout the

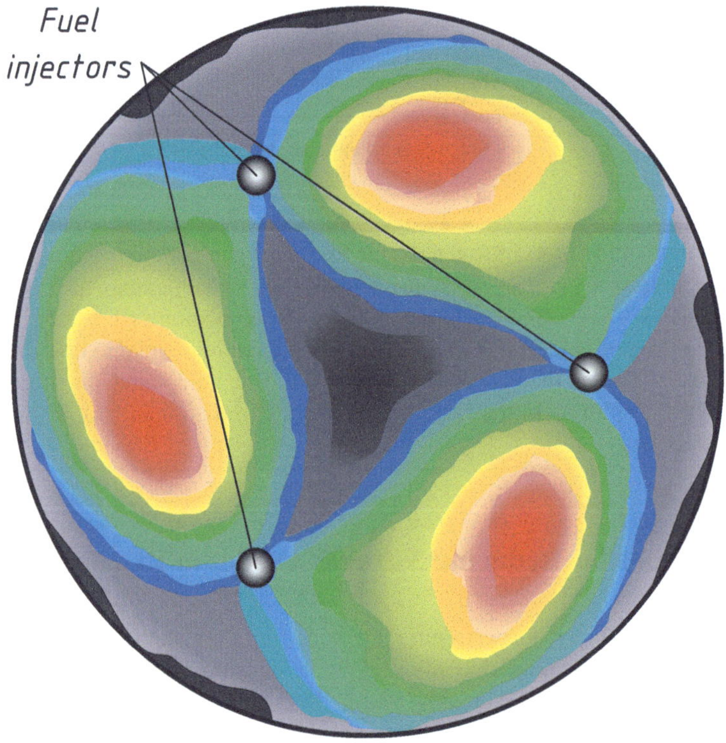

Fig. 2.5 Temperatures distribution in the combustion chamber of a two-stroke low-speed engine with three injectors per cylinder (adapted from [5, 6])

combustion chamber to create the most favorable conditions for its combustion. The peculiarities of mixture formation in diesel engines lead to the fact that it is practically impossible to achieve uniform distribution of fuel in the short periods allotted for the entire process. In this regard, mixture formation and combustion in these engines characterized by pronounced zonality. As a result, areas with temperatures significantly exceeding the average formed in the combustion chamber, and there areas in which the temperatures remain at a low level (Fig. 2.5). In high temperature zones, conditions created for the launch of the "thermal" mechanism of NO_x formation; in low temperature zones, reactions slow down significantly, which leads to the formation of incomplete combustion products.

Thus, a set of measures aimed at a uniform distribution of fuel throughout the combustion chamber volume allows to significantly improving the environmental performance of engines. These measures themselves come down to two main tasks:

- optimization of the processes of fuel atomization and its distribution throughout the combustion chamber volume;
- imparting mobility to the charge to intensify mixing of fuel with air.

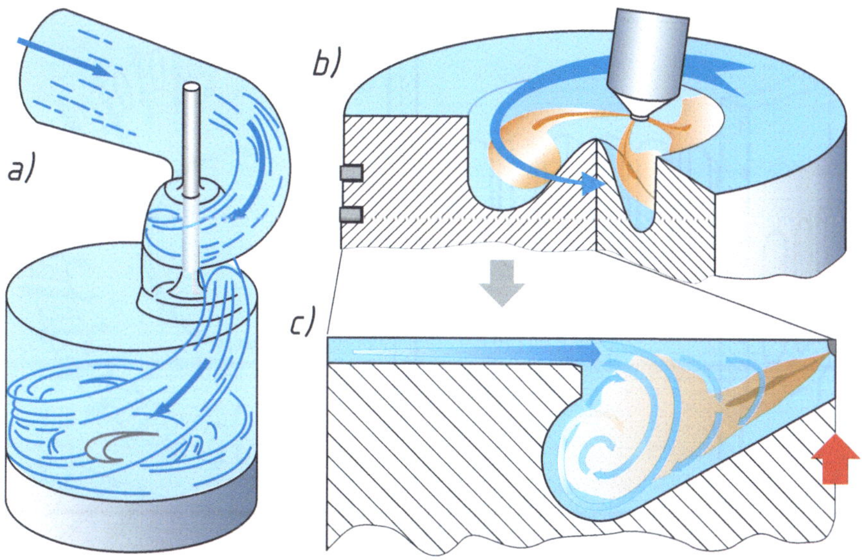

Fig. 2.6 Formation of a tangential vortex during the process of charge intake into the cylinder (**a**), deflection of spray plumes and distribution of their contents throughout the combustion chamber volume (**b**); charge turbulence during the compression process due to the displacement of air from the narrowest areas of the above-piston space (**c**)

The first achieved by increasing the fuel atomization pressure, selecting the geometry of the nozzle holes and their location, optimizing the shape of the combustion chamber, etc. The second achieved by organizing the gas exchange process, in which the charge given a directed or turbulent movement, which facilitates the transfer of droplets and vapors of fuel to areas not covered by spray plumes.

In four-stroke engines, directed movement is imparted to the charge, as a rule, during the process of filling the cylinder by profiling the inlet channels in such a way as to impart rotational motion to the flow due to its tangential entry into the working cylinder (Fig. 2.6a).

The rotational motion created in this way continues in the cylinder during the compression process, and at its final stage, when fuel injection begins, the airflow deflects the spray plumes, distributing their contents throghout the combustion chamber (Fig. 2.6b). In addition to the tangential direction of the flow, at the final stage of compression, the charge begins to displace from the gap between the piston bottom and the cylinder cover into the main volume of the combustion chamber. As a result, additional flows are created, which, twisting, deform the spray plume and facilitate the distribution of fuel throughout the combustion chamber (Fig. 2.6c). These same flows facilitate the mixing of fuel vapors evaporating from the walls of the combustion chamber during volumetric-film mixture formation [4].

In two-stroke engines, the tangential inclination of the scavenging ports used to impart rotational motion to the charge (Fig. 2.7).

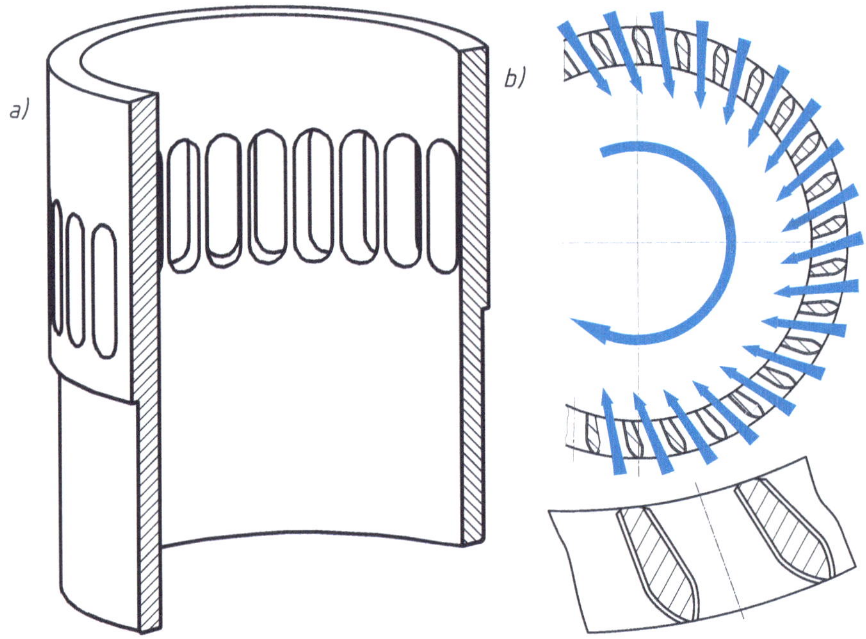

Fig. 2.7 Imparting rotational motion to the air charge in two-stroke low-speed engines: (**a**) location of the purge ports in the lower part of the cylinder liner; (**b**) section of the cylinder liner in the plane of the purge ports (adapted from [4])

The mobility of the charge effectively promotes a uniform distribution of fuel throughout the combustion chamber, but incorrect selection of charge turbulence parameters can also have negative consequences. Thus, at high flow rotation speeds, fuel from one spray plume can transfer to the zone of another plume, creating over-enriched areas in which high-quality combustion of the fuel becomes impossible. In two-stroke engines, during scavenging, too high a charge rotation speed can press the bulk of the charge against the walls, forming a zone saturated with exhaust combustion products along the cylinder axis. As a result, the quality of cylinder cleaning will deteriorate [5].

An important tool for improving the quality of mixture formation in modern engines is the introduction of accumulator fuel injection systems (Common Rail). These systems are capable of significantly improving the quality of fuel atomization, regardless of the engine-operating mode, as well as implementing complex, including multi-stage algorithms for its supply to the combustion chamber.

Figure 2.8 shows the options for changing the pressure before the injector nozzle for three different options for the operation of the Common Rail control valve of the medium-speed L32/40 engine from MAN Diesel & Turbo. For comparison, the injection pressure curve for the mechanical injection system shown. It is evident that the injection pressure at the initial stage, which most strongly affects the formation of conditions favorable for the formation of NO_x and particulate matter, is

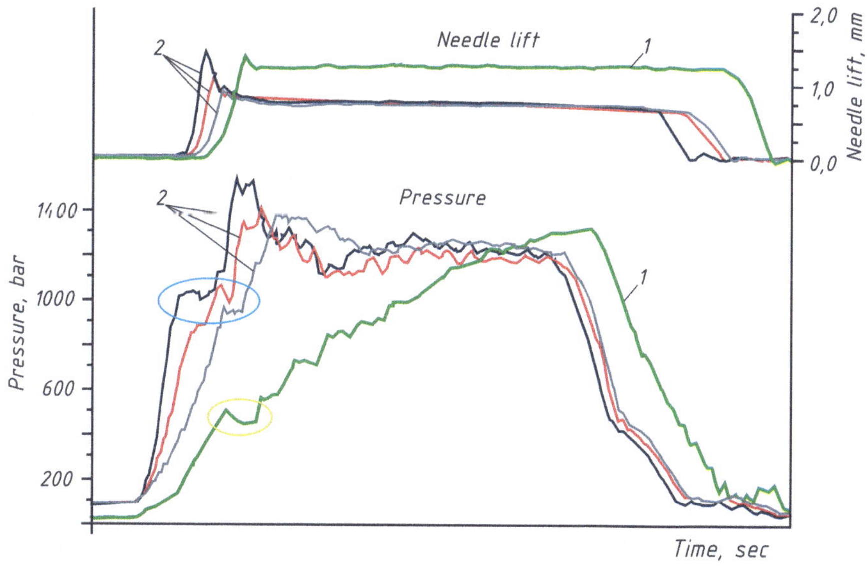

Fig. 2.8 Injection law variants implemented in the Common Rail system of the medium-speed L32/40 engine from MAN Diesel & Turbo: (1) injector needle lift and pressure before the nozzle for the mechanical injection system; (2) also for the Common Rail system (adapted from [6])

significantly higher (zone highlighted in blue) than for the mechanical system (zone highlighted in yellow).

High-pressure now of opening of the needle valve of the injector promotes finer atomization of fuel and its uniform distribution throughout the volume of the combustion chamber. As a result, the first portions well atomized, distributed throughout the entire volume of the combustion chamber, pre-flame processes occur faster, and the delay time of autoignition reduced.

Ignition and combustion of the first portions leads to the fact that subsequent fuel enters an environment with increased temperature and pressure, in which there are already centers of open flame. This promotes rapid evaporation of subsequent portions of fuel and intensive combustion. As a result, combustion efficiency increases and the likelihood of forming local zones with higher temperatures, in which NO_x formed decreases. By changing the operating mode of the control valve, the supply law can optimize in a wide range of engine operating modes. In most modern electronically controlled fuel systems, such as: hydraulic fuel injection pump drive system of ME engines from MAN Diesel & Turbo, Common Rail system, electrically controlled pump-injectors, in addition to single-stage supply of a cyclic portion of fuel, a fractionated supply can be implemented. With this type of supply, the cyclic portion enters the combustion chamber in the form of several successive injections.

The use of two-phase supply allows a significant reduction in the content of NO_x in the engine exhaust gases, which significantly reduces the rate of pressure increase in the initial stage and achieves lower temperatures of the working cycle (Fig. 2.9).

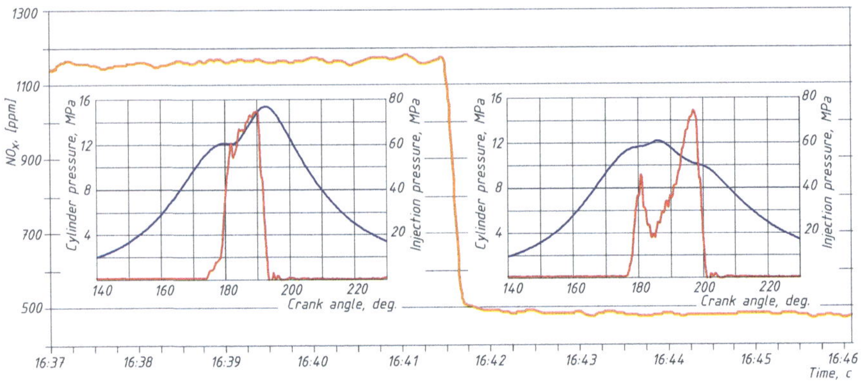

Fig. 2.9 Transition of the ME series engine from the economical operating mode to the mode that ensures the minimum content of harmful emissions with exhaust gases (adapted from [7])

The injection law consists of a pilot injection of a small amount of fuel into the combustion chamber and the main fuel injection following the pilot injection. To avoid deterioration in efficiency, the pilot portion of fuel in such a process should be as small as possible. Therefore, an important requirement for fuel equipment is its ability to ensure a minimum stable supply of the pilot portion of fuel and the time interval between it and the main supply. The presence of pre-injection creates conditions for reducing the dynamics of the process and more complete combustion, since the main supply occurs now when a flame appears in the cylinder from the burning pre-injected portion of fuel. Therefore, ignition of the main mass occurs almost instantly and the combustion process occurs at significantly lower speeds, determined by the speed of fuel supply to the cylinder. The nature of the working process with two-phase fuel injection in the ME series engine from MAN equipped with high-pressure pumps with hydraulic drive and electronic control is shown in Fig. 2.9 (right). In such engines, control of the injection law allows changing the nature of the working process during engine operation depending on the task. If the task is to achieve maximum engine efficiency, the supply law shown in Fig. 2.9 (left) can implemented. If it is necessary to minimize the content of harmful emissions, the supply law shown in Fig. 2.9 (right) implemented. The transition from one mode to another is carried out by changing the control algorithm, which requires a time of 0.25 s, i.e. commensurate with the interval between individual cycles. Figure 2.9 shows the change in the NO_x emission level when the ME series engine switches from the maximum economy mode to the maximum environmental friendliness mode. In this case, the maximum cycle pressure decreases from 15.7 MPa to 12.2 MPa. At the same time, a decrease in NO_x emissions noted when switching to the mode with preliminary injection from 1200 to 480 ppm.

In low-speed engines from Mitsubishi, a gentle law of fuel injection pressure increase at the initial stage of the supply process been applied (Fig. 2.10). The character of the change in fuel pressure before the nozzle shown in the figure realized in the UEC-Eco series engines with hydraulically driven fuel pumps. Two

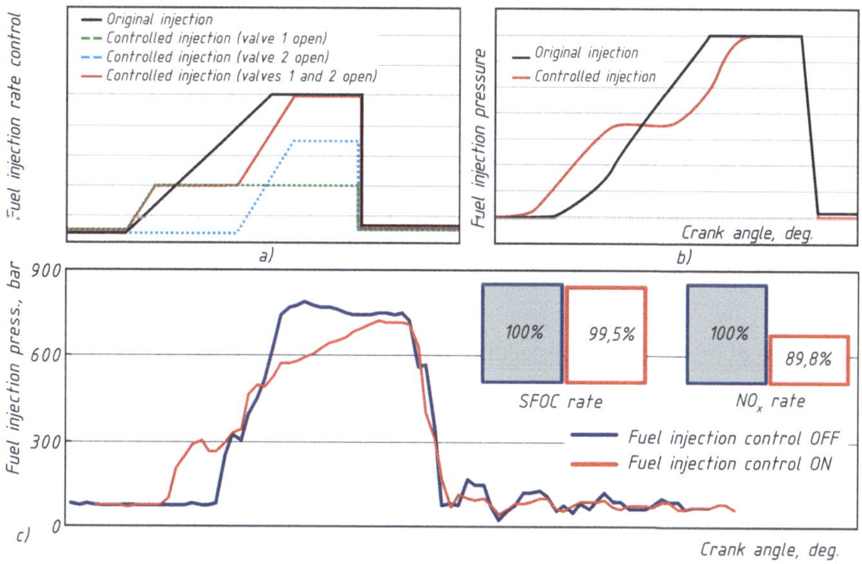

Fig. 2.10 Fuel supply law with a gentle pressure increase in the UEC50LSH-Eco engine compared to the classic supply law: (**a**) combinations of opening of the fuel pump hydraulic drive control valves; (**b**) formation of a stepwise supply law; (**c**) fuel pressure in front of the injector nozzle and its effect on specific fuel consumption and NO_x emissions [8, 9]

electromagnetic valves regulate the oil supply for the plunger drive. Various combinations of their activation (Fig. 2.10a) allow changing the plunger lift speed and, accordingly, the nature of the pressure change at the outlet of the fuel pump and the injector nozzle (Fig. 2.10b, c). The slow supply of the first portions of fuel to the combustion chamber does not allow a large amount of it to accumulate during the self-ignition delay period, which prevents a sharp increase in pressure and temperature at the initial stage of combustion. The character of the change in fuel pressure before the nozzle shown in the figure realized in the UEC-Eco series engines with hydraulically driven fuel pumps. Two electromagnetic valves regulate the oil supply for the plunger drive. Various combinations of their activation (Fig. 2.10a) allow changing the plunger lift speed and, accordingly, the nature of the pressure change at the outlet of the fuel pump and the injector nozzle (Fig. 2.10b, c).

The slow supply of the first portions of fuel to the combustion chamber does not allow a large amount of it to accumulate during the self-ignition delay period, which prevents a sharp increase in pressure and temperature at the initial stage of combustion. After ignition of a limited portion of fuel supplied in the initial phase, subsequent fuel enters the already prepared environment with an open flame source. Therefore, in the second phase of injection, the incoming fuel burns more intensively as it enters the combustion chamber.

This process not accompanied by a sharp increase in pressures and temperatures, so it considered a compromise solution between the efficient and environmental performance of the engine when operating at low speeds, but with a high load. As a result, with virtually unchanged fuel efficiency, NO_x emissions can be reduced more than 10% (Fig. 2.10c).

2.2.2 *Methods Based on Changing the Properties and Composition of the Working Fluid at the Entrance to the Working Cylinder or Immediately Before Combustion*

This group of methods includes adding water aerosol or steam to the incoming air or directly feeding water into the combustion chamber together or separately with the fuel.

This method also includes regulating the amount of oxygen in the cylinder charge by bypassing part of the exhaust gases into the intake system. This method called Exhaust Gas Recirculation (EGR) and, depending on the specific approaches to its implementation, has several modifications, some of which we will consider in more detail below.

Methods of using water to regulate NO_x emissions in exhaust gases discussed in sufficient detail in our book [4]. However, we will dwell on some of these methods in more detail.

2.2.2.1 Methods of Saturating Incoming Air with Water Vapor

The mechanism of reducing NO_x formation due to the increased content of water vapor in the working fluid based on smoothing out temperature peaks in the engine combustion chamber during the fuel combustion process (Fig. 2.11). Water vapor, being a triatomic gas, has a much higher heat capacity and thermal conductivity than atmospheric air. For conditions typical at the start of the combustion process in modern diesel engines, the average mass heat capacity of water is 2.0754 kJ/(kg × K), against 1.0710 kJ/(kg × K) for air, and thermal conductivity is 78.65 W/(m × K) and 57.3 W/(m × K), respectively. As a result, the heat released in local high-temperature zones distributed more evenly throughout the combustion chamber.

An example of NO_x emission reduction technologies by humidifying incoming air is the Humid Air Motor (HAM) system developed by Munters Euroform, which increases air humidity to 99% (Fig. 2.12). Air from the turbocharger with a temperature of 120–130 °C passes through a humidification column, where it cooled and humidified.

As part of research programs aimed at finding new technologies for controlling emissions of harmful substances with exhaust gases, MAN B&W conducted a set of studies on a laboratory two-stroke low-speed engine 4T50ME-X, including HAM technology.

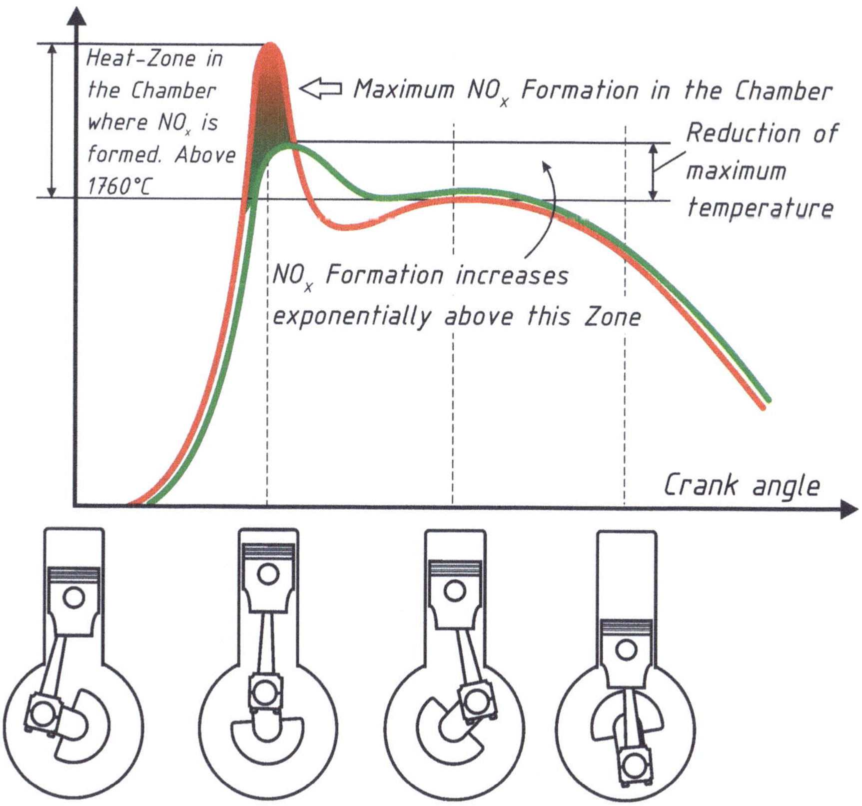

Fig. 2.11 Reduction of peak temperatures in the combustion chamber by saturating the air with water vapor to reduce NO_x formation (adapted from [10])

The results of measurements for four main types of pollution obtained during engine tests at 100% load, shown in Fig. 2.13.

The tests showed that this technology has little effect on the level of emissions of particulate matter and hydrocarbons, while carbon monoxide emissions increase significantly. This increase explained by the fact that, with an increase in the proportion of water vapor in the charge, large local areas with a lack of oxygen are formed in the combustion chamber. The reduction in NO_x emissions was about 35%, confirming the high efficiency of this method, developed primarily to control the emissions of this harmful component.

A disadvantage that noted based on the test results can consider the increased temperature of the steam-air mixture at the engine inlet. To ensure that the dew point exceeded, it had to maintain at 70 °C, which led to increased thermal stress in the engine's structural elements. Otherwise, no serious deviations in engine operation recorded.

The HAM system first installed on the "Viking Line" vessel "Mariella" in 1997. The upgrade was carried out on a vessel that was already in service, which made the

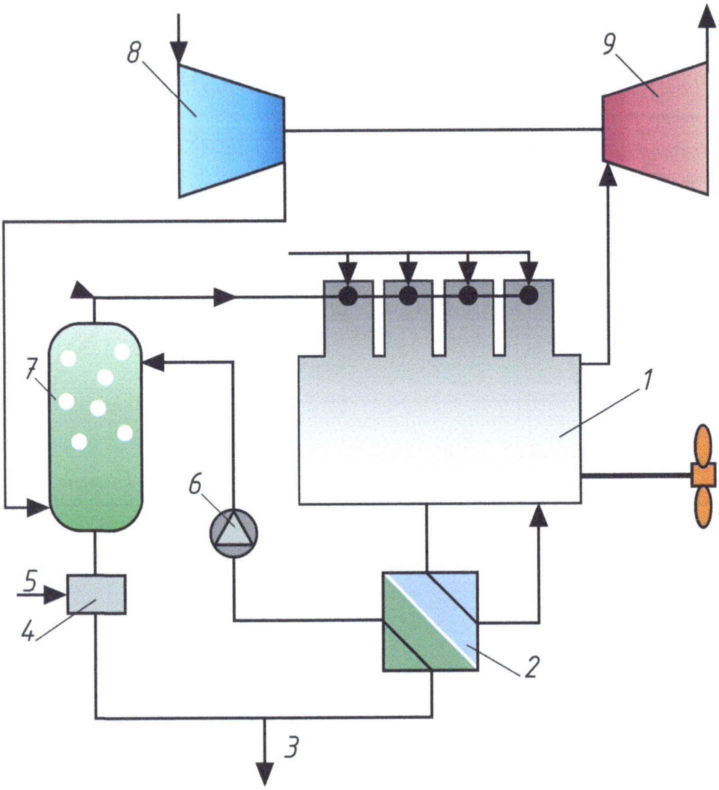

Fig. 2.12 Schematic diagram of the air humidification system at the engine inlet by Munters Euroform: (1) engine; (2) heat exchanger; (3) water drain; (4) water tank; (5) water supply; (6) circulation pump; (7) humidification column; (8) supercharger; (9) turbocharger turbine (adapted from [4])

task somewhat more challenging given that the vessel's propulsion system included four SEMT Pielstick 12VPC2–6 four-stroke medium-speed engines. For this reason, the upgrade carried out in several stages and finally completed in 2001.

The vessel uses water from the Baltic Sea. The amount of water required for the voyage between Stockholm and Helsinki is approximately 60 tons. Of this, 95% of the water is recycled and 5% discharged into the sea to control the salt content in the humidifier column. The water supplied to compensate for the discharges filtered to separate algae and other solid contaminants. The vessel is equipped with two parallel filters with a pollution alarm system. Testing of the system showed that in the operating mode, the NO_x content was reduced by 70–80%. Water consumption through the engine is approximately 3 times higher than fuel consumption; however, this technology has minimal requirements for the quality of the water used, including the use of seawater with its minimal preparation.

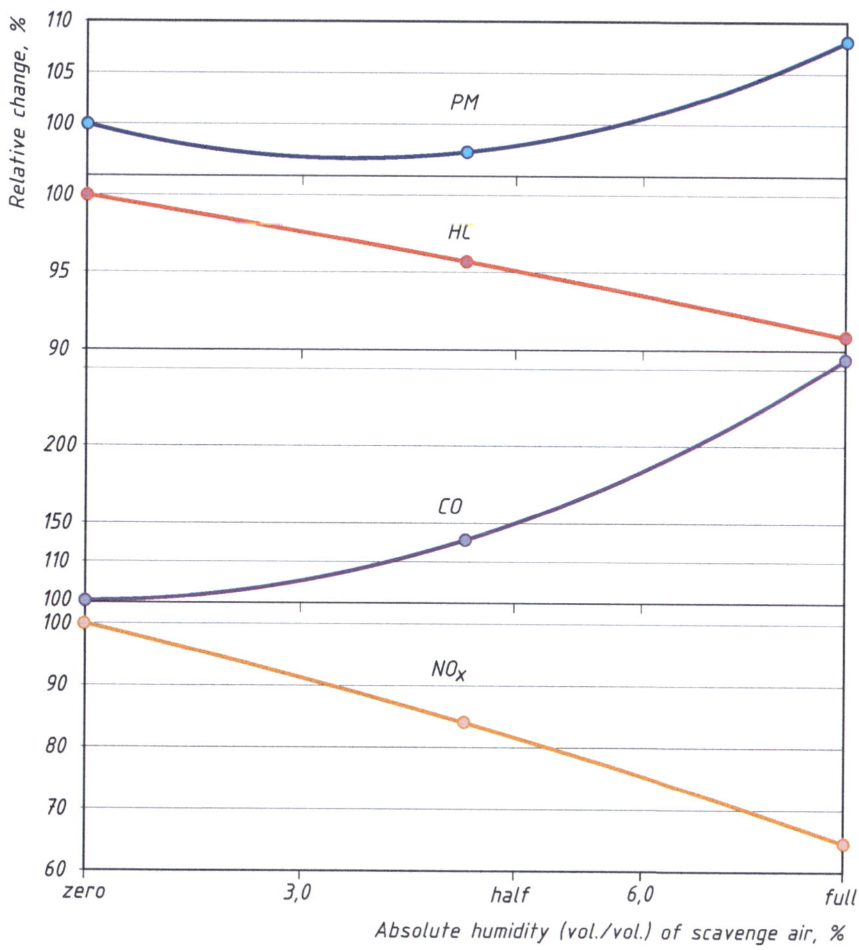

Fig. 2.13 Effect of saturation of scavenging air with water vapor on emissions of harmful substances with exhaust gases based on the results of tests of the 4T50ME-X laboratory engine equipped with the HAM scavenging air humidification system (adapted from [11])

2.2.2.2 Methods of Saturating Incoming Air with Water Aerosol

Marioff Oy, together with Wärtsilä, developed a technology for saturating the air entering the combustion zone with water vapor (Combustion Air Saturation System–CASS). Water aerosol was injected through nozzles directly into the airflow of the supercharged air immediately after the turbocharger (Fig. 2.14a).

In airflow with a temperature of over 473 K and a speed of 75 m/s, the bulk of the aerosol droplets, several micrometers in diameter, evaporate very quickly. The remaining droplets evaporated in the air cooler, which switched to heater mode. As a result, each kilogram of air entering the engine saturated with 60 g of water.

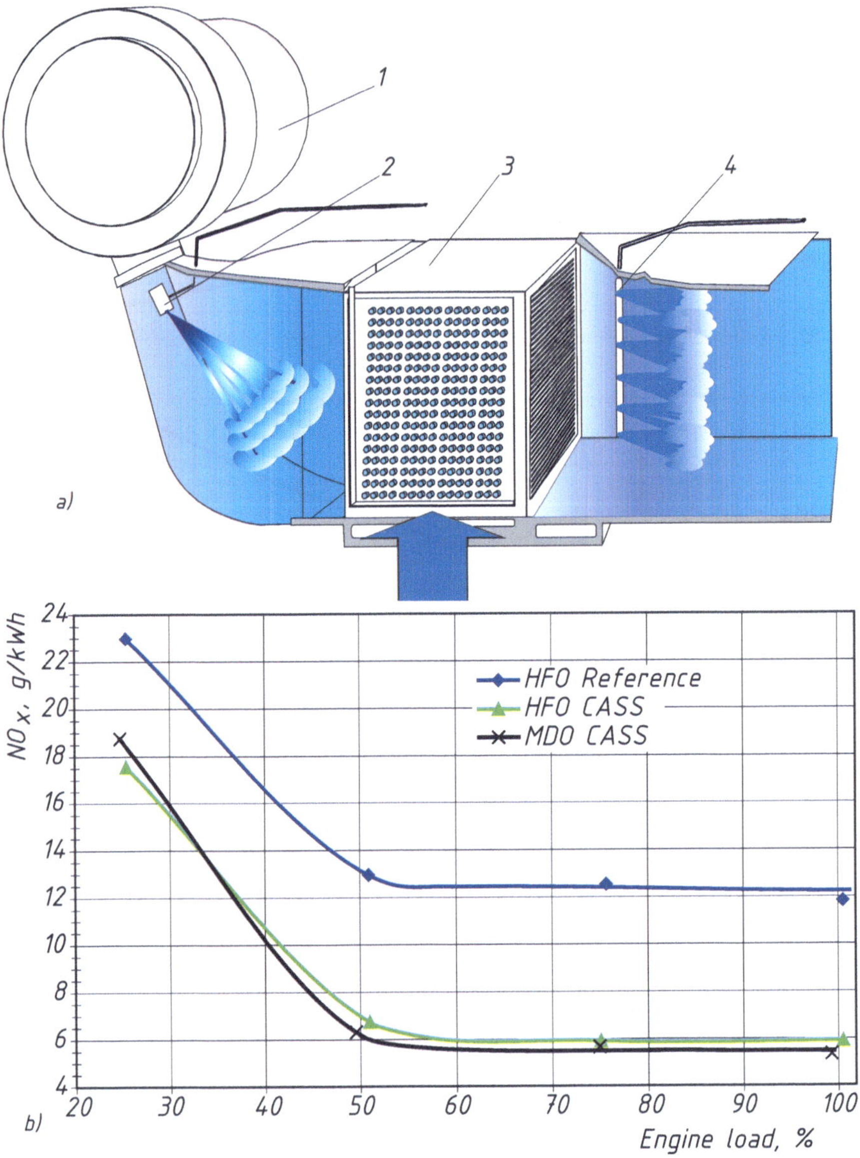

Fig. 2.14 Wärtsilä air humidification unit (**a**): (1) turbocharger; (2) water injection nozzle; (3) air cooler; (4) additional water injection. Effect of CASS on NO_x emissions of the W6L46CR engine, standard nozzle: 12 × 0.64 × 160°, fuel: HFO and MDO (**b**) (adapted from [12])

According to Wärtsilä reports, CASS technology allows reducing NO_x emissions with exhaust gases to 5.0–6.0 g/(kW × h) (Fig. 2.14b).

In case of insufficient humidity, the possibility of additional spraying of water aerosol after the air cooler considered.

In general, methods based on humidification of air entering the engine allow to obtain a reduction of NO_x emissions by 30–50% depending on the limitations of a specific technical solution, and water consumption is usually two to three times higher than fuel consumption. When using air humidification methods, a slight increase in fuel consumption is noted at a level of 1.5–3.0 g/(kW × h).

2.2.2.3 Direct Injection Methods of Water with Fuel

Direct injection of water into the engine's working space immediately before fuel supply or simultaneously with fuel has an additional effect on the charge. In the first case, water injected by specially installed injection valves, in the second case, water can supplied to the combustion chamber as part of a water-fuel emulsion or in the form of a stratified fuel-water plume.

Water-Fuel Emulsions Methods associated with the use of Water-Fuel Emulsions (WFE) are the most studied. According to numerous studies, the use of WFE allows to reduce nitrogen oxide emissions by 50–80%, reduce soot emissions by 35%, and increase the efficiency of a diesel engine by 3.0–5.0%.

An emulsion is a two-phase liquid system consisting of coarse dispersions of one liquid in another in the form of droplets with a diameter exceeding 0.1 μm. In an emulsion, droplets of one phase (dispersed or internal phase) encapsulated within the volume of another phase (continuous or external phase) [17].

Water-fuel emulsions contain water, hydrocarbon fuel, and emulsifiers. The ratio between individual components of the suspensions qualitatively and quantitatively affects the processes of their storage on board, preparation for combustion, supply to the combustion chamber, mixture formation, and combustion in the engine. Generally, there are two different types of emulsions: water-in-fuel and fuel-in-water. A system consisting of water droplets dispersed in a fuel phase is called a water-in-fuel emulsion (Fig. 2.15a), while a system consisting of fuel droplets dispersed in an aqueous phase is called a fuel-in-water emulsion (Fig. 2.15b). Multiphase emulsions consist of three phases: an internal phase and an external phase, separated by a dispersed phase (Fig. 2.15c) [17].

Water-in-fuel emulsions (Fig. 2.15a) are most suitable for diesel engines, due to the phenomenon of microexplosion of water droplets, which causes greater fragmentation of the fuel. These suspensions characterized by a smaller change in viscosity depending on the water content. The droplet size in the emulsion is in the range of 1.0–10.0 μm, although there are microemulsions in which the droplet size can be less than 0.2 μm.

A mixture of water and petroleum fuel is an unstable system, especially at rest. Droplets of homogeneous liquids tend to merge and then become larger. This leads

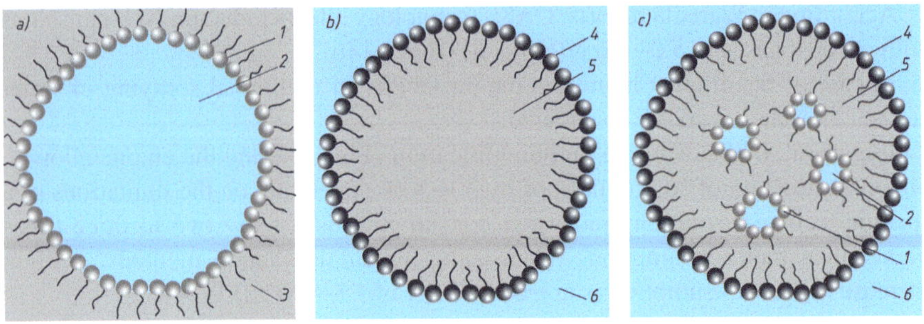

Fig. 2.15 Schematic representation of different types of emulsions: (**a**) "water in fuel"; (**b**) "fuel in water"; (**c**) multiphase "water-fuel-water": (1) hydrophobic emulsifier; (2) internal water phase; (3) fuel phase; (4) hydrophilic emulsifier; (5) external fuel phase; (6) external water phase (adapted from [13])

to gradual stratification of the emulsion due to an increase in surface tension at the phase boundary.

To prevent this process, a boundary layer created on the surface of the droplets by adding various surfactants–emulsifiers.

Emulsifiers that are not soluble in water but are readily soluble in oil called hydrophobic. Their presence in the fuel promotes the formation of an emulsion of the "water in fuel" type.

Hydrophilic emulsifiers, on the contrary, are readily soluble in water and insoluble in fuel. Their presence in the petroleum product promotes the formation of emulsions of the "fuel in water" type. In an emulsion of the "water in fuel" type, the surfactant suspends water droplets in the fuel, preventing its direct contact with the engine design elements. In addition to solving the problem of emulsion stabilization, additional additives can introduce into it to improve lubricating ability, inhibit corrosion and protect against freezing, and so on.

Emulsion stability characterized by kinetic and aggregate stability. Kinetic stability is the ability of dispersed droplets to remain suspended without settling under the action of Brownian motion. Kinetic stability increases with increasing emulsion dispersion. Aggregate stability understood as the ability of an emulsion to resist the merging of the dispersed phase (aggregation). Aggregate stability achieved by introducing emulsifiers. The main mechanism of the emulsion's influence on the diesel engine operating process is the peculiarities of heat and mass exchange of its droplets with the hot air environment in the engine combustion chamber. The combustion of a drop of emulsion accompanied by a micro-explosion, which occurs due to the difference in volatility of water and the base fuel (Fig. 2.16). Under the action of convective and radiant heat transfer, the droplet temperature quickly reaches the boiling point of water. This leads to the formation and rapid, explosive expansion of vapor bubbles inside the droplet, accompanied by the disintegration of the superheated emulsion droplet. The micro explosion causes secondary disintegration, during which a large number of much smaller droplets formed. These small droplets evaporate very quickly, and the additional impulse they receive because of the micro explosion improves the

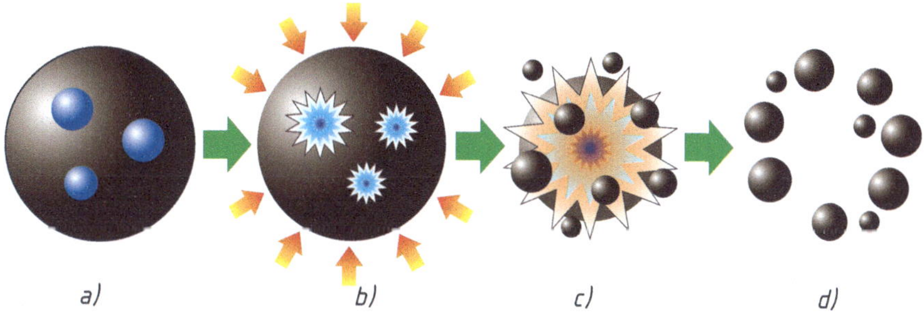

Fig. 2.16 Phenomenon of micro-explosion of a drop of water-fuel emulsion: (**a**) drop of emulsion with water inclusions; (**b**) action of high temperature leads to rapid boiling of water; (**c**) rapid evaporation of water and destruction of the drop of fuel; (**d**) fragmentation of the drop of emulsion into smaller fragments (adapted from [14])

mixing of fuel with air. Because of secondary disintegration, the droplet divided into smaller fragments and water vapor released into the combustion chamber space, reducing local temperatures, distributing heat more evenly throughout its volume. These features of the micro explosion help to increase combustion efficiency and, not only reduce NO_x emissions, but also suppress the formation of soot and unburned hydrocarbons. In combination, the action of the factors considered leads to an improvement in the environmental and economic indicators of the engine.

The cooling effect of water in the engine combustion chamber lasts for a very short time, coinciding with the achievement of the peak temperature of the cycle. However, this is enough to reduce the formation of NO_x approximately proportional to the proportion of water contained in the fuel (Fig. 2.17). In addition to reducing nitrogen oxide emissions, the presence of water in the combustion chamber allows for a significant reduction for particles in the exhaust gases. The effect of high temperatures in the combustion chamber leads to the decomposition of water molecules into hydrogen and oxygen, increasing their concentration in the charge. The participation of these additional components in the combustion process leads to its intensification and contributes to an increase in the completeness of combustion. As a result, the amount of incomplete combustion products, primarily soot, in the exhaust gases significantly reduced.

The main factor limiting the increase in the water content in the suspension is the performance of the high-pressure fuel pumps. Obviously, to obtain the same power, as when running on pure fuel, the cyclic supply of the suspension must increase proportionally to the water content in it. Of course, standard fuel pumps not designed for such supplies. For this reason, to increase the effect of using suspensions, it is necessary to either reduce the engine power or limit the water content in the suspension to 20–25%.

As can be seen in Fig. 2.17, during tests of a low-speed engine equipped with standard fuel pumps, the reserve of their performance in the main operating modes is sufficient to use emulsions with a water content of no more than 60% of the fuel volume. At full operating load, this figure decreases to 25%, which corresponds to a decrease in NO_x emissions by 21%.

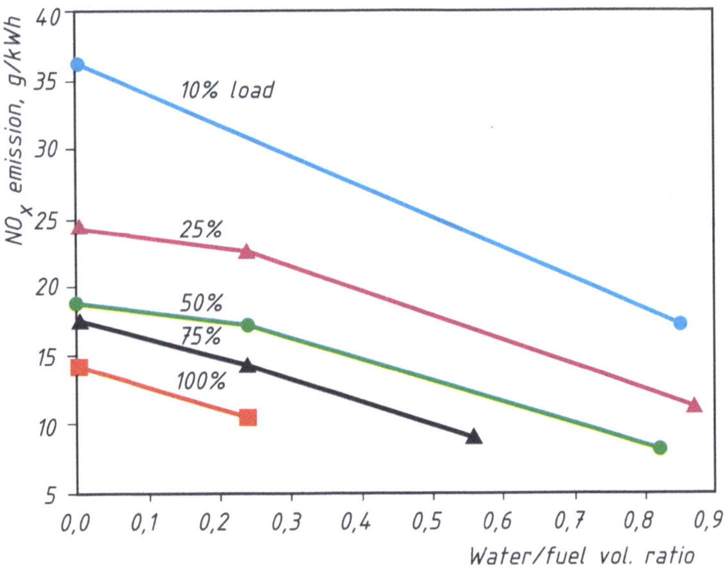

Fig. 2.17 Specific NO_x emissions depending on the water content in the fuel at various operating modes of the laboratory low-speed two-stroke engine 4T50ME-X (adapted from [14])

To achieve the maximum reduction in NO_x emissions at full load, it is necessary to redesign not only the fuel system, but also its drive mechanisms. The proportion of added water also limited by the viscosity of the emulsion and the degree of heating required to reduce its viscosity to the required level. Another aspect of the efficient use of water-fuel emulsions is that the design of the nozzle tip (orifice diameter, etc.) must be adapted to the increased amount of liquid injected. However, a spray nozzle optimized in this way will be extremely inefficient when running on pure fuel.

It is much easier to adapt engines with common rail injection systems or with hydraulically driven fuel pumps to use water-fuel emulsions. These engines have sufficient flexibility in choosing the fuel injection parameters, which is a great advantage when using water-fuel emulsions. The electronic control system can easily be adapted to various optimized injection control characteristics for both water-fuel emulsions and pure fuel. In addition, the engine settings can optimized according to the engine load and speed. As an effective measure aimed at reducing NO_x emissions in conditions of reduced commercial ship speeds, a system for preparing a water-fuel emulsion with different water content directly on board the ship can be considered. The amount of water will vary depending on the engine load.

Thus, the method of reducing NO_x emissions by using water-fuel emulsions still has enough unused reserves.

2.2.2.4 Methods of Direct Injection of Water into the Engine Working Space

Water can injected into the combustion chamber through a separate nozzle tip mounted in a single injector, through a separate nozzle mounted in the cylinder cover, or by stratified injection of water and fuel through nozzle tip of a single injection [11]

The Direct Water Injection (DWI) system, developed by Wärtsilä, shown in Fig. 2.18 [19].

Water and fuel supplied to the engine combustion chamber through a single injector, which has two nozzle tips. The volume of water supplied varies within the range of 15–70% of the cyclic fuel supply.

For water injection, an accumulator supply system with electromagnetic control of the hydraulic drive of the spray needle valve is used. Water supplied to the nozzle through a maximum supply limiter under a pressure of 21.0 MPa. Fuel supplied to the nozzle by a standard volumetric fuel pump.

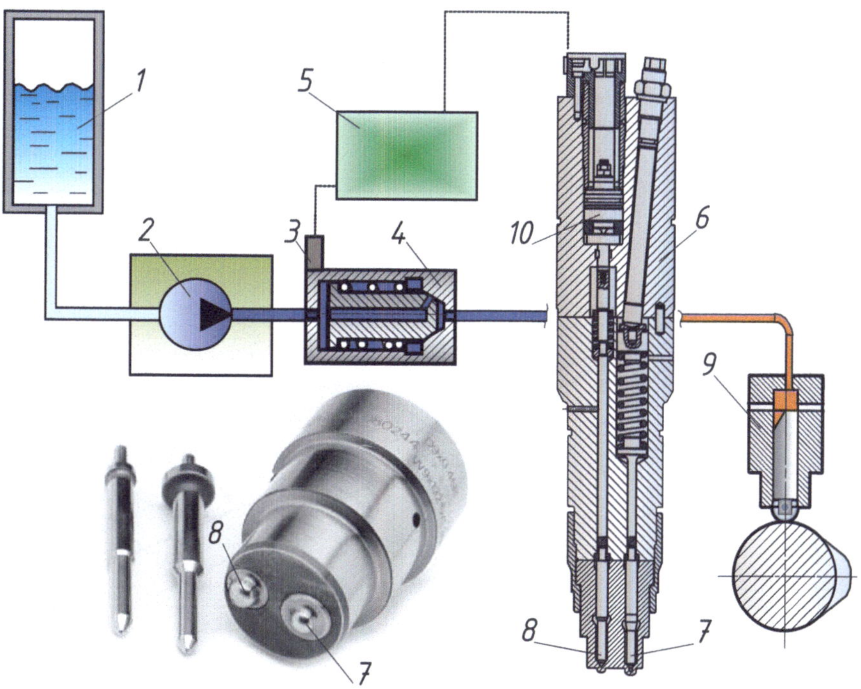

Fig. 2.18 Schematic diagram of the water injection system into the combustion chamber through an injector with a double nozzle tip: (1) water tank; (2) pressurized water supply pump; (3) maximum flow sensor; (4) maximum water flow limiter; (5) electronic control unit; (6) injector; (7) liquid fuel spray tip; (8) water spray tip; (9) high-pressure fuel pump used to supply liquid fuel; (10) electromagnetic-hydraulic drive of the water injection valve (adapted from [15])

The electronic control system allows you to adjust the water supply depending on the engine-operating mode within a wide range. The maximum supply limiter automatically turns off the water supply if the spray needle hangs. In this case, the valve actuation sensor transmits a signal to the electronic control unit and it turns off the water supply pump, which has an electric drive.

It should be noted that the two-nozzle injection valve is interchangeable with the standard nozzle of serial engines with a cylinder diameter of 20 cm (W20), which allows them to be modernized without significant costs, equipping only with a water supply system, while maintaining the standard fuel injection system unchanged.

To reduce the temperature in the combustion chamber accordingly, water injected shortly before the start of fuel injection. The start and time of water injection are of decisive importance for the efficiency of NO_x emission reduction. For this reason, water supply controlled by the electronic control unit.

Tests of the system have shown that its use makes it possible to reduce NO_x emissions by 50–60% without a noticeable reduction in power (Fig. 2.19).

From Fig. 2.19 it is evident that the direct water injection method allows obtaining the same reduction of NO_x emissions as the charge air humidification method at significantly lower water consumption (approximately 2 times). For the case under consideration, as for the case of air humidification, the reduction in harmful emissions is directly proportional to the amount of water supplied. For the 46 series engines (with a cylinder diameter of 46 cm), Wärtsilä developed a water injection system through an additional injection valve installed peripherally to the axis of the

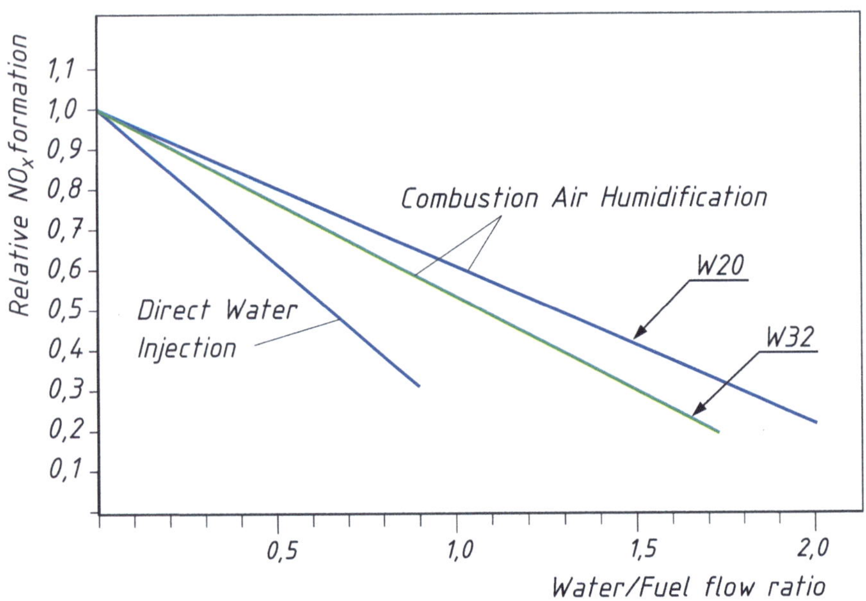

Fig. 2.19 Comparison of charge air humidification and direct water injection methods for Wärtsilä W20 and W32 engines (adapted from [16])

working cylinder (Fig. 2.20). Fuel injection, as in the case of small-sized engines, is carried out by a standard volumetric system through a injection valve installed along the cylinder axis. Thus, the water supply system completely separated from the fuel system. Disabling this system will not lead to the need to stop the engine.

When the NO_x emission reduction mode switched on, water injection begins 45° before TDC and ends immediately before the start of fuel injection, 10° before TDC. The injection parameters controlled by an electronic control system that responds flexibly to changes in the engine-operating mode. The nature of the working process of an engine with water injection, shown in Fig. 2.21. In 1999, the company equipped seven Ro-Ro vessels for the "Transfennica" company with separate water injection systems. The vessels with a deadweight of 7100 to 7250 tons were built at the German shipyard J.J. Sietas. They are equipped with 12- and 16-cylinder 46 series engines from Wärtsilä. Direct water injection provided an optimal combination of good efficiency and reduced harmful emissions with exhaust gases. This achieved with a one-time, relatively low investment and a minor increase in operating costs, without any reduction in reliability and, in fact, without any additional requirements for equipment placement.

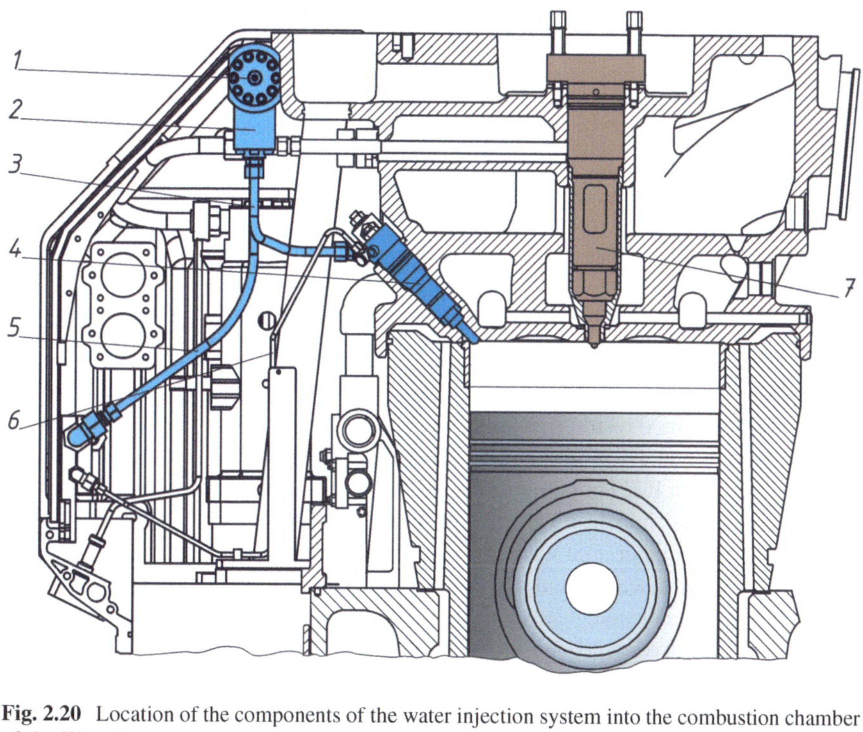

Fig. 2.20 Location of the components of the water injection system into the combustion chamber of the Wärtsilä 46 engine: (1) water accumulator; (2) maximum water supply limiter; (3) high-pressure water supply pipe to the injection; (4) water injection; (5) high-pressure water supply pipe to the accumulator; (6) injector control oil line; (7) fuel injector (adapted from [16])

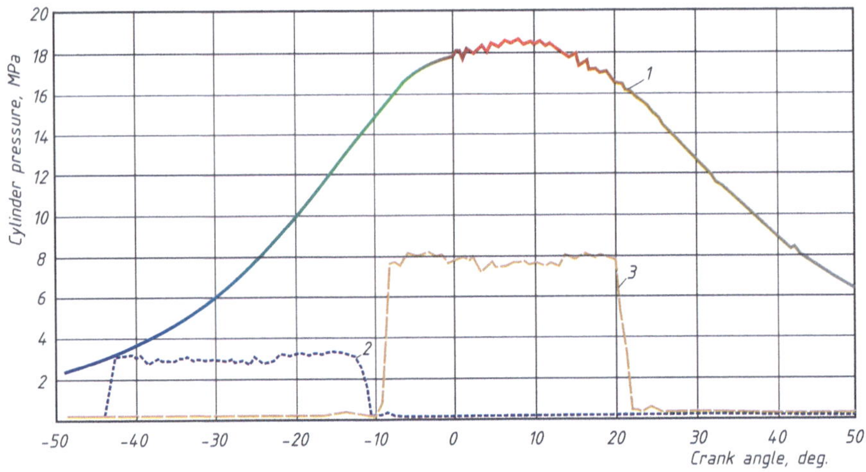

Fig. 2.21 Display of the working process in the Wärtsilä 46 series engine with separate water and fuel injection into the combustion chamber: (1) cylinder pressure; (2) water injector needle lift; (3) fuel injector needle lift (adapted from [12])

Long-term operation of vessels has shown that this technology has reduced NO_x emissions by 50–60%, while the loss of efficiency did not exceed 2.0–3.0 g/(kW × h).

Similar results obtained using direct injection in two-stroke low-speed engines of the RT-flex series, equipped with a common rail fuel system with electronic control. For water injection, the engine was additionally equipped with an independent water supply system, which also had electronic control. Tests have shown that direct injection is capable of reducing NO_x emissions by approximately 50% with a water-to-fuel ratio of 70% (Fig. 2.22).

In all the cases considered, the use of separate water injection systems allows, firstly, to inject large volumes of water without the need to reduce engine power, and secondly, to inject water and fuel at different stages of the cycle. Water can be supplied in parallel with fuel and/or during the compression stroke, which allows injection timing to be optimized taking into account fuel and water consumption. To achieve a programmed result, fuel and water injection systems can be equipped with various settings that allow the engine to operate with water injection on or off, ensuring maximum fuel economy or minimum NO_x emissions. In addition, installing a separate water injection valves allows the jets they create to direct directly to the combustion chamber area where the highest local temperatures formed.

The conducted research and the existing operating experience show that the reduction of NO_x emissions leads to a slight increase in the emission of other components, such as hydrocarbons and carbon monoxide. In some cases, a slight increase in fuel consumption is noted.

Unlike the use of water-fuel emulsions, direct injection technology has no restrictions on the amount of water supplied to the engine. However, this leads to the

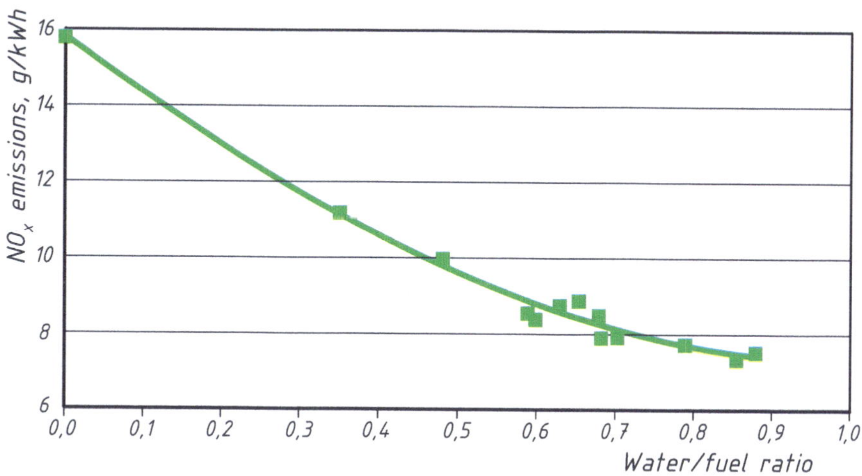

Fig. 2.22 NO_x emissions obtained on the Sulzer RT-flex research engine using direct water injection (adapted from [12])

problem of providing the power plant with fresh water, since with commensurate fuel and water consumption, the need for the latter will be significant.

2.2.2.5 Methods of Stratified Water and Fuel Injection

With this method, water and fuel enter the combustion chamber through one nozzle alternately, forming separate layers in the spray plume. An example of such technology is the Stratified Fuel-Water Injection (SFWI) system developed by Mitsubishi Heavy Industries and used on two-stroke low-speed engines of the UEC LSH-Eco series. The standard fuel injection system is equipped with a special valve device that allows part of the internal cavity of the high-pressure pipeline to be filled with water between injections (Fig. 2.23a).

During injection, fuel and water alternately displaced, resulting in distinct layers formed in the spray plume. The reduction in NO_x emissions achieved with this technology is directly proportional to the amount of water introduced (Fig. 2.23b), with a slight deterioration in the engine's fuel economy. The diagram showing the water supply to the high-speed engine's injection valve for stratified injection depicted in Fig. 2.24. The nozzle is equipped with a special spool valve device that allows part of the internal cavity adjacent to the injector needle to fill with water between injections.

When the needle valve is closed, the water supply channel in the nozzle body coincides with the channel in the needle valve guide. Water supplied under pressure slightly higher than the pressure in the pressure line displaces the fuel between injections, forming a water layer in the above-needle chamber. When fuel supplied, the check valve closes the water supply channel. When the needle valve lifted, the

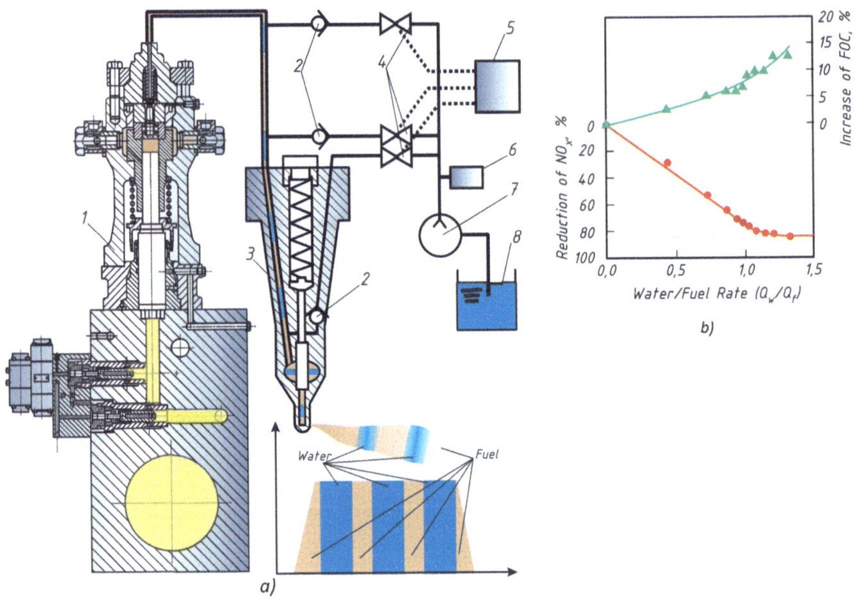

Fig. 2.23 Water injection system of the low-speed diesel engine UEC LSH-Eco from Mitsubishi (**a**) and its effect on NO_x emissions and fuel consumption (**b**) (1) high-pressure fuel pump; (2) check valves for supplying water to the injector and high-pressure line; (3) injector; (4) electrically controlled shut-off devices; (5) electronic control unit; (6) pressure accumulator; (7) booster pump; (8) water tank (adapted from [9])

water and fuel cavities of its guide, which functions as a spool, separated. The contents of the above-needle chamber forced out through the spray holes, forming clearly defined zones of fuel and water in the spray plume.

2.2.2.6 Comparison of Methods for Reducing Nitrogen Oxide Emissions Using Water

All the above methods of water supply to the engine working space lead to a significant reduction in NO_x emissions. The reduction is almost directly proportional to the amount of water introduced (Fig. 2.25a), but the amount required to achieve the same result will be different. The highest consumption observed for methods associated with humidification of the supercharged air, the lowest for methods associated with direct injection into the combustion chamber.

This can explain by the fact that when humidifying the air, only the temperature reduction mechanism associated with an increase in the heat capacity and thermal conductivity of the charge operates. With direct injection, this mechanism supplemented by a temperature reduction associated with the absorption of part of the heat for the phase transition of water. When injecting a water-fuel emulsion, a factor of improving mixture formation added due to the microexplosion effect. The latter

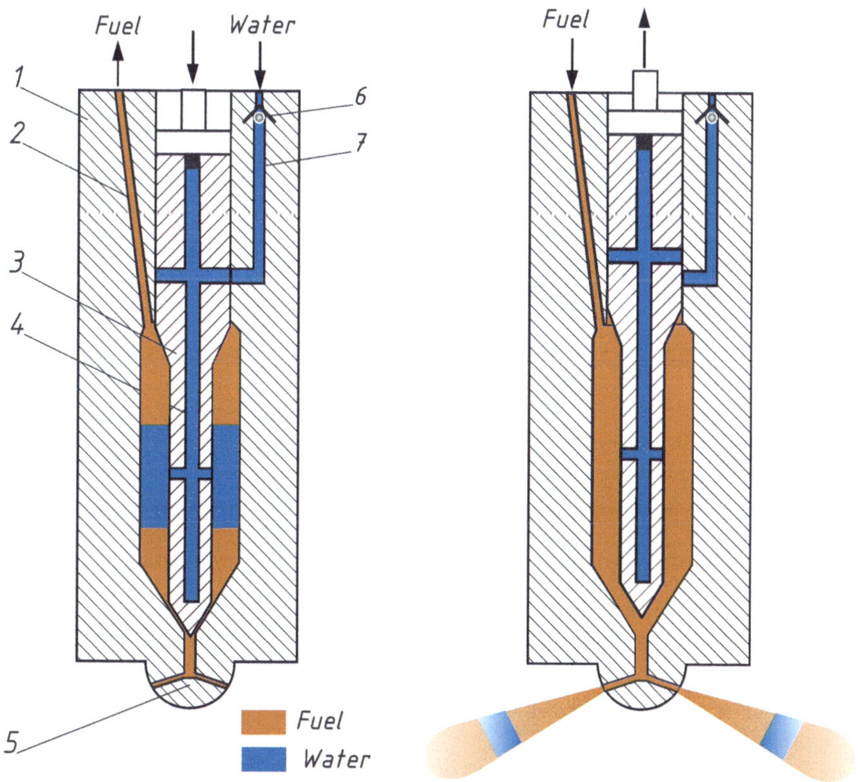

Fig. 2.24 Spray nozzle for stratified water and fuel injection of a high-speed diesel engine: (1) spray nozzle body; (2) fuel supply channel; (3) needle valve; (4) water supply channel to the above-needle chamber; (5) nozzle tip; (6) water line check valve; (7) water supply channel (adapted from [17])

circumstance leads to an increase in the quality of combustion, as a result of which not only NO_x emissions are reduced, but also soot (Fig. 2.25b), and the specific fuel consumption is reduced. In cases of direct water injection and air humidification, soot emissions, although slightly, increase, as does the engine's fuel consumption.

Various methods of using water to reduce NO_x emissions can successfully combine with each other, as well as with other known methods, both in-cylinder action and external exhaust gas treatment.

2.2.3 *Exhaust Gas Recirculation*

Exhaust Gas Recirculation (EGR) is another technology for influencing the engine operating process based on changing the composition and properties of the working fluid. The method based on adding a portion of the exhaust gases from previous operating cycles to the incoming air.

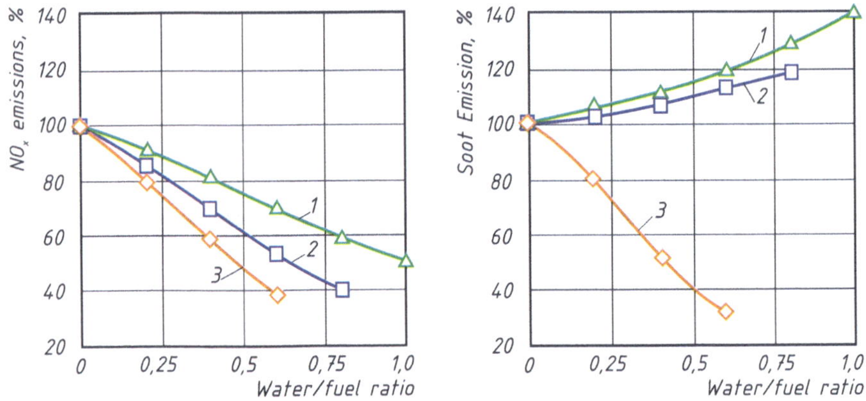

Fig. 2.25 Effect of different water supply methods on NO_x and soot emissions in a medium-speed engine: (1) charge air humidification; (2) direct water injection; (3) fuel-water emulsion (adapted from [18])

The impact of this method on NO_x emissions based on three main mechanisms:

- reducing the concentration of oxygen O_2 in the gas mixture used as the working fluid reduces the likelihood of its reactions with nitrogen N_2, during which nitrogen oxides, NO_x, are formed (see Sect. 2.1). Thus, recirculation of 15% of exhaust gases allows reducing the oxygen concentration in the supplied air from approximately 21 to 18%;
- reduced oxygen concentration slows down the combustion reactions of fuel components, which, in turn, does not contribute to the occurrence of local high-temperature zones in the combustion chamber;
- increasing the concentration of carbon dioxide CO_2 and water vapor H_2O, which are triatomic gases, increases the heat capacity and thermal conductivity of the gas mixture, which contributes to a more uniform temperature distribution throughout the combustion chamber.

The method is highly efficient and has proven itself well on small-sized engines, allowing to significantly reducing NO_x emissions without significantly complicating the power plant.

Conventionally, all known methods of exhaust gas recirculation can divide into two groups—internal and external.

With internal recirculation, a mixture of fresh air and exhaust gases is prepared directly in the engine's working cylinder, with external recirculation, outside the working cylinder, and is already in a finished form supplied to the engine.

2.2.3.1 Internal Recirculation Methods

The simplest way to implement internal recirculation is in two-stroke low-speed engines with uniflow scavenging. To do this, it is sufficient to reduce the amount of scavenging air entering the cylinder. As a result, some of the exhaust gases from the previous cycle will not removed from the cylinder during the time allocated for scavenging, and, mixing with the fresh charge, will form a gas-air mixture with a reduced oxygen concentration (Fig. 2.26).

This method of recirculation developed by Wärtsilä and called Water-Cooled Residual Gas (WaCoReG). Mixing combustion products with air directly in the working cylinder leads to a significant increase in the charge temperature. For this reason, cooling of exhaust gases directly in the cylinder by water injection provided. Special injection valves installed in the cylinder cover injected water at the end of the purging process–the beginning of the compression process into the upper part of the cylinder, where, in the main, the exhaust gases that not removed during the purging process were, located (Fig. 2.26). Evaporating, water, due to the heat of phase transition, cools the gases and additionally saturates the charge with water vapor, which, as has been shown earlier, has a positive effect on reducing NO_x emissions. As additional advantages of the method under consideration, the possibility of using a turbocharger with a lower capacity on the engine noted. This, in turn, would make it possible to reduce the height of the purge ports, which would make it possible to increase the piston stroke on the expansion stroke, increasing the efficiency of the working process.

It assumed that internal recirculation could use together with direct water injection to increase the efficiency of reducing NO_x emissions. It was expected that this combined technology would reduce NO_x emissions to 5 g/(kW × h), i.e. by approximately 70%. However, this technology did not go beyond laboratory tests. The risks associated with the possibility of water getting on the cylinder walls, which could potentially increase their wear and wear of the piston rings, turned out to be unreasonably high compared to the expected benefits. The implementation of other benefits associated with a decrease in the height of the purge ports and a decrease in the performance of the turbocharger required significant redesign of the engine. In addition, such a redesign will lead to a significant decrease in engine efficiency when the internal recirculation system is disabled. Four-stroke marine engines sometimes use internal recirculation based on delayed closing of the exhaust valve. An example of the implementation of this method in Yanmar 6AY high-speed engines is shown in Fig. 2.27. At the end of the exhaust stroke, the exhaust valve does not close completely, leaving a small gap through which, at the beginning of the intake stroke, some of the exhaust gases sucked back into the cylinder from the exhaust receiver. This method is simple to implement, but the use of uncooled gases greatly limits the amount that can returned to the cylinder without significantly deteriorating the operating process parameters. Because of these limitations, this method allows achieving NO_x emissions only in accordance with Tier II.

Internal recirculation methods are relatively simple, but the risks associated with their use today are significant, which prevents their widespread use in marine engines.

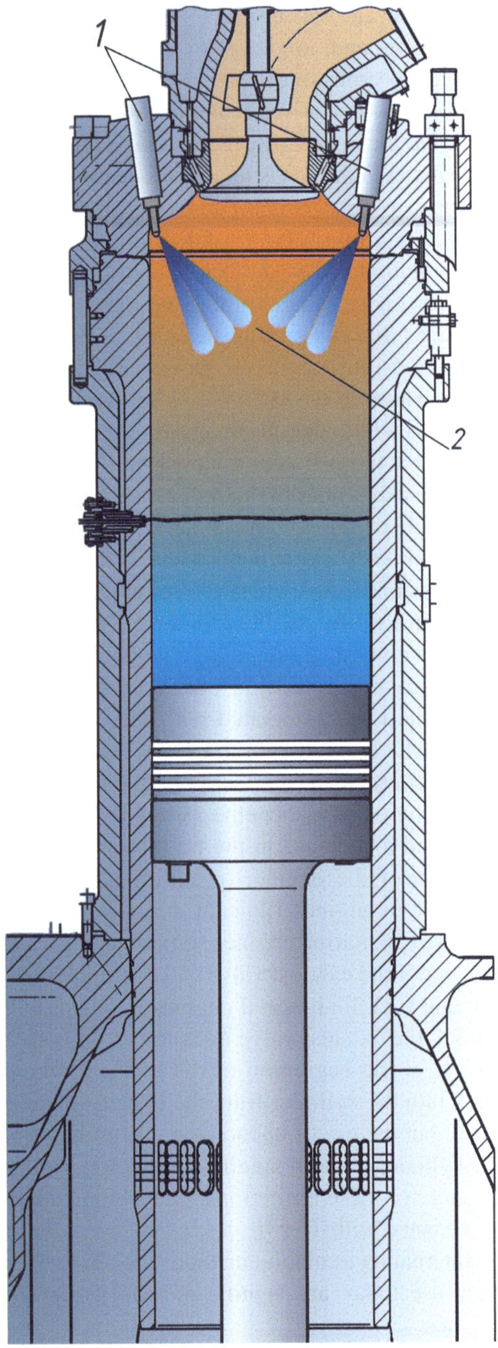

Fig. 2.26 Water-Cooled Residual Gas internal exhaust gas recirculation systems for two-stroke engines: (1) water injection valve; (2) residual exhaust gases (adapted from [19])

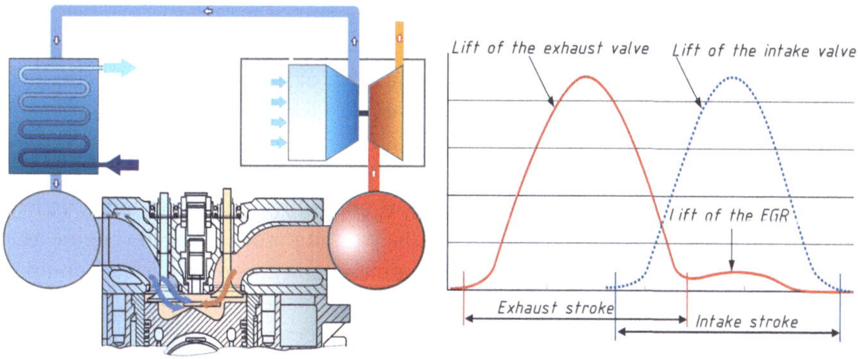

Fig. 2.27 Internal EGR system of Yanmar 6AYM four-stroke engines and valve timing graphs for internal recirculation (adapted from [20])

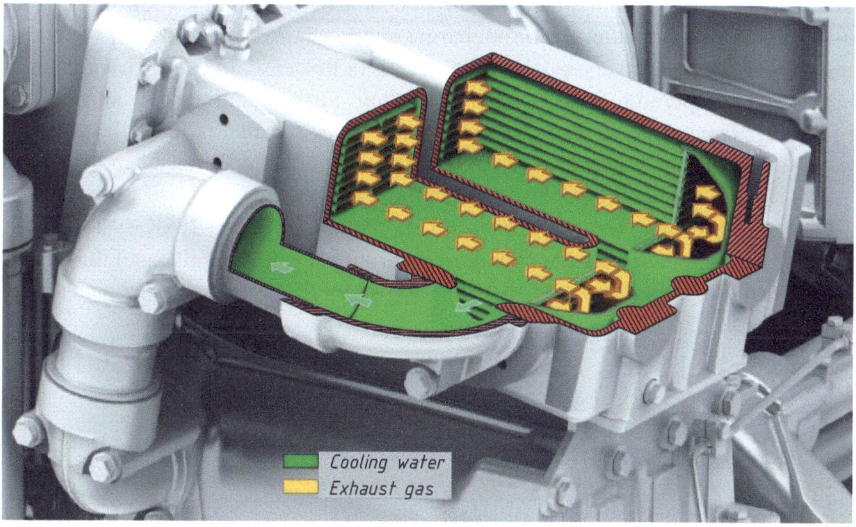

Fig. 2.28 External EGR exhaust gas cooling system of four-stroke 16 V 4000 R43 engines from MTU (adapted from [21])

2.2.3.2 External Recirculation Methods

In high-speed marine engines, which mainly operate on light fuels with low sulfur content, a simple internal recirculation scheme without exhaust gas cleaning is used. Exhaust gases discharged from the exhaust manifold with a temperature of about 650 °C enter the heat exchanger, where their temperature reduced to about 120 °C. The heat exchanger usually integrated into the engine design and uses coolant from the engine cooling system as a cooling medium (Fig. 2.28). To regulate the flow of gases through the cooler, a special damper with an automatic drive installed in the supply channel.

For other types of marine engines, a significant part of the risks associated with exhaust gas recirculation is associated with the fact that they operate mainly on heavy grades of fuel containing a significant number of impurities. For the effective implementation of this method, it is necessary to take into account a number of conditions that can only realized with external recirculation. First, the exhaust gases must cool significantly to avoid an increase in the charge temperature at the inlet. Otherwise, this will lead to a decrease in the weight filling of the working cylinders and will increase the heat intensity of both the working process and the engine as a whole. In addition, the exhaust gases of marine engines contain harmful components such as sulfur compounds SO_x and particulate matter, from which they must, cleaned. Otherwise, this will lead to rapid wear of the engine.

2.2.3.3 Low-Pressure Recirculation Systems

The simplest way to implement the method of external recirculation of exhaust gases is their selection after the turbocharger and in some cases even after the economizer of the heat recovery system (Fig. 2.29). In this case, the gases have a sufficiently low temperature, which simplifies their further cooling and subsequent processing. The relatively Low-pressure of the gases selected reduces the requirements for additional equipment necessary for the implementation of the method. Systems with gas selection after the turbocharger called low-pressure systems.

Figure 2.29 shows the basic diagram of the low-pressure EGR system. The exhaust gases selected after the economizer through a shut-off valve and then go to cooling and cleaning. The shut-off valve allows the EGR system to connected or disconnected.

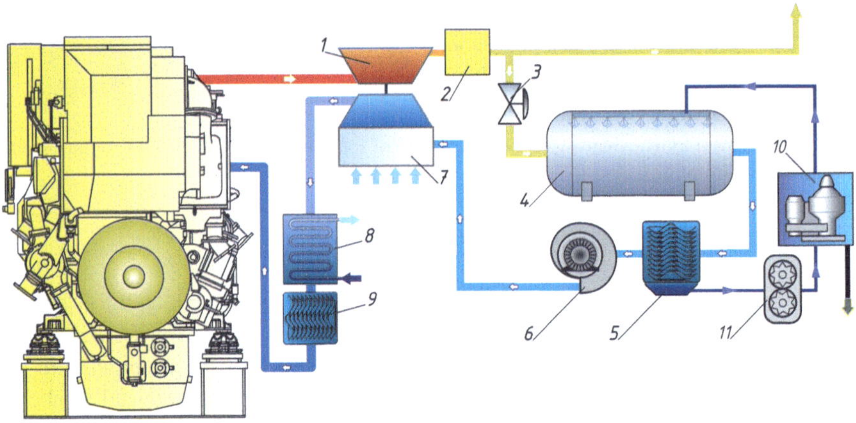

Fig. 2.29 Low-pressure EGR system: (1) turbocharger; (2) economizer; (3) shut-off valve; (4) scrubber; (5) moisture separator; (6) electric EGR supercharger; (7) mixer; (8) air cooler; (9) moisture separator; (10) water treatment module; (11) water supply pump (adapted from [22])

Scrubbers mainly used as a device for cleaning and cooling gases in ship systems, in which the gas flow interacts with a flow of sprayed water. As a result, water effectively removes heat from the gases, partly through the heat transfer mechanism, partly due to its absorption during the phase transition. In addition, water dissolves sulfur compounds SO_x and nitrogen NO_x, and absorbs particulate matter.

The gases purified in this way additionally saturated with water vapor, after which excess moisture removed in a special separator (moisture separator).

Then, using an electric supercharger, the gases fed to the turbocharger inlet, where they mixed with the flow of sucked-in air. The amount of gases fed to the engine inlet regulated by the position of the shut-off valve and/or the rotation speed of the electric supercharger. After the scrubber, the water fed to the processing module by a special pump. Here it purified, and the acids formed in it neutralized. After that, the water returns to the scrubber.

Low-pressure recirculation systems are quite simple and reliable in operation. They allow maximum use of the energy of exhaust gases for engine supercharging, and the passage of recirculated gases through the compressor promotes their effective mixing with the incoming air. However, if the efficiency of gas purification decreases, this can lead to accelerated clogging and wear of the compressor wheel.

The most significant disadvantage of these systems is the significant dimensions of the additional equipment, since large volumes of gases must processed.

As an example of such an installation, Fig. 2.30 shows a general view of a WinGD X-DF 2.0 series gas-diesel engine with a gas recirculation installation called Intelligent Control by Exhaust Recycling–iCER. This installation allows to significantly improving the environmental performance of an engine running on gas fuel.

An additional idea of the volumes of gases that pass through installations of this type is given by the general view of the shut-off valve through which gases are discharged (Fig. 2.31).

2.2.3.4 High-Pressure Recirculation Systems

Another circuit solution is recirculation systems, in which gas extracted before the turbocharger, and after processing, the gases enter the engine after the compressor. Such installations called high-pressure installations. They are more compact, but at the same time more complex both structurally and in terms of operation. Since the elements of such installations operate under significantly higher pressures and temperatures, the requirements for their manufacture and the materials used are much higher than for low-pressure installations.

A significant advantage of high-pressure systems is that gases that can potentially contain particulate matter and other contaminants do not pass through the compressor, but directly enter the engine (Fig. 2.32). This has a positive effect on the reliability of the turbocharger, allowing it to operate at rated capacity for a long time.

The basic diagram of the high-pressure system shown in Fig. 2.32.

Fig. 2.30 Low-pressure unit implemented using Intelligent Control by Exhaust Recycling (iCER) technology: (1) exhaust gas discharge channel after turbocharger; (2) shut-off valve; (3) water injection point for pre-cooling; (4) scrubber; (5) circulation tank; (6) moisture separator; (7) shut-off valve (Fig. 2.31) (adapted from [23])

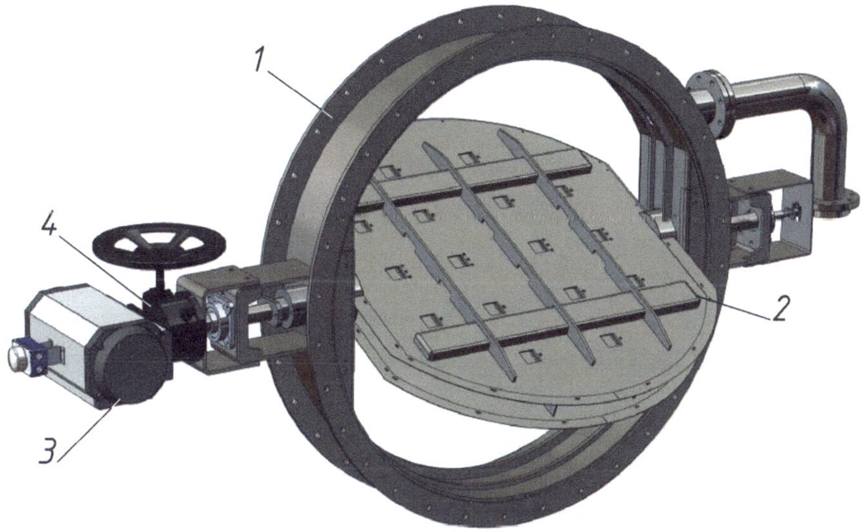

Fig. 2.31 Exhaust gas shut-off valve: (1) valve body; (2) damper; (3) electric drive mechanism; (4) manual drive mechanism (adapted from [24])

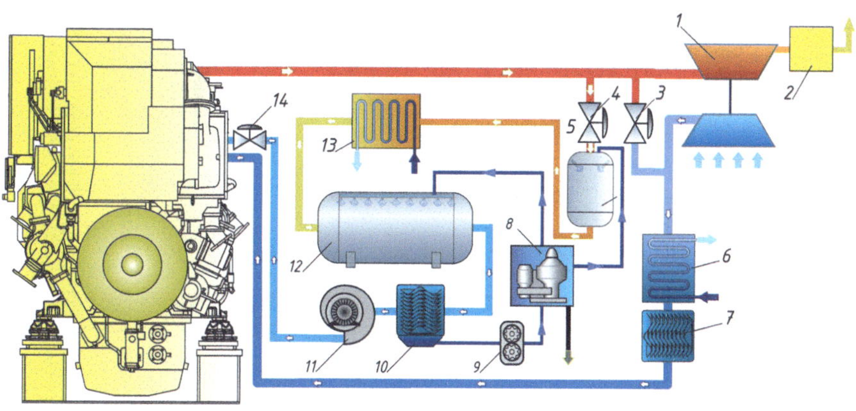

Fig. 2.32 High-pressure EGR system: (1) turbocharger; (2) economizer; (3) turbocharger balancing valve; (4, 14) shut-off valve; (5) EGR pre-scrubber; (6) air cooler; (7) moisture separator; (8) water treatment module; (9) water supply pump; (10) EGR moisture separator; (11) EGR electric supercharger; (12) scrubber; (13) EGR cooler (adapted from [22])

Considering the higher gas temperatures, a pre-scrubber and/or flow-through gas cooler usually installed before the main scrubber. The gas extraction before the turbine, which can change depending on the operating mode, requires the ability to balance the inflator, for which the system provides a bypass with a balancing valve. Higher pressures require a more powerful electric supercharger capable of creating pressure exceeding the boost pressure.

Fig. 2.33 Location of EGR system components on the 6S80ME-C9.2 engine (1) disconnectable turbocharger; (2) gas outlet to the EGR system; (3) water pre-spray; (4) EGR cooler; (5) EGR electric supercharger; (6) moisture separator; (7) EGR outlet pipe (8) mixing chamber (adapted from [25])

The general arrangement of the main elements of the high-pressure EGR system on the low-speed two-stroke engine 6S80ME-C9.2 shown in Fig. 2.33 (highlighted in orange).

The main area of application of EGR technologies in marine power engineering are installations with low-speed two-stroke engines. At present, it believed that on four-stroke engines, the disadvantages of recirculation systems offset the benefits of their use.

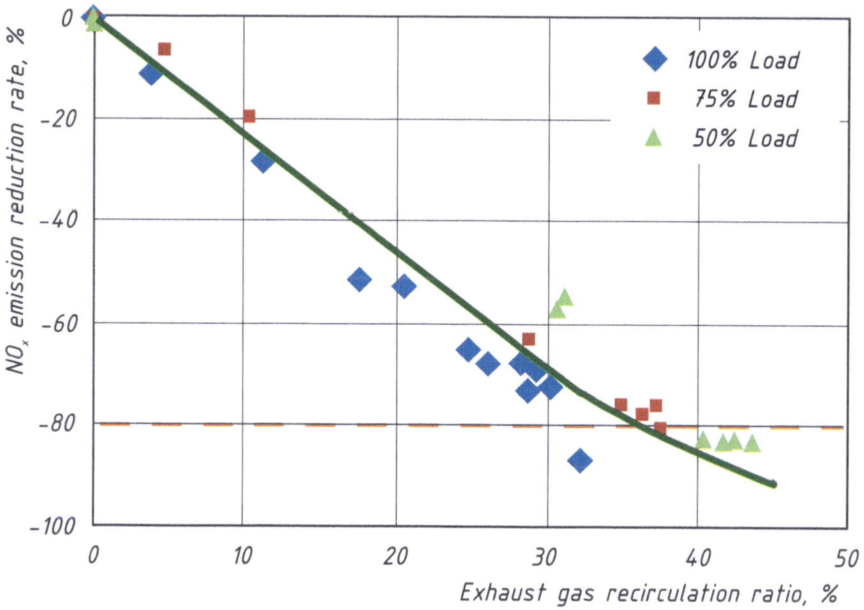

Fig. 2.34 Effect of exhaust gas recirculation ratio on NO_x emissions for a two-stroke low-speed engine 6UEC45LSE-Eco-B2 (adapted from [22])

This explained by the fact that low-pressure systems unreasonably increase the dimensions of the installation, while the main advantage of installations with four-stroke engines is their compactness. High-pressure installations unreasonably complicate the installation and its operation.

Given the availability of other alternative solutions for reducing NO_x emissions, the use of EGR systems on four-stroke marine engines considered insufficiently justified, although some manufacturers continue to search in this direction.

On two-stroke diesel engines, EGR systems used to achieve NO_x emissions up to Tier III requirements when the vessel operates in emission control areas.

Exhaust gas recirculation technology allows for an effective reduction in the content of nitrogen oxides, NO_x, in exhaust gases by up to 80% (Fig. 2.34). At the same time, their concentration decreases almost in direct proportion to the increase in the recirculation coefficient, which in some cases can reach 0.5.

The removal of a significant portion of exhaust gases before the turbocharger leads to a decrease in its performance and a reduction for air entering the engine, which also leads to a decrease in the oxygen concentration in the charge. Consequently, a decrease in the combustion rate associated with a decrease in the oxygen concentration in the combustion chamber leads to a decrease in the fuel efficiency of the engine by 0.5–3.0% depending on its operating mode (Fig. 2.35).

An example of such a scheme used on MAN Diesel&Turbo engines with a cylinder diameter of up to 70 cm, equipped with one turbocharger, shown in Fig. 2.36.

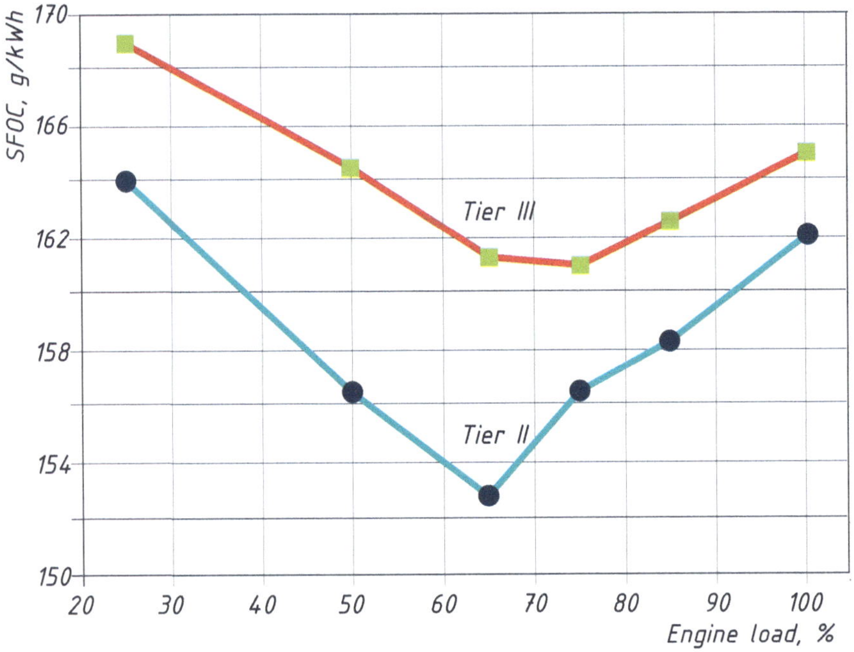

Fig. 2.35 Specific Fuel Oil Consumption (SFOC) of the low-speed 6G70ME-C9.5 engine with EGR system when operating in Tier II and Tier III modes (adapted from [26])

The scavenging air can enter the scavenging receiver through two lines, the main line and the EGR system line.

The main line designed for engine operation outside Tier III zones, when the EGR system line is disconnected. All air supplied by the turbocharger, passing through the air cooler and moisture separator, enters the scavenging receiver and then into the working cylinders.

The EGR line, capable of removing up to 40% of the exhaust gases from the outlet receiver, which pass through the scrubber, cooler and moisture separator to the EGR centrifugal supercharger, which feeds them into the scavenging receiver. Accordingly, the EGR line connected in parallel to the main line when the vessel enters the emission control area by opening the shut-off valves installed at the entry and exit points.

Since when the EGR line turned on, the extraction of gases leads to a decrease in the power of the turbocharger turbine, for its balancing the main line is equipped with bypasses of charge air and exhaust gases, on which balancing valves installed.

When operating outside Tier III zones, the bypass valve of charge air remains closed, and the bypass valve of exhaust gases is fully open at high loads and partially open at low loads.

When operating in the Tier III zone, the shut-off valves at the ends of the EGR line opened, connecting it in parallel with the main line. After cleaning and cooling,

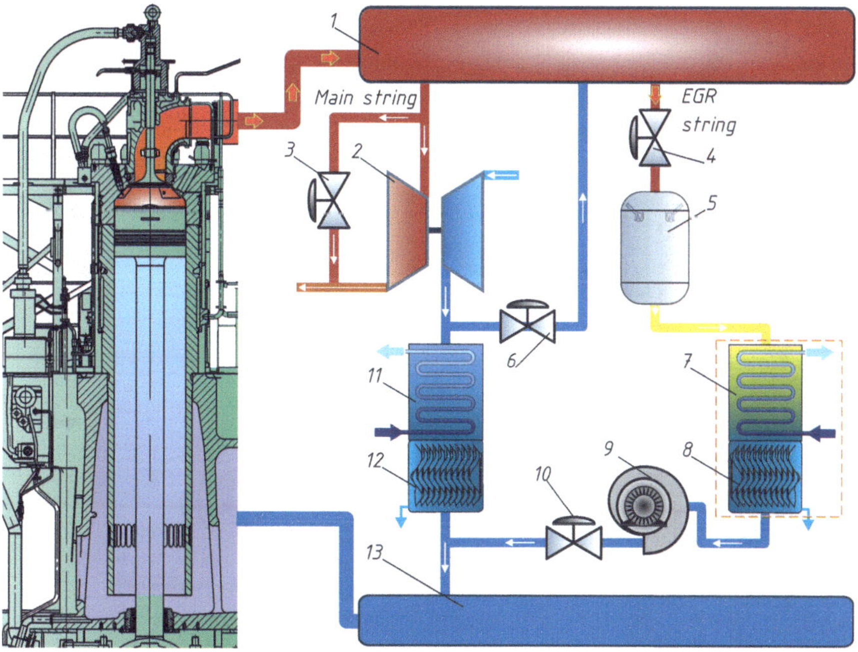

Fig. 2.36 EGR system for two-stroke engines equipped with one turbocharger: (1) outlet receiver; (2) turbocharger; (3) turbine bypass valve; (4) shut-off valve; (5) scrubber; (6) compressor bypass valve; (7) exhaust gas cooler; (8) moisture separator; (9) electric supercharger, (10) shut-off valve; (11) air cooler; (12) moisture separator; (13) inlet receiver (adapted from [26])

a centrifugal supercharger into the purge receiver pumps the gases, where they mixed with air supplied through the main line. The amount of gases supplied regulated by changing the flow rate of the EGR supercharger, the electric motor of which is connected via a frequency converter. In this mode, the compressor bypass valve is open and the exhaust gas bypass valve completely closed, which allows all remaining gases after extraction to direct to the turbine to increase the pressure of the scavenging air.

For engines with a working cylinder diameter of more than 70 cm, the diagram shown in Fig. 2.37 is used.

The diagram provides for three lines: the main line, the additional line, and the EGR line.

The main line includes a large turbocharger (hereinafter referred to as the main line), which supplies up to 70% of the scavenging air passing through the cooler and moisture separator to the scavenging receiver.

The additional line supplies up to 40% of the scavenging air through a smaller turbocharger (hereinafter referred to as the additional line). Before entering the scavenge receiver, the air passes through the EGR unit, consisting of a flow-through cooler and a moisture separator, which is built into the engine, similar to the scavenge air cooler (Fig. 2.32).

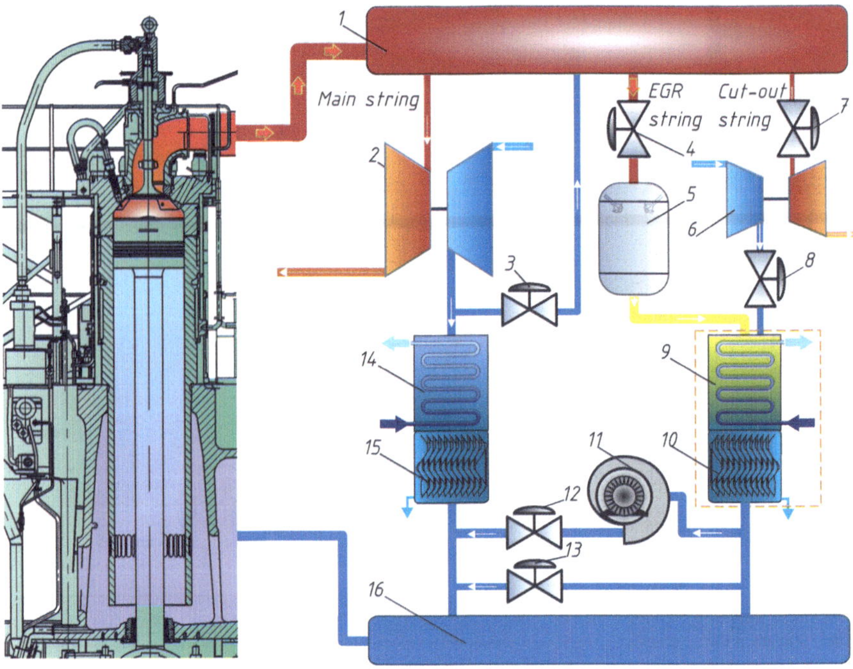

Fig. 2.37 EGR systems for two-stroke engines equipped with two or more turbochargers: (1) outlet receiver; (2) main turbocharger; (3) compressor bypass valve; (4) shut-off valve; (5) scrubber; (6) disconnectable turbocharger; (7, 8, 12) shut-off valve; (8) moisture separator; (9) exhaust gas cooler; (10) moisture separator; (11) electric EGR supercharger, (13) EGR supercharger bypass valve; (14) air cooler; (15) moisture separator; (16) inlet receiver (adapted from [26])

The EGR line directs up to 40% of the exhaust gases through the scrubber and the EGR unit. Then, the cleaned and cooled gases are fed by one or more superchargers to the mixing point in the main line before entering the scavenge air receiver. In this case, the shut-off valves turn off the additional compressor.

On large engines with a large number of cylinders, several main and additional turbochargers, as well as several EGR units, can install.

In the basic operating mode corresponding to Tier II, the main and auxiliary lines are in operation, and the EGR line closed using shut-off valves. In this case, the air supplied by the auxiliary turbocharger passes through the cooler and the moisture separator of the EGR unit, which perform the same functions as similar devices of the main line. The main line provides about 60% of the purge air, the remaining 40% falls on the auxiliary line. In this mode, the bypass of the main compressor remains closed.

At low load modes, this scheme allows you to turn off the auxiliary line, directing all exhaust gases through the main turbocharger. This makes it possible to increase the boost pressure and improve the fuel economy of the engine.

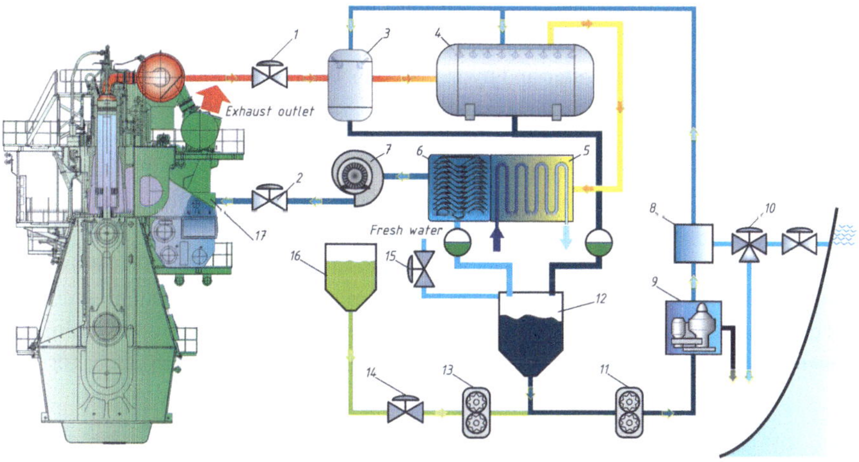

Fig. 2.38 Shematic diagram of the EGR system of a two-stroke low-speed engine with auxiliary subsystems: (1, 2) shut-off valve; (3) pre-scrubber; (4) scrubber; (5) cooler; (6) moisture separator; (7) supercharger; (8) filter; (9) water purification system; (10) discharge control valve; (11) scrubber pump; (12) buffer tank; (13) NaOH pump; (14) shut-off valve; (15) on/off valve; (16) NaOH tank; (17) mixing chamber (adapted from [27])

When switching to Tier III mode, the auxiliary line turned off, and the EGR line connected in parallel with the main line. The exhaust gases pass successively through the scrubber, cooler and moisture separator, after which they fed to the mixing point with the scavenging air by the EGR supercharger. The resulting mixture is fed to the scavenging receiver and then to the working cylinders. The composition of the mixture regulated changing the flow rate of the EGR supercharger. In this mode, the main compressor bypass valve partially opened to increase the scavenging air pressure.

From the above, it follows that modern EGR units are highly efficient, but quite complex and energy-intensive. They include a large number of servicing subsystems (Fig. 2.38), which require not only energy but also consumables for their operation. In addition to reducing the fuel efficiency of the engine, when assessing their operation, it is necessary to take into account the energy costs of additional equipment that ensures the operation of the EGR system. Only for the operation of the electric supercharger and water purification module, at the rated engine power, the energy consumption can reach 10 kW/MW. The water purification module, in addition to electricity, uses an aqueous solution of alkali (NaOH), the hourly consumption of which, at a 50% concentration and a sulfur content in the fuel of 0.1%, can reach 0.17 dm^3/MW.

All of the above, taking into account the more complex operation of the vessel's power plant, currently makes the use of the EGR system unprofitable outside the emission control areas where restrictions apply according to Tier II. Therefore, all implemented circuit solutions provide only for periodic commissioning of the system for a limited period.

It should noted that EGR systems not only have a positive effect on NO_x emissions, but also allow for a significant reduction in the emission of harmful components such as particulate matter and hydrocarbons. Figure 2.39 shows the test results of the laboratory two-stroke low-speed engine 4T50ME-X obtained by MAN B&W within the framework of research programs aimed at finding new technologies for controlling the emission of harmful substances with exhaust gases. The results are given for the mode corresponding to the engine load of 75% of the nominal. The engine is equipped with a high-pressure EGR system.

The reduction in particulate emissions was about 15% with up to 20% of exhaust gas recirculation. This is partly because the gases selected for recirculation cooled and cleaned in a scrubber before returning to the engine. The reduction in hydrocarbon emissions reached about 10%, while the content of another component formed

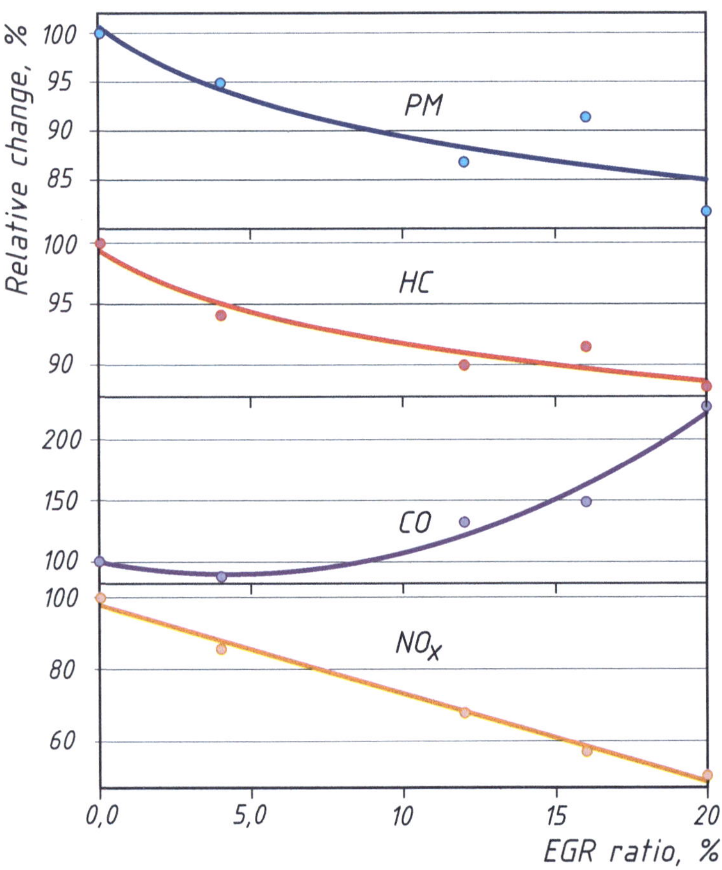

Fig. 2.39 Effect of the degree of exhaust gas recirculation on emissions of harmful substances with exhaust gases based on the results of tests of the 4T50ME-X laboratory engine equipped with the EGR system (adapted from [11])

because of incomplete combustion–carbon monoxide–increased more than twofold.

Such a generally contradictory effect can be the result of a complex interaction of various factors in the combustion chamber. The supply of exhaust gases to the engine leads to an increase in the charge temperature in general, especially in the final stages of combustion. This creates favorable conditions for the afterburning of hydrocarbon compounds that have not yet had time to react. The increase in the CO level is primarily due to a decrease in the oxygen concentration and a decrease in peak combustion temperatures. In addition, the afterburning of hydrocarbons in the final combustion phase, due to time limitations, may not be complete, with the formation of new products of incomplete combustion, which is carbon monoxide. Thus, the concentration of CO increases, due to a decrease in the concentration of hydrocarbons.

2.2.4 Selective Catalytic Reduction (SCR). Types of SCR Units

Selective Catalytic Reduction (SCR) is a method of off-cylinder control of nitrogen oxide emissions into the atmosphere. This technology involves processing gases generated in a marine engine in order to reduce NO_x to a level that meets Tier III requirements.

Gas processing performed in a special device, an SCR reactor, installed in the exhaust tract of the engine. Nitrogen oxides, interacting with the catalyst, reduced to nitrogen and water.

Rare earth metals such as platinum (Pt), palladium (Pd), rhodium (Rh), vanadium (V), as well as alloys and oxides based on them used as catalysts in SCR reactors.

Catalytic converters divided into two main types, passive selective catalytic converters and active catalytic converters. Passive catalysts are self-sufficient and do not require any additional reducing agents to reduce NO_x. Active catalysts require a reducing agent, which is ammonia or urea. Passive catalysts are quite expensive and very sensitive to fuel quality, so they are not widely used in marine installations.

SCR reactor Figure 2.40 shows the general structure of an active SCR reactor used in various types of marine engines.

The main element of the reactor is the catalyst blocks located in its body, which are a ceramic matrix (usually based on TiO_2) with a large number of channels on the surface of which metals used as a catalyst (usually V_2O_5) are dispersed. This provides a large surface area on which the catalytic process occurs. The thickness of the catalyst layer usually does not exceed 100 μm. At the entrance to the catalyst body, there is a mixing chamber, into which the reducing agent fed and sprayed.

Anhydrous ammonia (NH_3), an aqueous solution of ammonia (25% NH_3) or an aqueous solution of urea (urea–$(NH_2)_2CO$) are used as a reducing agent in the SCR

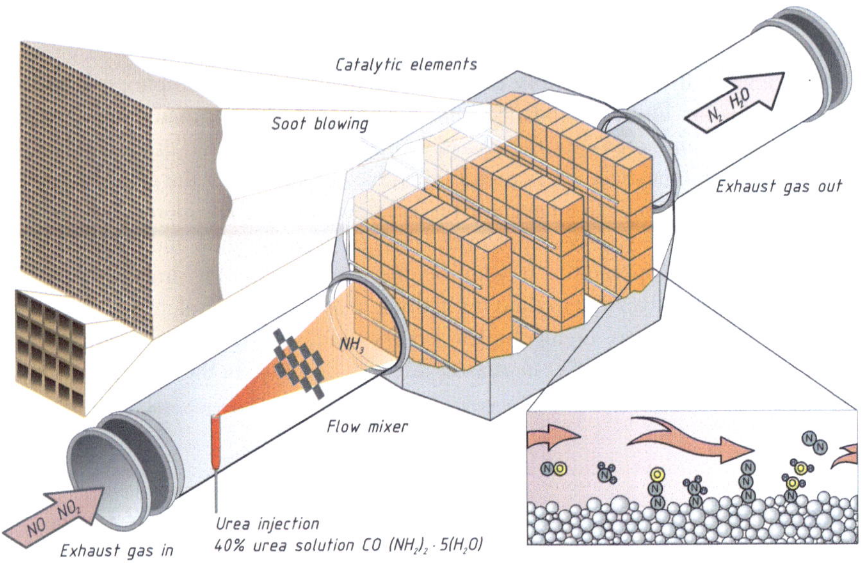

Fig. 2.40 General structure of horizontal type active SCR reactor [26, 28]

process. Considering the toxicity of the first two substances, an aqueous solution of urea in concentrations of 32.5% or 40.0%, which is sprayed in the mixing chamber, is most often used in ship installations. After mixing with engine exhaust gases, urea evaporates and decomposes into ammonia and carbon dioxide according to the reactions:

$$(NH_2)_2CO_{(aq)} \rightarrow (NH_2)_2CO_{(s)} + \times H_2O_{(g)}$$

$$(NH_2)_2CO_{(s)} \rightarrow NH_{3(g)} + HNCO_{(g)}$$

$$HNCO_{(g)} + H_2O_{(g)} \rightarrow NH_{3(g)} + CO_{2(g)}$$

For intensive mixing of the obtained ammonia with the exhaust gases of the engine, guide plates installed in the mixing chamber, which repeatedly change the direction of the flow. After the mixing chamber, the obtained mixture passes through the channels of the catalyst blocks, in which NO_x reduced according to the following reactions (Fig. 2.40):

$$4NO + 4NH_3 + O_2 \rightarrow 4N_2 + 6H_2O$$

$$2NO + 2NO_2 + 4NH_3 \rightarrow 4N_2 + 6H_2O$$

$$2NO_2 + 4NH_3 + O_2 \rightarrow 3N_2 + 6H_2O$$

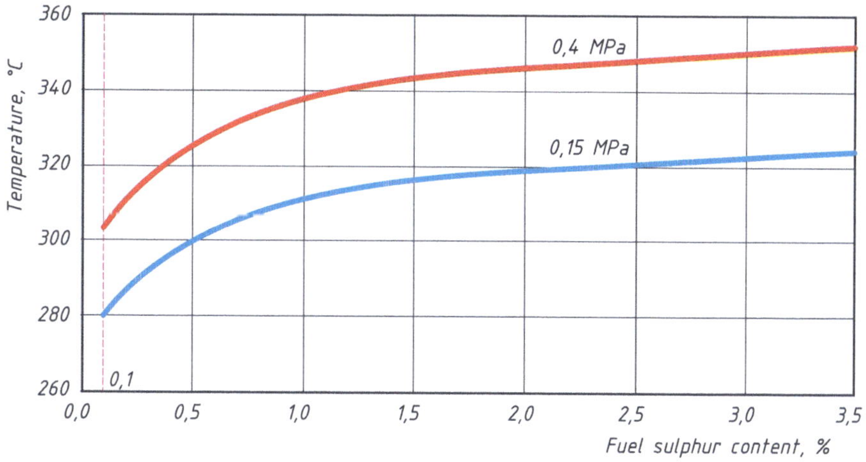

Fig. 2.41 Lower values of exhaust gas temperatures for SCR depending on the sulfur content in the fuel used at operating pressures of 0.15 and 0.4 MPa (adapted from [26])

To reduce one mole of NO, one mole of NH_3 must consume, and two moles must consume to reduce NO_2. Since NO accounts for about 90% of all NO_x in the exhaust gases, the first equation can considered the main one, which determines the consumption of ammonia. With the correct dosage of the reducing agent, up to 90% of nitrogen oxides decomposed during the catalytic reaction, forming molecular nitrogen (N_2) and water (H_2O). The NO_x reduction reaction is exothermic, but given that the concentration of nitrogen oxides in the exhaust gases is about 0.02% of their volume, the heat release is insignificant and has virtually no effect on the processes in the reactor.

Catalytic reactions of NO_x reduction have a high activation energy, which requires maintaining sufficiently high temperatures in the reactor in the range of 200–450 °C. At lower temperature ranges, the catalytic converter generates ammonium sulfates, and at higher temperatures, it generates NO. The maximum temperature when using ammonia as a reducing agent is approximately 550 °C. In other words, to ensure reliable operation of the SCR, it is extremely important to maintain the exhaust gas temperature within a certain temperature range. The lower limit of the operating temperature in the reactor determined the amount of sulfur in the fuel (Fig. 2.41). Its presence leads to the formation of sulfuric acid in the exhaust gases. When interacting with ammonia, sulfuric acid forms a by-product–ammonium bisulfate (ABS–NH_4HSO_4), which can accumulate in the reactor, disrupting its operation.

At higher temperatures, this reaction suppressed. For low-sulfur fuels with a sulfur content of less than 0.1%, this occurs at a temperature of about 310 °C at a pressure of 0.4 MPa. At a pressure of 0.15 MPa, the minimum temperature can reduced to 280 °C. The pressures for which the temperature ranges given approximately correspond to the pressures in the engine exhaust system at high and low engine loads. Excessively high temperatures lead to increased formation of SO_3 in the catalyst, which reacts with water to form sulfuric acid. Another undesirable reaction, which also limits the upper temperature in the reactor, is the oxidation of NH_3. This reaction begins when the exhaust gas temperature approaches 500 °C and leads to increased NH_3 consumption. In addition, the catalyst material begins to sinter at temperatures above 500–550 °C.

Types of SCR installations Temperature limitations on the efficiency and reliability of the SCR reactor affect the circuit solutions of cleaning systems using them. Installations distinguished by the level of operating pressure. They divided into low-pressure and high-pressure installations.

Below we will consider these circuit solutions in more detail.

2.2.4.1 Low-Pressure Units for Four-Stroke Engines

In installations equipped with four-stroke engines, the exhaust gases directed into the reactor after the turbocharger, since they have a high enough temperature for the reactor to function effectively (Fig. 2.42).

Such installations called low-pressure systems (LP SCR), since by the time they enter the reactor, the exhaust gas pressure is 0.15–0.20 MPa.

Figure 2.42 shows a general view of the installation for reducing NO_x emissions based on SCR technology for a four-stroke marine medium-speed engine. In addition to the reactor, the installation includes a module for preparing an aqueous urea solution, a tank for storing it, and a feed control system.

The module for preparing an aqueous urea solution produces it in the required quantity from crystalline powder. This eliminates problems associated with deterioration in quality caused by long-term storage of the solution. The use of urea powder reduces the cost of its purchase and allows for a reduction in the size of the tank for its aqueous solution.

The pump unit ensures the supply and spraying of a controlled amount of aqueous solution into the engine exhaust gas flow.

All processes of the installation and its subsystems controlled in a controller integrated with the general engine control system.

2.2.4.2 Low-Pressure SCR Systems for Two-Stroke Engines

In two-stroke engines, the temperature of the gases at the turbocharger outlet does not exceed 230–260 °C [34] at nominal conditions, and at partial load conditions it is even lower. This circumstance allows the use of low-pressure systems only in engines designed to operate on low-sulfur fuels. Low-pressure systems are simpler in design. They allow more efficient use of the free space in the engine room, since their elements can remove from the engine at large distances.

Given the existing restrictions on the use of fuel with high sulfur content, low-pressure technology installations increasingly being used for low-speed engines. A diagram of such a system installed in a ship's smoke ventilation shaft shown in Fig. 2.43.

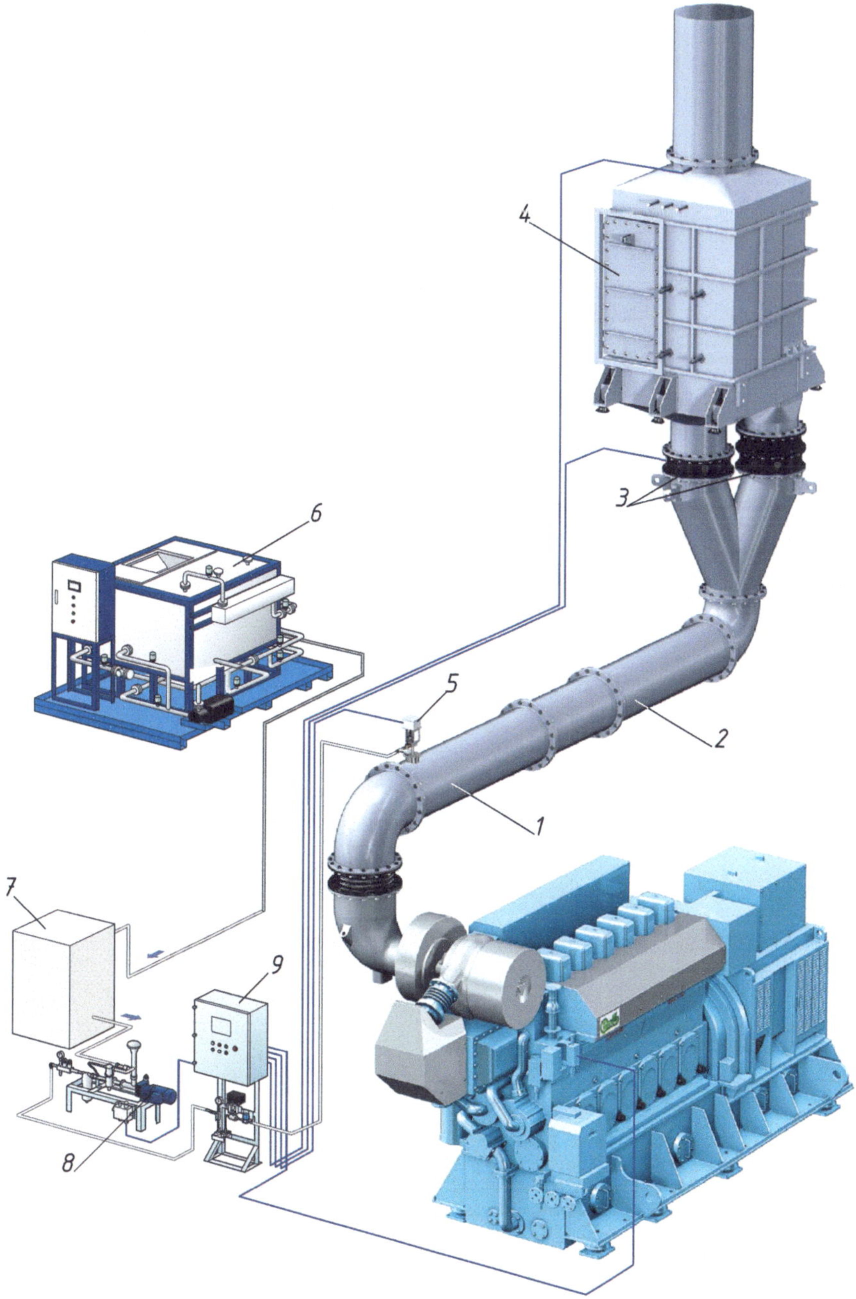

Fig. 2.42 General view of the installation for reducing NO_x emissions based on SCR technology for a four-stroke marine medium-speed engine: (1) outlet channel (aqueous urea sprayer); (2) outlet channel (with built-in mixer); (3) bypass damper; (4) reactor (with built-in catalyst); (5) nozzle for spraying an aqueous urea solution; (6) equipment for producing an aqueous urea solution; (7) tank for an aqueous urea solution; (8) pump unit for an aqueous urea solution; (9) controller and control panel (adapted from [29])

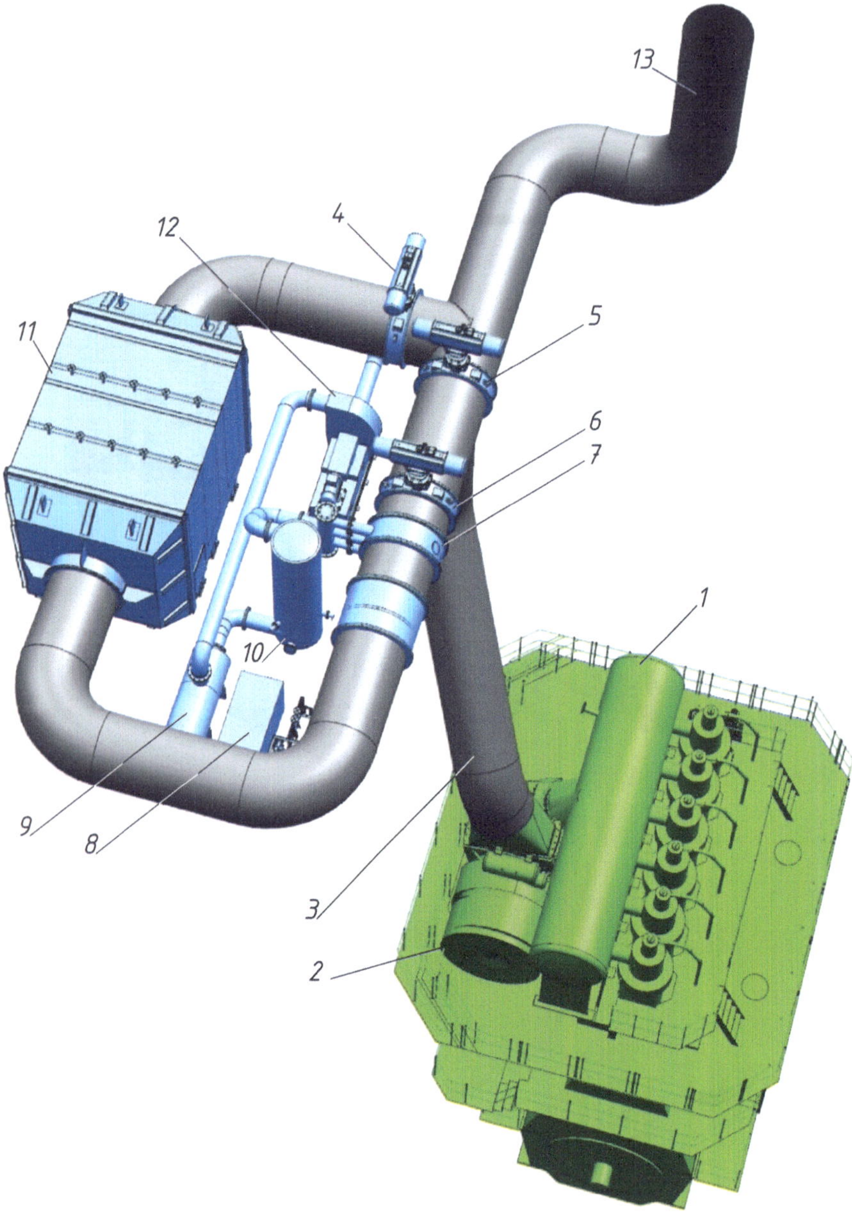

Fig. 2.43 Location of the main components of the low-pressure SCR system for two-stroke engines: (1) outlet receiver; (2) turbocharger; (3) gas pipe; (4) shut-off valve at the reactor outlet; (5) reactor bypass valve; (6) shut-off valve at the reactor inlet; (7) unit for feeding the reducing agent into the exhaust gas flow; (8) control panel; (9) burner; (10) evaporator; (11) SCR reactor; (12) air blower; (13) outlet pipe (adapted from [26])

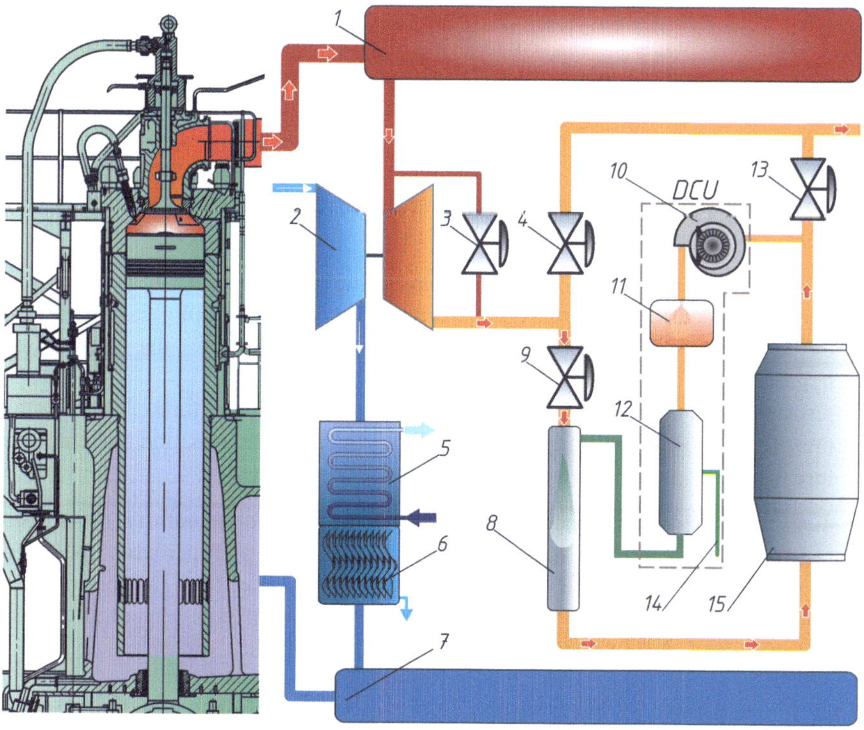

Fig. 2.44 Schematic diagram of low-pressure SCR system for two-stroke engines: (1) outlet receiver; (2) turbocharger; (3) turbine bypass valve; (4) shut-off valve; (5) air cooler; (6) moisture separator; (7) inlet receiver; (8) mixing chamber; (9, 13) shut-off valve; (10) supercharger; (11) burner; (12) evaporator; (13) reducing agent feed; (15) SCR reactor; *DCU* DeComposition Unit (adapted from [26])

Low-pressure SCR systems, the diagram of which is shown in Fig. 2.44, include three main components: the SCR reactor, the mixer (Ammonia Injection Grid—AIG) and the Decomposition Unit (DCU). In turn, the decomposition unit consists of a blower, a heater (burner) and an evaporator. The decomposition unit is located in the gas line between the reactor outlet and the mixer inlet. The reducing agent (aqueous ammonia solution) injected into the evaporator, forming a mixture of water vapor and ammonia. This mixture then enters the mixer and then the SCR reactor together with the flow of gases supplied by the blower.

Despite the use of low-sulfur fuels, it is impossible completely avoid the formation of ammonium bisulfate in the reactor. In order to minimize its formation, especially in partial load modes, it is necessary to increase the temperature in the reactor. For this purpose, the decomposition unit is equipped with a burner that uses liquid fuel to heat the exhaust gases before the evaporator. Increasing the temperature with the help of the burner allows not only to reduce the formation of ABS, but also to decompose the deposits already accumulated in it.

The burner also used to warm up the reactor before starting the engine from a cold state, for example, before leaving a port located in the emission control area.

In running modes, to maintain higher temperatures in the reactor, the bypass of exhaust gases past the turbocharger turbine used, using a bypass valve. The control system regulates the degree of valve opening based on the engine's operating mode, environmental conditions, and the required exhaust gas temperature. Bypassing a portion of the exhaust gases past the turbine increases the temperature of the gases entering the reactor by two main mechanisms:

- gases that have bypassed the turbine have a higher temperature than those that have exhausted their potential in the turbine. Because of mixing the two flows, the temperature of the gases increases proportionally to the amount of bypassed gas;
- gas bypass leads to a decrease in the power supplied to the turbocharger compressor, as a result, the weight filling of the cylinders decreases, and the temperature of the gas at the outlet of the cylinder increases.

It should understood that the need to maintain elevated temperatures before the reactor leads to an increase in specific fuel consumption, and increases operating costs due to the use of additional devices and consumables. For this reason, the operation of a low-pressure SCR system is advisable only in emission control areas.

2.2.4.3 High-Pressure SCR Systems for Two-Stroke Engines

If the use of fuels with sulfur content higher than 0.1% is expected, two-stroke engines are equipped with high-pressure systems (HP SCR). In such systems, the SCR reactor installed before the turbocharger turbine and operates under a pressure of 0.4–1.0 MPa (Fig. 2.45). When operating on heavy fuels, the minimum temperature of the exhaust gases entering the reactor must be 300–350 °C.

The high-pressure SCR system, the diagram of which shown in Fig. 2.45, includes two lines, the main line and the SCR line. The main line supplies gases from the outlet receiver directly to the turbine, and the SCR line supplies gases to the turbine through a combined mixer and SCR reactor. Given the high temperature of the gases entering the mixer, there is no need for a separate evaporator for the high-pressure system, which is present in low-pressure systems. Evaporation and decomposition of the reducing agent occurs directly in the mixing chamber.

The need to place the reactor on the high-pressure side imposes restrictions on the layout of the system. All gas channels must have a minimum length, which means that all elements of the system must compactly placed directly on the engine (Fig. 2.46).

When operating in modes in accordance with Tier II, the SCR line disconnected using shut-off valves. In this case, the turbine bypass valve is open, and the exhaust gases enter directly into the turbocharger. The main line includes a turbine bypass with a control valve, which allows controlling the turbocharger performance when operating at low loads.

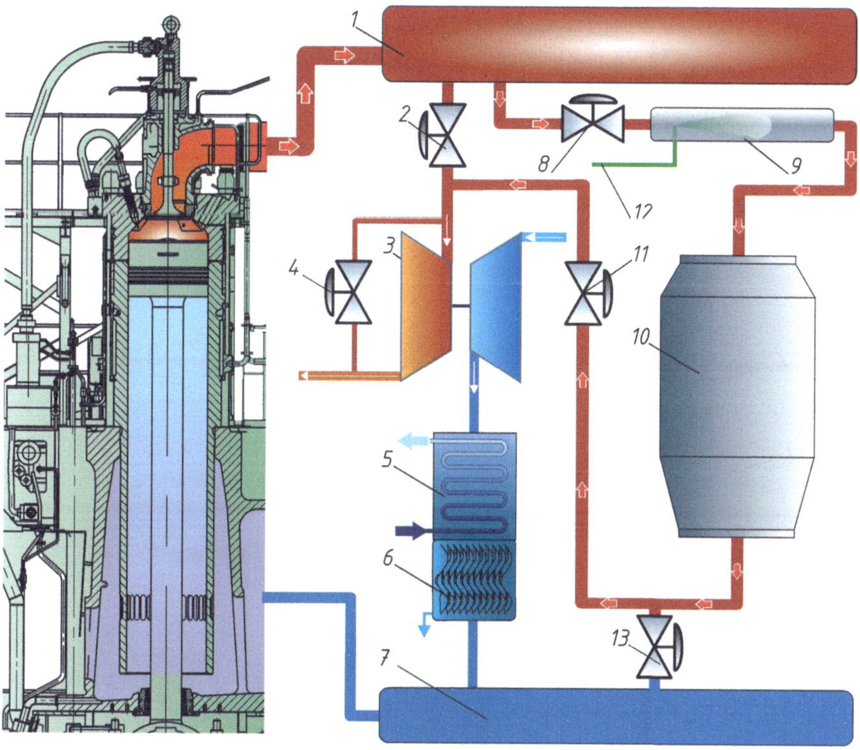

Fig. 2.45 Schematic diagram of high-pressure SCR system for two-stroke engines: (1) outlet receiver; (2) main line shut-off valve; (3) turbocharger; (4) turbine bypass valve; (5) air cooler; (6) moisture separator; (7) inlet receiver; (8, 11) SCR line shut-off valves; (9) mixing chamber; (10) SCR reactor; (12) reducing agent feed; (13) working cylinder bypass valve (adapted from [26])

When operating in Tier III modes, the SCR line activated. To do this, the SCR line shut-off valves opened, and the main line shut-off valve is closed.

Despite the location of the SCR reactor in front of the turbine, at low loads the exhaust gas temperature is insufficient for efficient reactor operation. To increase the temperature, a bypass channel provided from the purge air receiver to the turbine inlet. The channel is equipped with a control valve.

As the valve opens, the mass of air entering the cylinders will decrease, and the exhaust gas temperature will increase. In this way, the gas temperature can maintained at the required level (Fig. 2.47). However, air bypass increases the engine's fuel consumption almost proportionally to the level of temperature increase. In addition, this leads to an increase in thermal stress, both in the working process and in the engine as a whole.

At the same time, engines with lower temperatures of gases at the outlet of the working cylinders will require air bypass at sufficiently high loads. For engines with higher temperatures, such regulation will be required only at medium and low loads (Fig. 2.47). This circumstance largely neutralizes the advantages of using engines

Fig. 2.46 Location of the main components of the high-pressure SCR system for two-stroke engines: (1) turbocharger; (2) outlet receiver; (3) mixing chamber; (4) SCR reactor; (5) shut-off valves; (4) gas supply channel from the reactor to the turbine (adapted from [26])

with higher efficiency in the ship's power plant, at least in emission control areas. At low loads (approximately 15% MCR and below), depending on the engine type and the sulfur content in the fuel, the injection of the reducing agent into the SCR reactor is stopped to prevent ABS deposits in the system caused by insufficient temperature. Another significant problem associated with the installation of the SCR

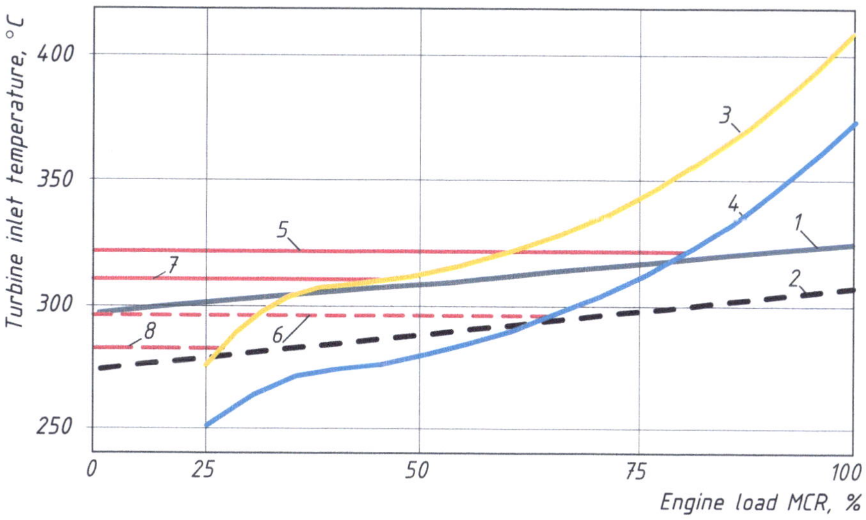

Fig. 2.47 Effect of the compressor bypass valve opening degree on ensuring minimum temperatures at the turbine inlet; (1, 2) minimum temperature in the reactor when using fuel with a sulfur content of 3 and 0.1%, respectively; (3, 4) temperature in front of the turbine for engines with different efficiency levels; (5, 6, 7, 8) temperatures in the reactor when regulated by gradually opening the working cylinder bypass valve (adapted from [26])

reactor directly before the turbine is its high thermal inertia. This especially affects transient modes, leading to imbalance of the engine and turbocharger unit. For example, when the load dropped, the exhaust gases passing through the reactor will heated by the heat accumulated by the catalytic blocks and will enter the turbine with a higher temperature than at the outlet of the exhaust receiver. As a result, the turbine will have excess power, which can lead to compressor surge. The resulting imbalance can eliminated by means of a controlled bypass of some of the gases past the turbine using a controlled bypass valve (Fig. 2.46) or by bypassing some of the air after the compressor into the outlet receiver.

The opposite effect occurs when the load increases, the temperature of the gases at the outlet of the cylinders increases, but the gases reach the turbine with a significantly lower temperature, since a significant part of the heat spent on heating the reactor. As a result, the turbine does not develop power, and the compressor does not supply the required amount of air.

The easiest way to eliminate the resulting imbalance is to supply additional air past the compressor. In practice, this achieved by switching on auxiliary blowers during the transition period. In this regard, auxiliary blowers must be able to operate in the entire range of engine loads. To do this, they must develop a pressure corresponding to the boost pressure at 100% load, which requires an increase in the power of their electric motors by approximately 1.6–2.2. Auxiliary blowers must operate during engine acceleration and throughout the entire time required to warm up the SCR reactor.

Since ammonia is a flammable and extremely toxic gas, it, and reducing agents based on, it supplied through a double-walled pipe system with appropriate ventilation. An ammonia leak detector installed at the outlet of the ventilation system.

To clean the surfaces of the catalyst blocks from soot, the reactors are equipped with soot blowing systems. Compressed air used for purging, which supplied to the reactor from the general ship compressed air system. Water vapor can also used. In some cases, the SCR system is equipped with its own air system. If a waste heat boiler installed after the SCR reactor, it is necessary to install its bypass, which bypasses the exhaust gases past the boiler during the operation of the SCR system. This is necessary to prevent the formation of ABS deposits in the boiler, since the temperature in it is significantly lower than in the reactor.

The control and management system of the SCR unit is equipped with the necessary sensors that allow monitoring both the NO_x and ammonia content in the exhaust gases. This system is integrated with the general engine management system, which allows optimizing the temperature in the SCR system under different load conditions. In addition, such integration allows optimizing the operation of the unit by accurately dosing the reducing agent, reducing its consumption and reducing the likelihood of the so-called ammonia "slip", that is, the carryover of unreacted ammonia with the exhaust gases.

In conclusion, it should note that the considered methods of NO_x emission control could combine with each other to one degree or another, allowing a comprehensive approach to the creation of power plants with minimal impact on the environment.

References

1. Okubo, M.: New Technologies for Emission Control in Marine Diesel Engines. Okubo, M., Kuwahara, T., Butterworth-Heinemann, Kidlington, Oxford, United Kingdom, Elsevier Inc. 2020. 284 p
2. Nielsen, K.V.: Exhaust Recirculation Control for Reduction of NO_x from Large Two-Stroke Diesel Engines. Nielsen, K.V.: PhD Thesis, November 2016 Technical University of Denmark Kgs. Lyngby, Denmark, 2016. 143 p
3. Білоусов, Є.В.: Теоретичні основи робочих процесів в суднових двигунах, що працюють на альтернативних паливах: монографія / Є.В. Білоусов. – Херсон: ОЛДІ-ПЛЮС, 2020. 444 с. ISBN 978-966-289-417-2
4. Bilousov, I.V.: Internal combustion engines of modern marine vessels. Textbook for higher maritime educational institutions. Bilousov, I.V., Bulgakov, M.P. Astroprint publishing house – Odesa, 2024. 512 p. ISBN 978-617-7797-54-7
5. Lalića, B.: Two-Stroke Low Speed Diesel Engine Simulation Model for NO_x Analysis. Lalića B., Račića, N., Radicab, G. : Transactions On Maritime Science. 2017. Pp. 14–23, doi:10.7225/toms.v06.n01.002
6. Common Rail-Design & Field Experience. MAN Diesel & Turbo. Augsburg, Germany. 20 p
7. Costs and Benefits of LNG as Ship Fuel for Container Vessels. MAN Diesel & Turbo. Copenhagen SV, Denmark. 2012. 18 p
8. Terata, I., Aota, A., Edo, K.: Introduction of the Latest Mitsubishi UE Engine Technologies. J. JIME Vol. 50. No.6, 2015. pp. 707–714

9. Matsuda, C.: The latest technologies for new concept "UEC-LSJ" engine. WMS-F440, 10th CIMAC Congress in Kobe, Japan. October 12.2018. 29 p
10. Bernodusson, J.: Removal of undesirable Gases and Soot-Particles from Marine Engines. Exhaust Gas. Icelandic Transport Authority. 2018. 36 p
11. Emission Control MAN B&W Two-stroke Diesel Engines. MAN B&W Diesel A/S, Copenhagen, Denmark. 22 p
12. Paro, D.: Technical improvements for ships. Wärtsilä Corporation, Finland. The Haagen-Smit Symposium April 18 to 21, 2005. 45 p
13. Scarpete, D.: Diesel-water emulsion, an alternative fuel to reduce diesel engine emissions. A review. Machines, Technologies, Materials. ISSN 1313-0226. ISSUE 7/2013. 4 p
14. Andreasen, A., Nyggard, K.B.: Water-in-fuel emulsion as marine engine fuel for reduced NO_x and particulate emissions. Environmental Project No. 1380, MAN Diesel & Turbo, 2011. 75 p
15. Steffens, D.: The Diesel Engine and the Environment. Wärtsilä North America, Inc. Houston, Texas. September 16–17, 2003, 36 p
16. Wärtsilä 46 Technology Review. Wärtsilä Corporation. 2008, 20 p
17. Bedford, F., Rutland, C.: Effects of Direct Water Injection on DI Diesel Engine Combustion. Society of Automotive Engineers, Inc. 2000. 10 p
18. Goldsworthy, L.: Design of Ship Engines for Reduced Oxides of Nitrogen Emissions Presented at "Engineering a Sustainable Future" Institution of Engineers, Australia Hobart July 2002. 39 p
19. Schmid, H., Weisser, G. Marine Technologies for Reduced Emissions. Wärtsilä Switzerland Ltd, Winterthur. April 2005. 12 p
20. Yanmar SolucIoneerIng Together. Marine Diesel Engine 6AYM-WST, 6AYM-WET. YANMAR CO., LTD. Marine Operations Division. Hyogo, Japan. 2 p
21. Exhaust Gas Recirculation: Internal engine technology for reducing nitrogen oxide emissions. Kech, J., Schmidt, G., Philipp C., Rall H. MTU Friedrichshafen GmbH, A Rolls-Royce Power Systems Company. 2012. 4 p
22. The World's First Onboard Verification Test of UE Engine with Low Pressure EGR complied with IMO's NO_x Tier III Regulations. Hiraoka, N., Miyanagi, A., Kuroda, K., Ito, K., Nakagawa, T., Ueda, T. Mitsubishi Heavy Ind. Tech. Rev. Vol. 53 No. 2 2016. – pp 40-47
23. Add more certainty to your future. iCER Delivering enhanced combustion control. WinGD's. Modified 10 August 2021. 13 p
24. ABS advisory on exhaust gas scrubber systems. American Bureau of Shipping. July 2018. 52 p
25. Tier III Two-Stroke Technology. MAN Diesel & Turbo, Copenhagen, Denmark. 2012, 36 p
26. Emission project guide. MAN B&W Two-stroke marine. MAN Energy Solutions, Copenhagen, Denmark. 2023, 96 p
27. Kaltoft J., Tier, III.: Emission technology. MAN Diesel & Turbo, Danmarks, København den 22. November 2011. 55 p
28. A practical guide to exhaust gas cleaning systems for the maritime industry. EGCSA Handbook. 2012, 145 p
29. Daihatsu Diesel SCR System. Selective Catalytic Reduction. Original Bypass-Integrated., Daihatsu Diesel MFG.CO., LTD. 8 p

Chapter 3
Reducing Sulfur Combustion Products Emissions from Marine Engines

3.1 Methods for Reducing Emissions of Sulfur Combustion Products from Exhaust Gases

As already noted, the main method of reducing sulfur oxide emissions in the exhaust gases of marine engines is their indirect regulation through standardization of the sulfur content in the fuel used. The sulfur content in marine fuel depends on the origin of crude oil and the process of its refining. In recent years, the tightening of environmental standards largely been aimed at reducing the sulfur content in marine fuels. Thus, for ships not equipped with exhaust gas cleaning systems, outside special control zones, the use of fuels with sulfur content by volume of no more than 0.5% permitted. Such fuels, known as Very Low Sulphur Fuel Oils (VLSFO). In Sulphur Emission Control Area (SECA), the use of fuels with a sulfur content of no more than 0.1% permitted. Such fuels, known as Ultra Low Sulphur Fuel Oils (ULSFO). The mass fraction of sulfur in the fuel must indicate in bunkering certificates. Ships using fuels with higher sulphur content must be equipped with exhaust gas cleaning systems. These systems must ensure the removal of sulphur combustion products to a level corresponding to the use of fuels with a specified sulphur content for a given water area. Currently, several different designs of cleaning systems using scrubbers have developed to remove SO_x from the exhaust gases of marine engines and boilers operating on high-sulphur fuels. The scrubber cleaning method consists of passing the exhaust gases from the engine through a cleaning medium. This can be seawater, fresh water with an appropriate additive (usually sodium hydroxide) or dry granules. Most of the components of the exhaust substances dissolve or chemically react with the ingredients of the water or granules and are removed from the exhaust gas stream. The main pollutants that scrubbers target are sulfur oxides, but solids also removed at the same time, thus reducing emissions of heavy metals, soot, and sulfur associated with particulate matter.

I. Bilousov, R. Varbanets, *Marine Internal Combustion Engines and Environmental Protection Issues*, Springer Series on Naval Architecture, Marine Engineering, Shipbuilding and Shipping 27,
https://doi.org/10.1007/978-3-032-20940-5_3

Depending on the working environment and operating mode, there are different types of scrubbing systems.

3.1.1 Dry Exhaust Gas Cleaning Systems

Dry scrubber systems rarely used on ships. As a rule, they use granulated hydrated (slaked) lime as a reagent.

This type of scrubber cleans by chemical adsorption, a phenomenon also known as chemisorption. In this process, a chemical reaction occurs to capture sulfur dioxide by converting it into a stable compound. The efficiency of a dry scrubber depends heavily on the characteristics of the adsorbent, and in particular on its specific surface area. The adsorbent, calcium hydroxide, is supplied in the form of granules with a size of 2–8 mm with a very high surface area to mass ratio. A large surface area combined with the use of only about 50% of the active chemisorption potential provides a cleaning efficiency of up to 99%.

The disadvantage of dry scrubbers is that during the dry cleaning process, the temperature of the exhaust gases in them does not decrease; the gas volume remains unchanged throughout the process. For effective operation, the scrubber volume must be large enough to reduce the gas flow rate to an acceptably low level.

3.1.2 Cleaning Systems Using Wet Scrubbers

The so-called "wet" scrubbers are the most common in the fleet (Fig. 3.1). In "wet scrubbers", the mixing of exhaust gases and washing water carried out by passing the gases through sprayed washing water or by passing the gases through a layer of packing wetted with water. The simplest units are hollow scrubbers (Fig. 3.1a). They have low hydraulic resistance, which allows simplifying the design scheme of the cleaning system. Hollow scrubbers can be direct-flow and counter-flow (Fig. 3.1a), they are a hollow cylindrical or square container, usually located vertically. Nozzles for spraying water installed in the center or along the perimeter of the device in the upper part so that their torches cover the entire cross-sectional area. Exhaust gases fed to the lower part of the unit and as they move upward, they cleaned, interacting with droplets of water moving downward under the action of gravity. The water that has passed through the scrubber collected in a special container and removed for disposal, further cleaning or processing.

The hollow scrubber shown in Fig. 3.1a has three levels of water supply (irrigation). At the lower level, the gases undergo intensive cooling, exchanging heat with a large number of large droplets with high heat capacity. Although water-soluble pollutants already captured at the lower level, this process is not intense enough due to the small contact surface between the media. Gas cooling is necessary to prevent their intensive evaporation upon contact with smaller droplets of

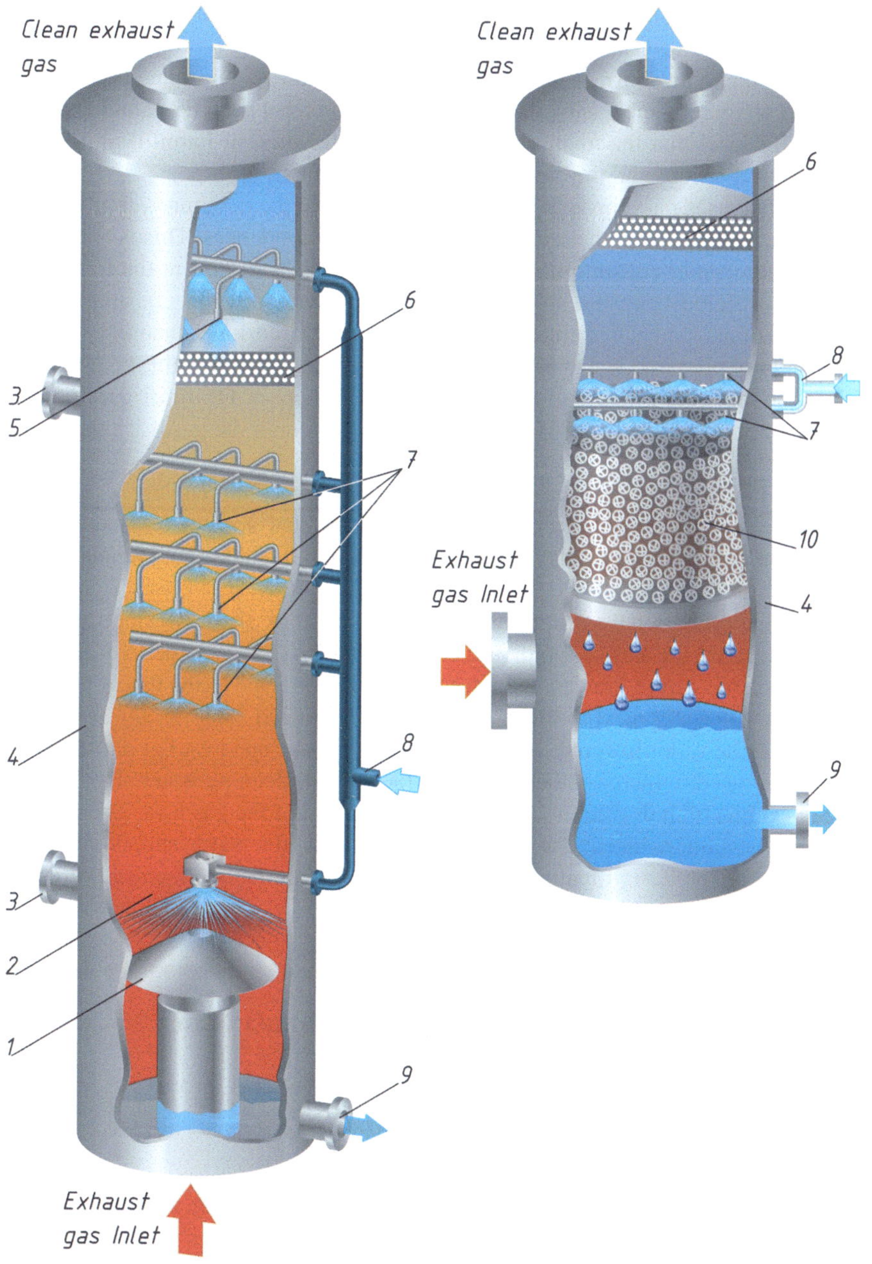

Fig. 3.1 Scrubbers for cleaning exhaust gases from water-soluble impurities and particulate matter: (**a**) hollow type; (**b**) packed type; (1) lower level irrigation drip collector; (2) lower irrigation level; (3) inspection window; (4) housing; (5) moisture separator irrigation; (6) moisture separator, (7) main (middle irrigation level) sprayers; (8) irrigation water supply; (9) contaminated water discharge for cleaning; (10) packing (adapted from [1])

the middle irrigation level, which can lead to the release of already captured pollutants.

At the middle level, irrigation is most intense; the sprayers installed here form a fine aerosol, thereby increasing the contact surface area between the liquid and gas. It is at this level that the main cleaning and neutralization of harmful impurities, such as sulfur oxides and the capture of particulate matter, occurs.

Then the gases pass through a moisture trap, where the moisture vapor formed during the heat exchange between the liquid and gas condenses. This is necessary to reduce the water consumption through the scrubber and prevent the formation of a steam plume from the ship's smokestack. Reducing water consumption is especially important when the scrubber used in a closed-loop cleaning system, and specially prepared water used as circulation water. To maintain the temperature in the moisture trap area below the dew point, there is a third irrigation level at the top, which cools the moisture trap and finally cools the gases leaving after cleaning.

Depending on the tasks that are set when designing the cleaning system, the scrubber can be equipped with additional elements that are not included in the description provided. It can also have a simpler design that does not include any of the elements considered.

Packed scrubbers have additional packing of various shapes inside the tank, which increases the efficiency of the cleaning process (Fig. 3.1b). Nozzles can have transverse, direct-flow and counter-flow irrigation. Due to the use of nozzles, the amount of water used to clean exhaust gases reduced by 40%, the operating pressure of the water lowered. Nozzles have the form of spheres, flat partitions or rings. They used to increase the contact surface of the cleaning water. This allows guaranteeing a high degree of cleaning at a low speed of the gas flow. The dimensions of the unit reduced; the power of the pumps for water supply reduced. Regardless of the design features, cleaning systems can be open, closed or combined cycle.

3.2 Schemes of Systems for Cleaning Exhaust Gases from Sulfur Combustion Products

3.2.1 Open Loop Exhaust Gas Cleaning Systems

Open loop exhaust gas cleaning systems use seawater. After the water has passed through the scrubber, it discharged back to the sea in accordance with IMO regulations. This is the simplest and least expensive installation with low operating costs, but its use may limit by low alkalinity of the water or local regulations on its discharge.

3.2.2 Closed Loop Systems

Closed loop systems can use both seawater and fresh water with alkaline additives. Water circulating in a closed loop purified as it becomes contaminated and reused to feed the scrubber. During the purification process, part of the circulating water becomes the so-called wash water, which contains contaminants. Such a system has no restrictions on the area of use, but its cost and operating costs are higher than those of open loop systems.

3.2.3 Combined Systems

Combined systems combine open and closed cycle modes. Allowing reducing operating costs where it is possible to work in an open cycle.

All existing cleaning systems allow achieving a reduction of emissions of sulfur compounds up to 98%, but the resulting waste and washing waters differ in quality and quantity, therefore, they require different methods of their disposal in ports.

Components included in the exhaust gases, such as carbon dioxide, nitrogen and nitrogen oxides, have very limited solubility in water; therefore, they not retained by wet cleaning systems, while sulfur oxides dissolve well in water.

Sulfur oxides in the exhaust gases of marine engines largely represented by sulfur dioxide—SO_2, and only a very small percentage of this substance further oxidized to sulfur trioxide—SO_3. When dissolved in seawater, reactions occur because of which:

- sulfur dioxide is ionized to bisulfite and sulfite, which is then oxidized to sulfate in seawater containing oxygen O_2:

$$SO_2 + H_2O \leftrightarrow H_2SO_3\left(\text{sulfurous acid}\right) \leftrightarrow H^+ + HSO_3^-\left(\text{bisulfite}\right)$$

$$HSO_3^- \leftrightarrow H^+ + SO_3^{2-}\left(\text{sulfite}\right)$$

$$SO_3^{2-} + ½O_2 \rightarrow SO_4^{2-}\left(\text{sulfate}\right);$$

- sulfur trioxide:

$$SO_3 + H_2O \rightarrow H_2SO_4\left(\text{sulfurous acid}\right)$$

$$H_2SO_4 + H_2O \rightarrow HSO_4 - \left(\text{hydrogen sulfate}\right) + H_3O^+\left(\text{hydroxonium}\right)$$

$$HSO_4^-\left(\text{hydrogen sulfate}\right) + H_2O \rightarrow SO_4^{2-}\left(\text{sulfate}\right) + H_3O^+\left(\text{hydroxonium}\right).$$

In open-loop systems, acids neutralized by seawater's natural alkalinity, which a result of its bicarbonate content. The neutralization reactions are vigorous until the drops to pH 3, after which the reactions slow down considerably and the scrubbing efficiency drops. For this reason, the feed water to the scrubber must be capable of maintaining the required alkalinity. In practice, for open-loop systems, the seawater flow rate is approximately 40–45 $m^3/(MW \times h)$. This figure depends heavily on the alkalinity of the water in a particular region, the sulphur content of the fuel, the water temperature and other factors that must have taken into account when designing such systems and during their operation.

In closed loop systems or when using fresh water for recirculation, the required alkalinity is achieved by adding a suitable alkaline reagent. Most marine exhaust gas cleaning systems use sodium hydroxide, also known as caustic soda, for this purpose, which usually sold as a 50% aqueous solution:

$$NaOH_{(s)} + H_2O \leftrightarrow Na^+{}_{(aq)} + OH^-{}_{(aq)} + H_2O$$

The process of neutralization of sulfur oxides occurs according to the following reactions:

- sulfite dioxide:

$$Na^+ + OH^- + SO_2 \rightarrow NaHSO_3 \left(\text{aqueous sodium bisulfite solution}\right)$$

$$2Na^+ + 2OH^- + SO_2 \rightarrow Na_2SO_3 \left(\text{aqueous sodium sulfite solution}\right) + H_2O$$

$$2Na + 2OH - +SO_2 + \tfrac{1}{2}O_2 \rightarrow Na_2SO_4 \left(\text{aqueous sodium sulfate solution}\right) + H_2O;$$

- sulfite trioxide:

$$SO_3 + H_2O \rightarrow H_2SO_4 \left(\text{sulfurous acid}\right)$$

$$2NaOH + H_2SO_4 \rightarrow Na_2SO_4 \left(\text{aqueous solution of sodium sulfate}\right) + 2H_2O.$$

Artificial alkalinity allows reducing water consumption in exhaust gas cleaning systems with the addition of caustic soda to approximately 20 $m^3/(MW \times h)$. This leads to a reduction in the size of the equipment, a decrease in energy costs for pumping water, but leads to an increase in operating costs for the purchase, storage and processing of the alkaline solution.

In addition to the chemical process of neutralizing sulfur compounds, scrubbers effectively capture particulate matter from 0.1 to 100.0 μm in size contained in exhaust gases. The particles captured by water droplets, after which the water fed to the cleaning, where all solid impurities removed from it.

In closed or combined cycle plants (Fig. 3.2), the used circulating water collected in a process tank, from which it fed back to the scrubber. In order to maintain the

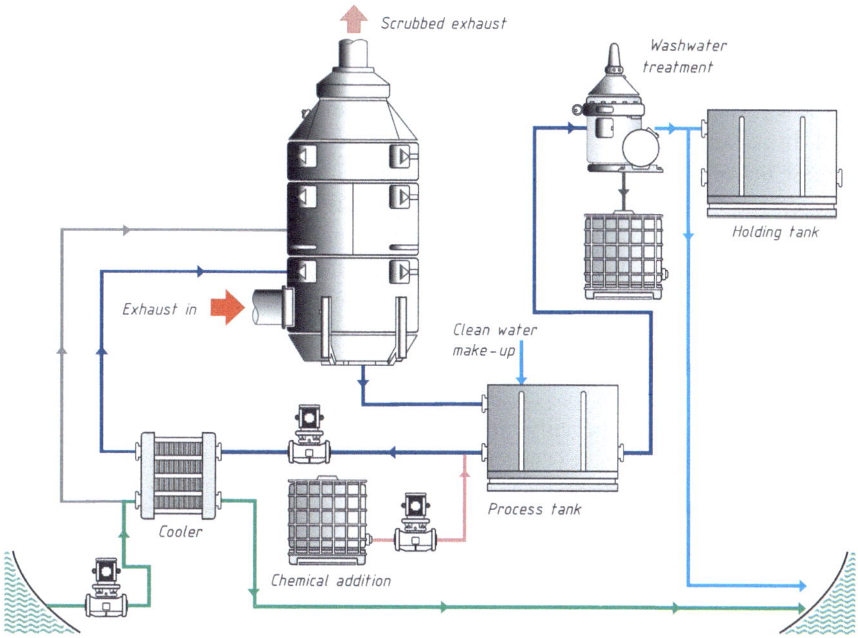

Fig. 3.2 Combined exhaust gas-cleaning system in closed cycle mode [2]

quality of cleaning, a small amount of circulating water collected from the bottom of the tank in the cleaning unit. The water collected in this way is called flushing water.

The wash water is stored in a tank, which allows gases to separate from it and facilitates further removal of solids. After cleaning, the liquid phase separated from the wash water pumped back into the circulation tank.

A small portion of the wash water during the cleaning process, together with mechanical pollution products, passes into wastewater, which after additional cleaning pumped overboard. During pumping, the wastewater carries sulfur compounds to the sea, where they effectively diluted in seawater as the vessel moves. Dilution of 1:2000 occurs already 50 m from the discharge point.

Sulfates formed because of exhaust gas cleaning are a natural component of seawater. Their concentration in open seawater remains constant and is about 2.65 g/l. The main suppliers of sulfates to seawater are volcanoes and river runoff. The increase in sulfate content because of exhaust gas cleaning will be insignificant compared to the amount already in the oceans. As an example, if all the sulphur in the world's oceans were removed, it would cover the Earth with a layer about 1.7 m thick. Sulfur extracted from oil already produced or yet to produce would only add another 10 μm to this layer [2].

When operating near ports, additional seawater may inject into the wastewater tank to dilute the water to pH 6 to meet pH requirements.

In ports, estuaries and rivers, local authorities may impose zero discharge requirements for wastewater, forcing it to be pumped into a tank and then into the public sewer (if the quality meets local acceptance criteria).

The separated solid phase is stored in sludge tanks for subsequent onshore disposal. In certain instances, an additional dewatering system may employ to further decrease the water content of the sludge, yielding a near-solid product.

When scrubbing in "wet" scrubbers, the exhaust gas temperature significantly reduced. For this reason, the cleaning system elements are located after the waste heat recovery systems. In addition, scrubbers are effective spark arrestors and noise suppressors, allowing abandoning the use of these elements in the ship's exhaust system. One scrubber can clean several exhaust gas flows from different ship equipment. In this case, means provided for isolating the exhaust systems to prevent the backflow of exhaust gases into engines or boilers that are not operating. Since different equipment, and especially boilers, can have different backpressure limits in different operating modes, it is necessary to provide the ability to maintain it at the required level. For this purpose, a fan with forced draft can installed at the gas outlet from the cleaning system. This is especially important when cleaning boiler flue gases, since boilers are more sensitive to increased backpressure than engines.

References

1. Your Quick Guide To Types Of Gas Scrubber. Wet Scrubbers. The Go-to People for Spray Nozzle Solutions. SNP. 3 p
2. A practical guide to exhaust gas cleaning systems for the maritime industry. EGCSA Handbook. 2012 145 p

Chapter 4
Reduction of Emissions of Products of Incomplete Combustion of Fuel

4.1 Hydrocarbons

Hydrocarbon compounds (hydrocarbons) are a general name for organic compounds based on hydrogen and carbon, differing from each other in the number of carbon atoms in the molecule, but they can also include atoms of other substances. At the same time, the composition of hydrocarbons present in exhaust gases differs from the original composition of hydrocarbons in fuel and oil, since they undergo thermal decomposition under the influence of high temperatures. This group of substances consists of many individual organic compounds, which present a wide range of chemical variations, many of which are present in extremely low concentrations. Individual components can be not only in the vapor phase, but also in the dispersed phase. Many of them can also distribute between the two phases with evaporation, condensation and polymerization reactions, leading to a constantly changing distribution. The diverse nature of the components of the hydrocarbon fraction makes it difficult to quantify their emissions. Hydrocarbon content typically expressed as the methane equivalent (CH_4). Getting into the atmosphere, some of the hydrocarbon compounds contribute to the formation of acid rain. Volatile organic compounds participate in photochemical reactions in the upper atmosphere, which leads to the formation of the troposphere ozone layer, thus affecting global climate change.

When these compounds enter the human body, their health effects can range from drowsiness and eye irritation to high toxicity, mutagenicity and carcinogenicity. Some hydrocarbons formed during diesel operation, such as benzopyrene, have strong carcinogenic properties. Therefore, despite the low concentration, the problem of reducing the emission of these components is quite relevant.

The hydrocarbon content in the exhaust gases of diesel engines depends on the type of fuel used, the design features of the engine, as well as its adjustments, operating mode and technical condition.

I. Bilousov, R. Varbanets, *Marine Internal Combustion Engines and Environmental Protection Issues*, Springer Series on Naval Architecture, Marine Engineering, Shipbuilding and Shipping 27,
https://doi.org/10.1007/978-3-032-20940-5_4

Hydrocarbons are products of incomplete combustion of fuel and lubricating oil. They enter the exhaust gases from the near-wall space, where, due to low temperatures, the reaction rates are greatly reduced (Fig. 4.1a, c). This is especially true for the space in the gap above the upper compression ring, between the piston and the cylinder (Fig. 4.1c). A stagnant zone formed in this space, into which the flame front from the main combustion chamber practically does not penetrate.

A significant flow of hydrocarbons into the engine working space after the combustion process is complete can occur through the nozzle openings of the injectors from the under-needle chamber in the tip of the nozzle (Fig. 4.1b). The fuel trapped there exposed to high temperatures after the needle valve closes, because of which light fractions boil out of it, creating excess pressure, under the action of which the fuel forced out through the nozzle openings into the working space of the cylinder.

In low-speed engines with volumetric fuel systems, for which the size of the under-needle space is quite large due to design features, this factor is the main source of hydrocarbons entering the exhaust gases.

To reduce hydrocarbon emissions, the design of the nozzle on the latest model engines has undergone significant changes. Earlier designs used a suspended valve that closed the under-needle space at the level of the upper nozzle opening, thus approximately halving the harmful volume (Fig. 4.2a).

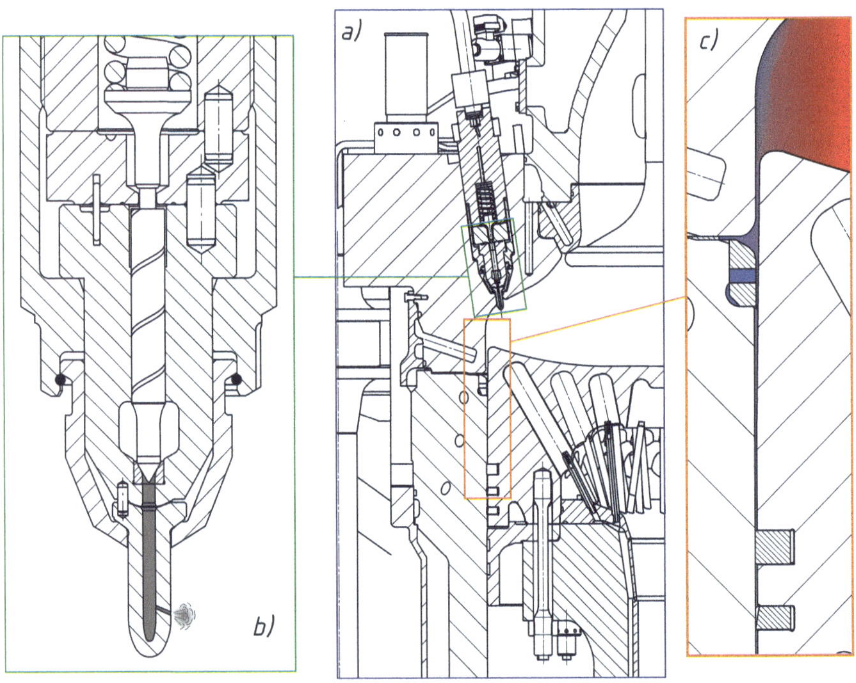

Fig. 4.1 Sources of incomplete combustion products and hydrocarbons entering the engine exhaust gases (**a**). The under-needle space of the nozzle (**b**) and the wall area, as well as the gap between the cylinder liner and the piston (**c**) (adapted from [1])

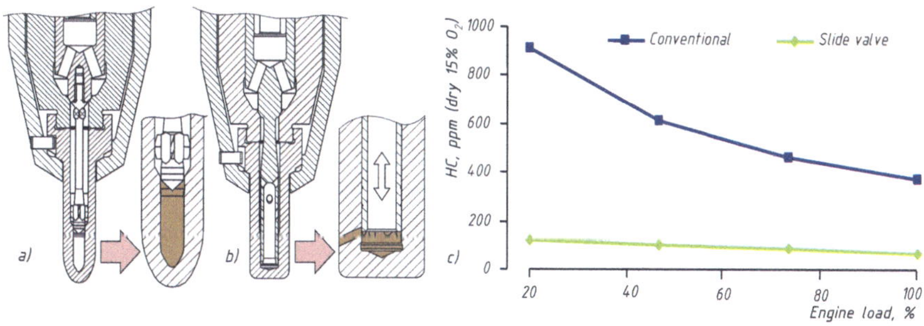

Fig. 4.2 Nozzle with a hanger valve (**a**) and a spool valve cutting off the volume of the underneedle chamber (**b**). Hydrocarbon emission levels of the S50MC-C engine, for a conventional nozzle and a nozzle with a spool valve (**c**) (adapted from [2])

With the advent of new injection systems, in which the pressure of the supplied fuel does not depend on the engine speed, the sprayers were equipped with an additional slide valve, closing the nozzle holes in the absence of fuel injection (Fig. 4.2b). The result was a significant reduction in hydrocarbon emissions, coupled with an increase in engine efficiency and a decrease in unproductive fuel consumption (Fig. 4.2c).

In general, the problem of reducing hydrocarbon emissions directly related to the efficiency of the fuel combustion process in the engine. Accordingly, all measures aimed at improving mixture formation and combustion automatically lead to a reduction in emissions of these substances. At present, the level of hydrocarbon emissions typical for all types of marine engines does not exceed 0.5 g/(kW × h).

4.2 Carbon Monoxide

Carbon monoxide (CO) is a product of incomplete combustion of carbon contained in hydrocarbon fuels. In its normal state, a colorless and odorless gas is naturally present in the atmosphere at a concentration of about 0.2 ppm. At such a low content, it is harmless, but at elevated concentrations, it causes dizziness and headaches, and at higher concentrations (more than 750 ppm) it can be fatal. Carbon monoxide very effectively binds hemoglobin in human and animal blood, forming carboxyhemoglobin. In this indicator, it is 200 times superior to oxygen—O_2. As a result, the ability of the blood to transport oxygen to organs and tissues sharply reduced even at low concentrations of this gas in the inhaled air. In addition, carbon monoxide contributes to the formation of smog and the ozone layer above the Earth's surface. Thus, it also contributes to global warming. Although, compared to carbon dioxide and other greenhouse gases, its role in this process is very limited.

As a rule, CO emissions from marine diesel engines are very insignificant, which is associated with high efficiency of the working process and a large excess of air, which helps to increase the completeness of combustion. At high and medium loads,

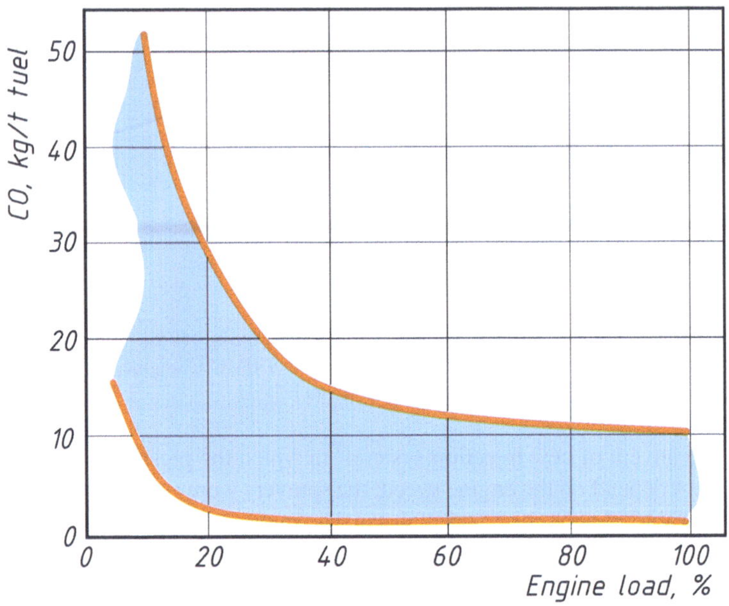

Fig. 4.3 Range of carbon monoxide emissions depending on the operating mode of a diesel engine

they are more or less constant (Fig. 4.3), but can increase significantly in transient modes, when the fuel-air ratio violated, as well as at low loads or at idle. In these modes, the fuel-air mixture is often over-enriched, and the temperature in the combustion chamber is relatively low, which leads to incomplete combustion. Fuel entering the combustion chamber because of post-injection [3] or from the under-needle chamber of the nozzle also contributes to an increase in CO emissions.

The technical condition of the engine and its systems has a significant impact on CO emissions. This primarily concerns the fuel supply and turbocharging systems. Like hydrocarbons, carbon monoxide formed primarily in localized areas of the combustion chamber with insufficient oxygen concentration, in wall zones with slow reaction rates, or in stagnant zones of a semi-closed space.

Improving the quality of mixture formation and combustion is an effective means of reducing CO emissions. During operation, increased carbon monoxide levels may caused by faulty fuel equipment. This primarily concerns injectors, as well as any faults related to the air supply to the working cylinders. However, the latter applies to all products of incomplete combustion.

Currently, the level of CO emissions typical for modern two-stroke low-speed engines does not exceed 0.35 g/(kW × h), for four-stroke engines—0.5 g/(kW × h).

The previously considered emission control methods, although not aimed at regulating the concentration of CO, also affect its content in the final gas mixture entering the atmosphere. For some cases, we have already considered such an effect in the relevant chapters.

4.3 Particulate Matter

Particulate matter is a complex mixture of inorganic and organic substances. This fraction consists of elemental carbon, ash minerals and heavy metals. It also includes various unburned or partially burned hydrocarbon components of fuel and lubricating oil. As a rule, these are particles with a diameter of 10 μm or fewer, suspended in the atmosphere. Particulate matter also includes accumulated fragments of deposits carried out from the exhaust system, formed because of their destruction. The latter can be quite large.

The bulk of particulate matter is easily carried by air currents and has a low settling rate, as a result of which, potentially harmful effects can manifest themselves at a significant distance from the source of exhaust gases. Increased levels of suspended particles in the atmosphere contribute to its warming, and their deposition on ice and further changes in albedo can lead to catastrophic consequences. This is especially true for the Arctic, where shipping activity has recently increased. Particulate matter easily deposited in the respiratory system of humans and animals, which can potentially provoke general respiratory problems, as well more serious toxic, mutagenic and carcinogenic effects.

Despite the significant impact of particulate matter on the environment and human health, IMO does not currently define clear, globally applicable limits on emissions of this type of pollutant.

This is due to a number of reasons, one of which is the difficulty of developing universal control methods applicable to different types of vessels and types of fuel. For different types of engines, the efficiency of combustion processes in them, the types of fuel used, and the methods of after-treatment of exhaust gases, particulate matter can vary significantly in size, composition and quantity. At the same time, the toxicity and environmental impact of various components can vary greatly.

It should noted that indirect measures to limit particulate matter emissions include tightening by IMO of standards for the sulfur content of the fuel used, which is the main source of sulfate particles, which make up a significant part of the solid products contained in exhaust gases. In addition, most measures aimed at meeting IMO NO_x emission standards, especially Tier III, lead to increased combustion efficiency, which can also significantly reduce the formation of particulate matter.

Particulate matter formed from components and compounds of incomplete combustion of the fuel and, to a lesser extent, the oil used to lubricate the cylinder liner. The fuel may contain metals, especially vanadium (V) and nickel (Ni), but in low concentrations. The oil may contain magnesium (Mg), calcium (Ca), zinc (Zn) and phosphorus (P), mainly metals used in modern additive technology.

The initial stage of particulate matter formation is the formation of soot during combustion of hydrocarbon fuel in the combustion chamber of the engine. Its amount depends on the conditions of mixture formation and combustion, which determined by such factors as the operating mode of the diesel engine, the chemical and physical properties of the fuel, the design features of the combustion chamber, the parameters of the fuel equipment, etc. The process of formation of primary carbon particles strongly depends on the molecular weight of the fuel, the higher it is, the higher the rate of particle formation.

During the formation of carbon particles, several sequential-parallel physical and chemical processes can distinguish:

- pyrolysis (cracking) in which the hydrocarbon fuel injected into the combustion chamber begins to decompose under the influence of high temperature into smaller hydrocarbon fragments and radicals. This process occurs under conditions of oxygen deficiency;
- formation of polycyclic compounds (Polymerization). Hydrocarbon fragments react, forming larger Polycyclic Aromatic Hydrocarbons (PAH);
- nucleation or the origin of primary particles. PAH molecules and other heavy hydrocarbons collide and combine, forming tiny particulate matter—soot nuclei. The process of formation of nuclei can occur either from the gas phase (homogeneous nucleation) or by deposition on existing ions or other small particles (heterogeneous nucleation).
- surface growth. Hydrocarbons and radicals from the gas phase are adsorbed on the surface of the already formed soot particles and undergo further chemical transformations, increasing the mass and size of the particles (Fig. 4.4a);
- coagulation and agglomeration. Small soot particles collide with each other under the action of gas flow turbulence and Brownian motion. In this case, they

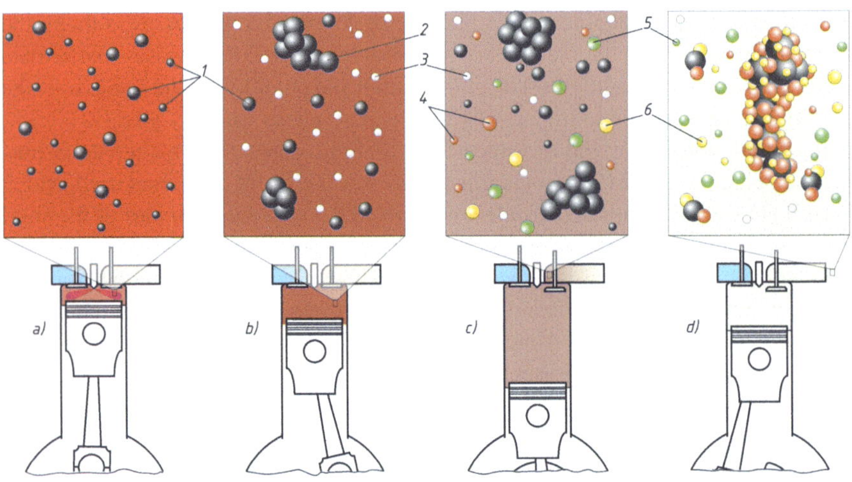

Fig. 4.4 Stages and locations of particulate matter formation: (**a**) formation and growth of soot particle nuclei in local areas of the spray plume; (**b**) coagulation and agglomeration of soot particles, condensation of metallic ash compounds during the process of decreasing charge temperature in the expansion section; (**c**) condensation of volatile compounds at the end of the expansion stroke, beginning of the exhaust stroke; (**d**) adsorption of volatile compounds on carbon arrays. (**1**) primary soot particles; (**2**) agglomerations of soot particles; (**3**) metallic ash; (**4**) hydrated sulfates; (**5**) sulfuric acid; (**6**) condensed hydrocarbons (adapted from [4])

stick together, forming larger agglomerates (clusters) of primary soot particles, which have a shape close to spherical (Fig. 4.4b);
- soot oxidation. The process of soot formation occurs in parallel with the processes of its oxidation under the influence of oxygen (O_2), hydroxyl radicals (OH) present in the combustion chamber. The amount of soot emitted with the exhaust gases of the engine is the result of the balance between the rates of its formation and oxidation;
- adsorption and condensation. Various substances such as unburned hydrocarbons, sulfur compounds, water, and ash particles from burnt oil and fuel additives can adsorbed and condensed on the surface of soot particles. This adsorbed and condensed fraction (Soluble Organic Fraction—SOF) is also part of the particulate matter (Fig. 4.4d).

Primary carbon structures (nuclei) are formed in the combustion chamber during the processes of pyrolysis, polymerization and nucleation (Fig. 4.4a). They are an agglomerate of crystals consisting of a set of individual networks of graphite hexagons. They formed in local zones of the combustion chamber under conditions of oxygen deficiency ($\alpha = 0.33–0.7$), the occurrence of which facilitated by the peculiarities of mixture formation in diesel engines [3]. The processes of formation of primary particles occur at temperatures above 1300–1800 K. At the same time, with an increase in temperature, the formation of soot increases sharply, shifting towards richer mixtures. An increase in pressure, on the contrary, shifts the process towards lean mixtures. The formed primary soot particles contain $10^5–10^6$ carbon atoms, have a diameter starting from 0.002 and up to 0.04–0.15 μm. A number of studies have shown that the formation of nuclei increases with increasing sulfur content in the fuel. The mechanism of this effect currently not well understood.

Under the action of van der Waals attractive forces arising from chemical reactions, primary soot particles coagulate and agglomerate when they collide (Fig. 4.4b). As a result, the primary particles combine and increase their surface area by adsorbing surrounding gas phase molecules, forming secondary structures. Coagulation and agglomeration reduce the number of particles, increasing their size. As a result, secondary soot structures formed in the combustion chamber, which are agglomerates of primary particles, the number of which can reach 4000, and the specific surface area of the agglomerates can reach 75 m^2/g.

After the start of expansion, when combustion is still ongoing during piston movement, the gases cool rapidly. In this case, some of the fuel hydrocarbons may not complete the oxidation process or may change the structure of their molecules, forming a whole series of new organic compounds. Inorganic products of the transformation of components included in the fuels begin to condense from the gas environment (Fig. 4.4c). Some of these compounds adsorbed on the surface of secondary carbon structures. With further expansion in the cylinder and then in the exhaust tract of the engine, ash particles, some wear products, metals, etc. settle on the surface of the carbon mass (Fig. 4.4d).

Sulfur contained in marine hydrocarbon fuel plays a special role in the formation of particulate matter.

As has already been shown above, during the combustion of sulfur in the engine, some of it oxidized to sulfur trioxide (SO_3), which reacts with water vapor, forming sulfuric acid (H_2SO_4). As the gases cool, some of the acid can adsorb on secondary soot particles (Fig. 4.4d). As a result, tertiary structures of particulate matter formed (Fig. 4.5), the formation of which occurs in not only the engine and its exhaust tract, but also continues after the exhaust gases enter the atmosphere (Fig. 4.4d). Most of the particles formed at this stage have sizes of 0.4–5.0 μm. At the same time, their density does not exceed 0.05 g/cm^3, and the specific surface reaches 300 m^2/g.

Sulfuric acid that is not adsorbed on the carbon arrays condenses in the form of submicron droplets (Fig. 4.4d), which are the main component of the inorganic fraction of particulate matter, making a significant contribution to the mass and number of particles, especially in the ultrafine range (particles less than 0.05 μm). These particles, having a small mass, can be very numerous, posing a serious danger to the environment and health.

For marine fuels containing a significant amount of sulfur, the latter is practically the main source of the formation of particulate matter, the content of which is directly proportional to the sulfur content in the fuel (Fig. 4.6).

In addition to the direct effect on the amount of emissions of particulate matter, increased sulfur content leads to the need to increase the supply of lubricating oil on the cylinder liner. Cylinder oils, as is known, contain a significant amount of alkaline additives necessary to neutralize the acids formed. During the neutralization process, particulate matter are formed containing compounds of the metals included in the additives. An example of how the amount of cylinder oil supplied affects the concentration of particulate matter shown in Fig. 4.7.

Measurements of the sizes of particulate matter for most diesel engines have shown that their size distribution, although corresponding to a wide range of diameters (varying within several orders of magnitude), has some regularities, which are

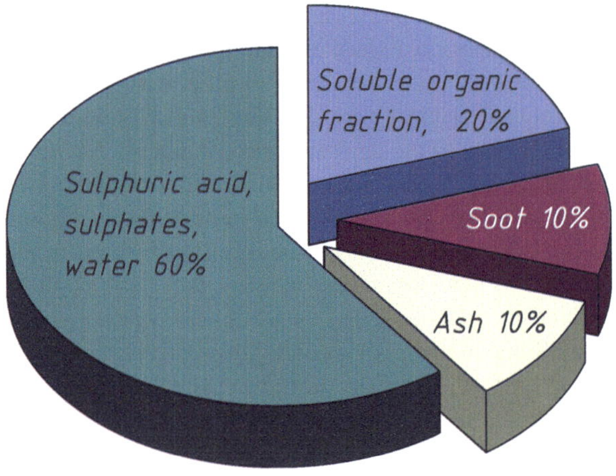

Fig. 4.5 Example of typical exhaust particulate matter composition at high engine loads when running on high-sulfur residual fuel (adapted from [4])

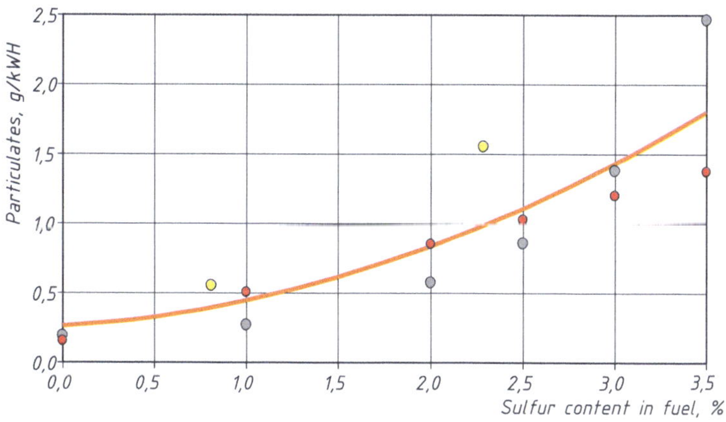

Fig. 4.6 Effect of fuel sulphur content on exhaust particulate matter concentration (adapted from [6])

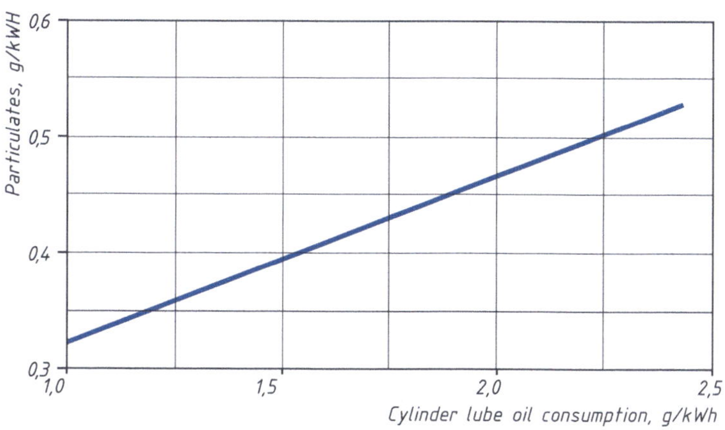

Fig. 4.7 Effect of oil supply for lubricating the cylinder liner of a two-stroke engine on particulate emissions with exhaust gases (adapted from [2])

characteristic to varying degrees for all types of engines. This distribution usually described as bimodal or even trimodal, with a predominance of particles in certain ranges.

For example, in the exhaust gases of high-speed diesel engines, three predominant particle size groups (modes) can identify, the distribution patterns of which shown in Fig. 4.8.

The first mode**,** called Nucleation Mode, combines particles with diameters in the range of 0.005–0.03 μm. It consists predominantly of volatile and semi-volatile compounds such as sulfates, nitrates, organic hydrocarbons, and water that condensed after leaving the cylinder and mixing with cold air. It may contain a small amount of solid nuclei.

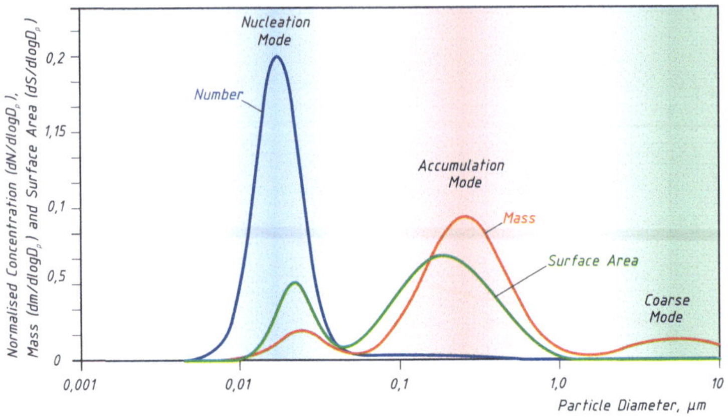

Fig. 4.8 Typical distribution of particle number ($dN/d\log D_p$) mass ($dm/d\log D_p$) and surface area ($dS/d\log D_p$) as a function of particle diameter (adapted from [4])

This mode contains the largest number of individual particles (high concentration by number), but due to its small size, its contribution to the total mass of particulate matter is insignificant.

The second mode, called Accumulation Mode, includes particles with sizes of 0.05–1.0 μm, with a peak concentration usually in the range of 0.01–0.03 μm. It consists mainly of agglomerates (clusters) of primary spherical soot particles (elementary carbon) with hydrocarbons, sulfates and other substances adsorbed on their surface (Soluble Organic Fraction—SOF).

The content of particles related to this mode is significantly less than in the nucleation mode, but the contribution to the total mass of particulate matter maximum due to their larger size.

The third mode, called coarse (Coarse Mode), contains larger particles (more than 1 μm), which can form as a result of engine wear, oil or fuel particles entering the exhaust system, and the agglomeration of smaller particles. Contains a small amount of particles that may vary significantly over time. It can also make a significant contribution to the total mass, but usually this contribution is less significant than that of the accumulation mode.

4.3.1 Particulate Matter Emission Control

Particulate matter, and, first, the soot included in their composition, lead to an increase in optical density and the appearance of a cloud of black smoke. The optical density of exhaust gases depends on the amount and size of soot particles. Visible smoking of diesel engines corresponds to a soot content in the exhaust gases of more than 1.1 g/m^3.

As shown earlier, the amount of emissions of particulate matter depends largely on the completeness of combustion, while monitoring the transparency of exhaust gases traditionally acts as a method of qualitative control of combustion efficiency. However, quantitative assessment of particulate matter is difficult due to the complex nature of their formation, which makes it difficult to develop any universal control methods. The formation of particulate matter begins directly in the engine, and the most toxic components may well form in the atmosphere at a sufficient distance from the end of the exhaust pipe. Thus, even the question of where to take samples is already difficult. To analyze the content of particulate matter in exhaust gases and their impact on the environment and living organisms, a number of methods have been developed based on sampling with subsequent dilution with atmospheric air and cooling to a certain temperature. Thus, to some extent, the processes occurring after the gases have left the exhaust tract of the engine, when the processes of condensation, polymerization, adsorption, etc. are completely completed, are simulated. Such methods have generally proven themselves quite good for engines of land-based vehicles that use fuels with extremely low sulfur content. For marine fuels, even low-sulfur ones, these methods have extremely low reproducibility, which makes their use in marine transport virtually impossible. To control emissions of particulate matter, marine engine manufacturers usually use the Filter Smoke Number (FSN) method. The essence of the method is to take samples of exhaust gases, filter them, and then optically measure the blackening of the filter paper before and after sampling using a reflectometer or photoelectric measuring head. The obtained results then correlated to FSN values using a standardized scale. The sampling procedure and their subsequent analysis are usually carried out in accordance with ISO 8178 (Reciprocating internal combustion engines—Measurement of exhaust emissions).

The result is a dimensionless FSN number, which varies from 0 to 10. The minimum value indicates a clean filter (no smoke), 10 interpreted as a very black filter (heavy smoke). The scale is logarithmic, i.e. equal increments of FSN represent increasing levels of smoke opacity.

In operational practice, a simplified method called the Bosch Smoke Number (BSN) is often used. In general, this method is very similar to the Filter Smoke Number method, but there are key differences. The Bosch Smoke Number is determined by passing a fixed volume of exhaust gases through filter paper of certain porosity. A special pump with a spring-loaded piston used to collect samples. The blackness of the soot stain on the filter paper visually compared with a calibrated 10-step gray scale (Bosch scale) ranging from 0 (white) to 9 (black). Thus, the determination of the BSN based on a visual assessment of the degree of blackness of the spot on the filter. The objectivity of this method is highly dependent on the qualifications of the operator. In some cases, portable devices called Bosch smoke testers may be used.

The lack of universal methods capable of producing comparable results does not yet allow IMO to directly monitor particulate emissions from marine engines. The traditional measure of combustion quality and the traditional way of monitoring its completeness is to monitor the optical transparency of engine exhaust gases.

4.3.2 Reducing Particulate Matter Emissions

Reduction of particulate emissions can achieve:

- by reducing the content of components in fuels that contribute to the formation of particles during combustion (primarily by removing sulfur, Fig. 4.6, metals and their salts);
- by improving engine operating processes aimed at increasing the completeness of fuel combustion (mainly aimed at reducing soot particles);
- by treating exhaust gases before they are released into the atmosphere.

The first group of methods not related to engines, but rather relates to fuel production and storage technologies.

The second group of methods aimed at improving mixture formation and combustion, for example, by introducing battery injection systems (Fig. 4.9) or systems with hydraulic plunger drive (Fig. 4.10) [3]. Such systems, in combination with electronic control, can significantly reduce particulate emissions, especially in partial load modes.

At the same time, measures to reduce other emissions can significantly reduce the concentration of particulates in the exhaust gases, although this does not apply to all methods. In some cases, the amount of particulate matter may actually increase.

Of the in-cylinder methods, humidification of the charge air and direct water injection, although insignificantly, still increase the content of particulate matter, primarily due to a decrease in the combustion temperature (Fig. 2.25). At the same time, the use of water-fuel emulsions, due to the micro-explosion effect, leads to a significant decrease in the content of particulate matter in the exhaust gases (Fig. 4.11).

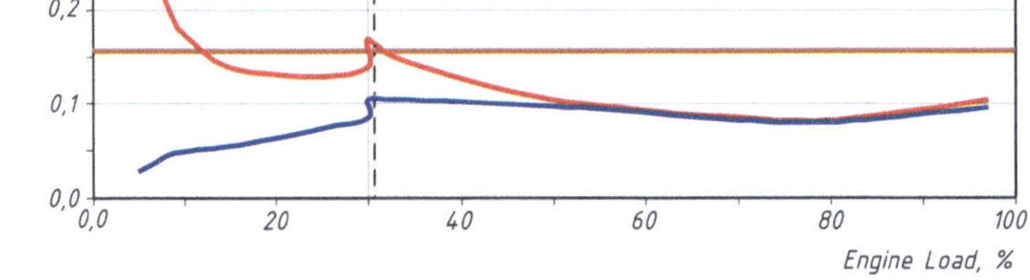

Fig. 4.9 Reduction of particulate matter emissions when using a common rail fuel injection system in low-speed engines of the 6RT-flex58T-B series compared to an engine equipped with a conventional fuel injection system (HFO fuel 380 cSt, 3% sulfur, 0.1% ash) (adapted from [5])

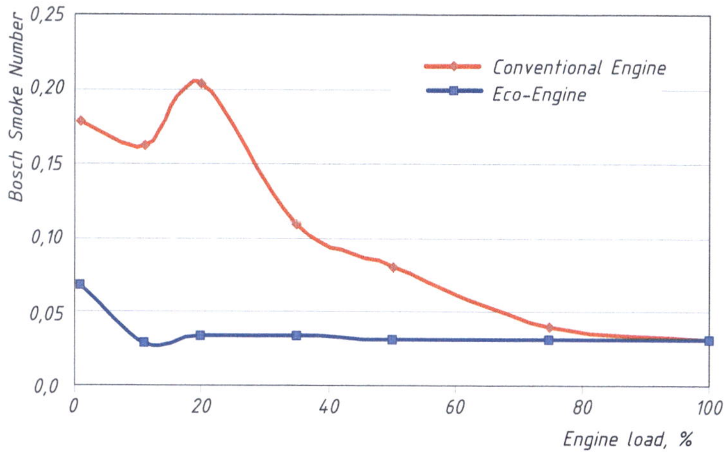

Fig. 4.10 Reduction in particulate emissions when using a fuel injection system with a hydraulically driven high-pressure pump plunger in low-speed UEC Eco-Engine series engines compared to an engine with a traditional (volumetric) fuel injection system (adapted from [7])

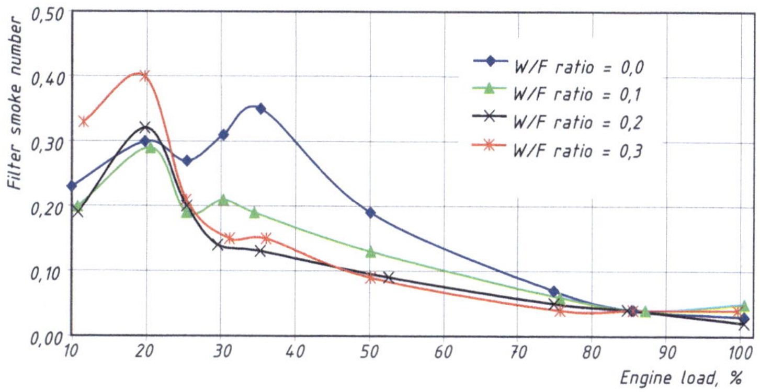

Fig. 4.11 Effect of water-fuel emulsion injection on particulate emissions of the W6L46CR engine, for different water-fuel ratios. Standard injector: 12 × 0.64 × 160°, fuel: HFO (adapted from [8])

A reduction in particulate emissions also noted with exhaust gas recirculation (Fig. 2.14).

The methods used on ships to clean exhaust gases from sulfur compounds, using scrubbers, very effectively capture particulates from the gas flow.

Combining different methods to reduce emissions of harmful substances allows modern ship power plants to eliminate, at least, visible smoke in almost the entire range of operating modes (Fig. 4.9).

References

1. Engine Documentation Wärtsilä RT-flex82C. Operating Manual "Marine". General. Explanations on the Use of the Operating Manual. Wärtsilä Switzerland Ltd. Winterthur, Switzerland. 2008. 3 p
2. Emission Control MAN B&W Two-stroke Diesel Engines. MAN B&W Diesel A/S, Copenhagen, Denmark. 22 p
3. Bilousov, I.V.: Internal combustion engines of modern marine vessels. Textbook for higher maritime educational institutions. Bilousov, I.V., Bulgakov, M.P.: Astroprint publishing house – Odesa, 2024. 512 p. ISBN 978–617-7797-54-7
4. A practical guide to exhaust gas cleaning systems for the maritime industry. EGCSA Handbook. 2012. 145 p
5. Schmid, H., Weisser, G.: Marine Technologies for Reduced Emissions. Wärtsilä Switzerland Ltd, Winterthur., April 2005. 12 r
6. Kristenen, H.O.: Energy demand and exhaust gas emissions of marine engines. Project no. 2014–122: Mitigating and reversing the side-effects of environmental legislation on Ro-Ro shipping in Northern Europe Work Package 2.3, The Technical University of Denmark. Report No. 03 September 2015. 30 p
7. Sakaguchi, K., Sugihara, M.: Development of Electronically Controlled Engine "Mitsubishi UEC Eco-Engine" Paper No.: 221. Mitsubishi Heavy Industries Ltd. CIMAC Congress 2004, Kyoto. 11 p
8. Steffens, D.: The Diesel Engine and the Environment. Wärtsilä North America, Inc. Houston, Texas. September 16–17, 2003. 36 p

Chapter 5
Reducing Greenhouse Gas Emissions

5.1 Greenhouse Gases and Their Impact on Global Climate Change

The Kyoto Protocol defined the main gases responsible for global climate change on the planet, which enter the atmosphere, including because of human economic activity. According to this protocol, the main greenhouse gases include water vapor (H_2O), carbon dioxide (CO_2), methane (CH_4), nitrous oxide (N_2O), ozone (in the troposphere) (O_3) and fluorinated gases (CFCs). The contribution of each gas to the greenhouse effect is different and depends on its concentration in the atmosphere and the ability to absorb infrared radiation (Global Warming Potential—GWP). All these compounds, with the possible exception of fluorinated gases, are, in one way or another, products formed because of the working processes in internal combustion engines. As can be seen from Fig. 5.1, the greatest contribution to the global process of increasing the temperature of the Earth's atmosphere is made by carbon dioxide, which is formed in large quantities as a result of burning carbon-containing fuels, including hydrocarbon fuels of petroleum origin used in marine energy. In recent decades, anthropogenic carbon dioxide has been accumulating in the atmosphere faster than at any other time in human history. Thus, in 2004, the level of carbon dioxide was 377.1 ppm, and by 2023 it had already increased to 420.0 ppm. This is an increase of 42.9 ppm, or 11.4% in just 20 years. The current level of CO_2 in the atmosphere is already 51% higher than the level of the pre-industrial era (before 1750) (Fig. 5.2).

I. Bilousov, R. Varbanets, *Marine Internal Combustion Engines and Environmental Protection Issues*, Springer Series on Naval Architecture, Marine Engineering, Shipbuilding and Shipping 27,
https://doi.org/10.1007/978-3-032-20940-5_5

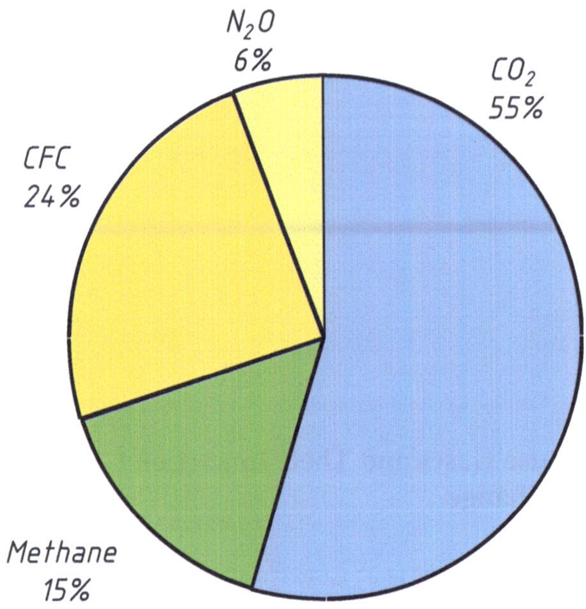

Fig. 5.1 Major greenhouse gases and their contribution to global warming (adapted from [1])

It should understood that the increase in CO_2 levels is the result of more complex processes, which, in addition to fuel combustion, include a reduction in the absorption of carbon dioxide by biosystems, which is associated with deforestation, fires in ecosystems, and much more. Volcanic activity also makes a significant contribution, but it human economic activity that disrupts the fragile balance that nature maintained for a long time before the advent of the industrial era.

The share of CO_2 emissions from international maritime shipping, according to IMO's own research, is approximately 2–3% of the total global anthropogenic emissions of this gas into the atmosphere, which in absolute terms corresponds to 1.05 million tons.

At the same time, IMO is one of the most effective regulators in this area, since almost 70% of the merchant fleet registered in countries not covered by Annex I of the Kyoto Protocol. Thus, ship-owners, regardless of the decisions of their countries to ratify the Kyoto Protocol, covered by IMO rules for reducing CO_2 emissions. Through its legislative framework, the IMO stimulates the development of technologies aimed at reducing CO_2 emissions. Because of such regulation, shipping is already the most efficient mode of transport in terms of carbon dioxide emissions

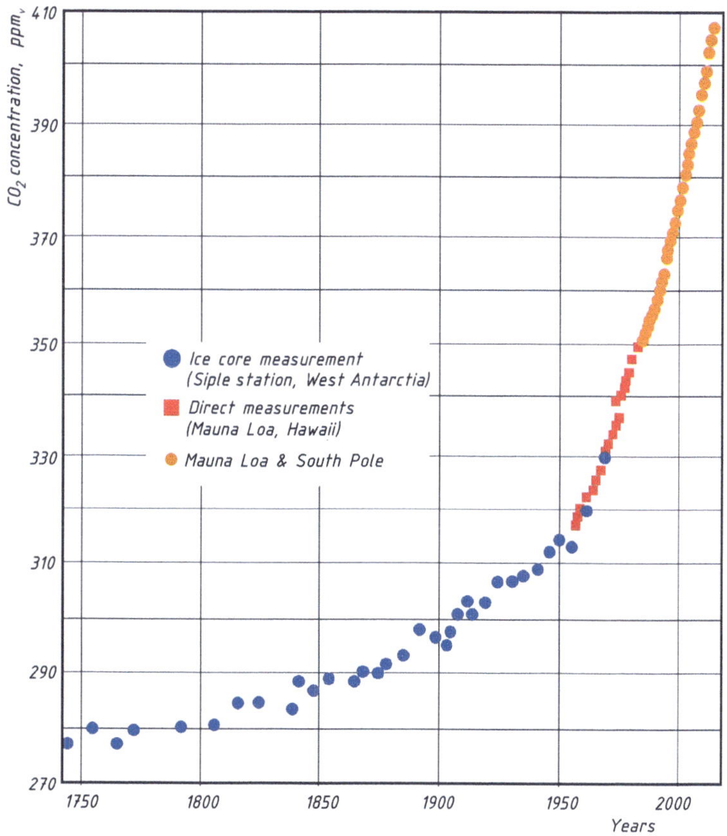

Fig. 5.2 The increase in the concentration of carbon dioxide in the earth's atmosphere since 1750 (the beginning of the industrial era) (adapted from [1])

(Fig. 5.3). Thus, increasing the share of sea transport is in itself an effective way to reduce anthropogenic CO_2 emissions. Given the IMO strategy aimed at the complete decarburizations of the industry, it can expected that in the coming years, the use of sea transport will become an even more effective measure to prevent global warming.

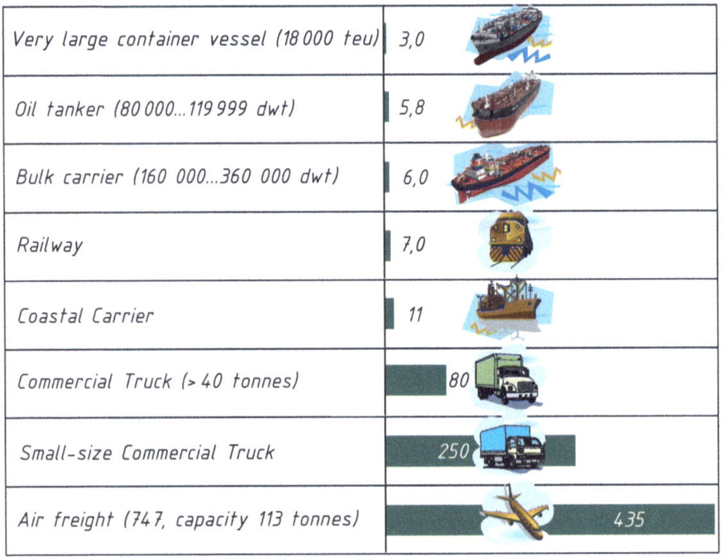

Fig. 5.3 Specific CO_2 emissions in grams per tonne-kilometer by main types of transport (adapted from [2, 3])

5.2 Measures Aimed at Reducing Carbon Dioxide Emissions in the Short Term

This set of measures includes increasing the efficiency of both the working process of ship engines and the entire power plant as a whole. It is obvious that increasing the efficiency of the power plant leads to a reduction in fuel consumption per unit of transport work, and, consequently, emissions of combustion products associated with the combustion of organic fuel. Now, ship engines are the most efficient heat engines available to humankind. In recent years, ship engines of all types been significantly improved. The introduction of electronic control systems, both for working processes and for the power plant as a whole, has practically exhausted the possibilities for further efficiency improvement. Today, the actual efficiency of engines has almost approached the theoretically possible, and therefore, each subsequent step aimed at increasing efficiency will require more and more significant complications of the design and financial investments that will be disproportionate to the effect obtained. A certain reserve for increasing the efficiency of a ship's power plant is the improvement of deep heat recovery systems, but these reserves are also very limited. In this direction, significant progress been achieved recently, which has largely exhausted the possibilities of this direction.

Another approach that allows, in the short term, if not to solve the problem of anthropogenic carbon dioxide emissions, then at least to significantly reduce them, is the transition to the use of alternative fossil fuels with lower carbon content. First, such fuels include natural gas, which consists of more than 95% methane (CH_4). Of considerable interest are propane (C_3H_8) and butane (C_4H_{10}), as well as their mixtures.

Table 5.1 Hydrogen-carbon ratio for gaseous and liquid fuels

CH_4	C_3H_8	C_4H_{10}	MGO/MDO	HFO
4.0	2.67	2.50	2.10–2.22	2.04–2.10

As a criterion for assessing the carbon content in fuel, the dimensionless parameter hydrogen-carbon ratio (H/C) is used. For the gases under consideration, as well as for traditional types of fuel, the value of the carbon-hydrogen ratio, presented in Table 5.1.

As can be seen from Table 5.1, the most interesting from the point of view of reducing CO_2 emissions is methane, when burning 1 kg of which approximately 2.74 kg of carbon dioxide is formed, while when burning heavy marine fuel, about 3.15 kg of carbon dioxide is formed. Thus, replacing heavy fuel with methane allows you to reduce CO_2 emissions into the atmosphere by 13%. The real figure, due to the higher heat of combustion, can reach 20–25%.

The use of gases as motor fuel can significantly improve other environmental indicators of engines. The absence of sulfur in gas fuels almost completely solves the problem of sulfur compounds, and the combustion features of these fuels in engines can significantly reduce NO_x and CO emissions.

The risks of methane and other hydrocarbons, which are also greenhouse gases, entering the atmosphere increase significantly. At the same time, the risks of entry exist at all stages of use: production, transportation, bunkering and directly in the engine operating process. In the following chapters, we will consider the features of using gas fuels in marine engines of various designs and purposes, where issues of improving the environmental performance of these engines will cover in more detail.

5.3 Measures Aimed at Reducing Carbon Dioxide Emissions in the Medium and Long Term

This set of measures provides for a complete transition to carbon-neutral or carbon-free technologies.

The basis of such a transition is the idea of using renewable energy sources, which must be learned how to use correctly.

Theoretically, such a transition is quite feasible. Every year, up to 86,000 TW of solar energy hits the Earth; of which 48,000 TW reach its surface. About 90 TW spent on photosynthesis processes; another 870 TW spent on the processes of moving air masses in the Earth's atmosphere [4]. For comparison, in 2018, 19 TW were for human economic activity, of which shipping costs amounted to 0.332 TW.

Thus, the potential for switching to alternative energy sources is enormous, but requires solving several fundamental problems. We will dwell only on those that directly or indirectly relate to the topic of this book—marine internal combustion engines.

Today, the problem of converting solar radiation energy into electrical energy being actively addressed. The main disadvantage of modern technologies is the cyclic nature of such a transformation associated with the periodicity of the impact of a particular natural factor on generating devices (tides, wind, solar activity, etc.). That is, even for stationary energy, the problem of accumulating energy when it is in excess and using it when it is in short supply is relevant. At the same time, electric energy is the most convenient for storage, since it allows you to avoid a number of transformations during which a significant part of the energy can be lost. The rapid development of electric energy accumulation technologies already today makes its use as the main driving force in land transport, primarily in cars, quite profitable.

The transfer of water transport to electric traction is, in principle, also possible for technical fleet vessels that do not depart far from service bases, which allows for the rapid replacement of discharged batteries or their charging from shore power sources. And such technologies are already being implemented and successfully used, which significantly improves the environmental situation in port waters and coastal regions.

For transport vessels, it is obviously impossible to abandon internal combustion engines in favor of electric traction in the near future. The operating conditions of vessels, both for reasons of profitability and navigation safety, imply significant autonomy, which cannot achieved using existing or prospective technical solutions. Figure 5.4 compares the mass and volume ratios for obtaining the same amount of

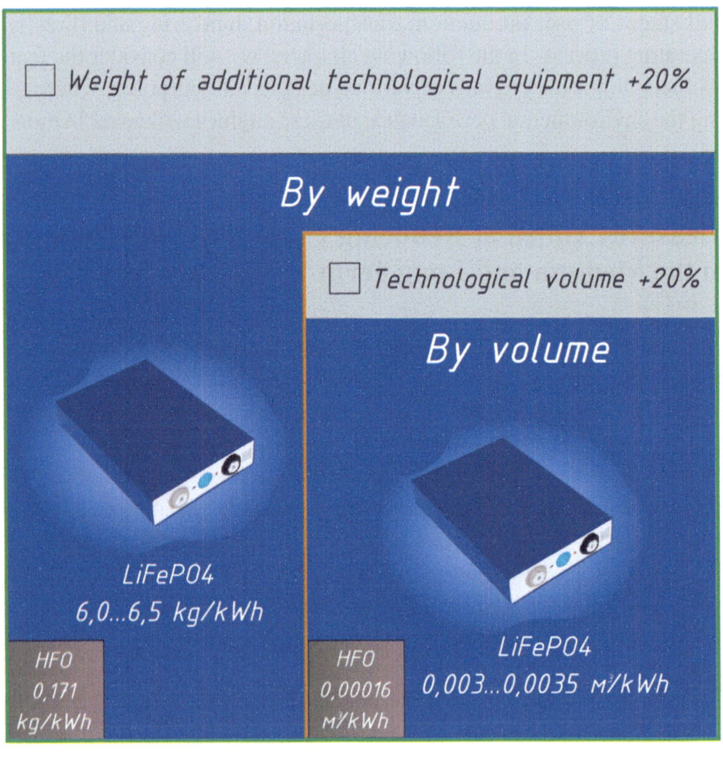

Fig. 5.4 Comparison of mass and volume ratios for obtaining the same amount of energy from organic fuel and stored using $LiFePO_4$ batteries

energy on board a vessel from organic fuel (HFO) and stored using lithium-iron-phosphate batteries ($LiFePO_4$). Simple calculations show that to obtain a unit of power, the mass of batteries should be 35–38 times greater than the mass of hydrocarbon fuel. At the same time, the increase in volume will be 18.5–22.0 times, and this does not take into account the technological volume and the mass of additional technological equipment (battery racks, charge/discharge controllers, conductive buses, etc.).

It should be noted that $LiFePO_4$ batteries are among the most energy-intensive among commercially available batteries (energy capacity up to 165 (W × h)/kg). The best battery samples available on the market have a mass energy capacity of up to 260 (W × h)/kg, but this figure is more than 22 times worse than that of petroleum fuel.

In order for the use of electrical energy to power ships to become feasible, it is necessary to increase the energy capacity of batteries by at least an order of magnitude. However, even this is unlikely to solve the problem, since it will be necessary to create port infrastructure that allows transmitting a huge amount of electrical energy on board to charge the batteries in a commercially acceptable period. Considering that the whole point of electrifying the transport fleet aimed at reducing anthropogenic carbon dioxide emissions, the electricity transmitted to ships must obtain from renewable sources. In this case, infrastructure for generating a huge amount of green energy must create in the immediate vicinity of the ports. The number of places where this can implement in practice is very limited.

Taking into account the arguments presented, it should assume that in the near future, there is no alternative to internal combustion engines in the merchant fleet. In this case, the only solution to the problem of completely stopping anthropogenic CO_2 emissions can only be a transition to carbon-neutral or carbon-free fuels of biological or artificial origin.

Fuels of biological origin considered carbon-neutral, such as biodiesel (mainly in the form of methyl esters of fatty acids), Hydrated Vegetable Oil (HVO), bioethanol, biomethanol. When these types of fuel burned, CO_2 emissions into the atmosphere offset by the consumption of the same amount of carbon dioxide in biological processes of formation of carbon-containing organic compounds in plants, living organisms or algae.

The advantage of using biological fuels is that their use does not require significant modification of existing engines. This is especially true for biodiesel fuel, which differs little in its characteristics from traditional fossil fuels. The use of alcohols in engines based on marine diesel engines is also not a technically difficult task.

The disadvantage of biodiesel fuels, which mainly obtained from plant products, is the competition of their raw material base with the raw material base of food products. For humanity, where every seventh inhabitant of the planet is undernourished, such a situation can hardly consider normal.

Perhaps this problem will solve when mass production of second-generation biofuels obtained from agricultural waste, forestry, etc. is established. Work is also underway to create third-generation fuels, the raw material for which will be algae.

Many countries, by law, regulate the use of mixed fuels on their territories, which contain additives of organic fuels to traditional hydrocarbon fuels. The amount of such additives is 10–20%, which helps to reduce the harm caused to the environment.

Mixed fuels also used in the fleet. In 2017, additional grades of marine motor fuel containing Fatty Acid Methyl Esters (FAME) introduced in the sixth edition of the ISO 8217 standard.

Biological fuels are products of direct conversion of solar energy into hydrocarbon compounds, although their production requires additional energy costs.

Another approach to solving the problem of accumulating renewable energy is to obtain so-called "electric" fuels. These are compounds synthesized from carbon contained in biomass, using energy (primarily electrical) obtained from renewable sources. Electrical energy used primarily to produce hydrogen by hydrolysis of water and to create the necessary conditions for the synthesis. Examples of such fuels can be methanol (CH_3OH) or ethanol (C_2H_5OH) synthesized in this way.

The raw material for obtaining methanol can also be CO_2 obtained by capturing it from flue gases of industrial processes associated with the combustion of fossil fuels. Nuclear power can use as a source of relatively clean energy for obtaining synthetic methanol. Such technologies do not fully solve the problem of CO_2 emissions, but they allow for a significant reduction in their volumes.

Carbon-free fuels do not contain carbon in their structure, and therefore the combustion products formed during their combustion do not affect the overall balance of CO_2 in the atmosphere. Examples of such fuels are ammonia (NH_3) and hydrogen (H_2). "Electric" ammonia can be synthesized using renewable electrical energy from atmospheric nitrogen (N_2) and hydrogen obtained by hydrolysis of water using the same energy. Such ammonia classified as "green" and assigned the role of one of the main types of fuel for future shipping. The production of such ammonia requires virtually no raw material base, so production is possible anywhere where sufficient renewable energy can obtained. Despite its toxicity, ammonia is easily accumulated, transported, and can be used as a fuel, including for marine internal combustion engines.

Hydrogen can produce by electrolysis of water using electricity generated from renewable sources, with virtually no greenhouse gas emissions. Its use on ships is associated with the need to solve a larger number of problems than, for example, when using ammonia or methanol. Despite the difficulties, the problems associated with the use of hydrogen now being successfully solved.

References

1. Wärtsilä 46. Technology Review. Wärtsilä NSD Finland Oy. Vaasa, Finland. 28 p
2. Exhaust Gas Emission Control Today and Tomorrow. MAN B&W Two-stroke. Marine Diesel Engines. MAN Diesel. Copenhagen, Denmark. – 31 p
3. Delivering CO_2 Emission Reductions. International Chamber of Shipping. Published in 2015, London, 6 p
4. Methanol as a Marine Fuel – Availability and Sea Trial Considerations Supported by Maritime Energy and Sustainable Development (MESD). Singapore, January 2021. 36 p

Chapter 6
Alternative Marine Fuels and Prospects for Their Use on Ships

The following considered as promising fuels, the use of which can significantly reduce or eliminate emissions of anthropogenic carbon dioxide, for marine power engineering: natural and petroleum gases, methanol, ethanol, ammonia and hydrogen. The use of natural and petroleum gases are the most proven technologies that can significantly reduce emissions of carbon dioxide and harmful substances into the atmosphere. However, the most promising are technologies using alternative artificial fuels, which in the future can make shipping carbon-neutral or even carbon-free [1, 2].

6.1 Properties of Artificial Alternative Fuels

The properties of some artificially produced alternative fuels that being consistently introduced into the merchant fleet in the context of reducing or eliminating emissions of anthropogenic carbon dioxide, presented in Table 6.1.

For ships carrying ethanol, its use as fuel also considered a promising direction. The use of ethanol is similar to the use of methanol with some modified engine settings. Therefore, we will not consider this type of fuel separately. The following chapters will be devoted to the issues of using individual types of alternative fuels in marine engines for various purposes.

I. Bilousov, R. Varbanets, *Marine Internal Combustion Engines and Environmental Protection Issues*, Springer Series on Naval Architecture, Marine Engineering, Shipbuilding and Shipping 27,
https://doi.org/10.1007/978-3-032-20940-5_6

Table 6.1 Properties of the most promising synthetic fuels [2–4]

Fuel	DF	Methanol	Ethanol	Ammonia	Hydrogen
Chemical formula	C_8-C_{25}	CH_3-OH	C_2H_5-OH	NH_3	H_2
Proportion of carbon, % wt.	85	38	52	0	0
Proportion of hydrogen, % wt.	15	12	13	17.78	100
Proportion of oxygen, % wt.	0	50	35	0	0
Molar mass, kg/kmol	183	32	46	17.031	2.016
Density in liquid state, kg/m^3	840	798	794	730	70.8
Lower heat of combustion, MJ/kg	42.7	20.1	27.0	18.6	120
Boiling point, °C**	180–360	65	78	−33.34	−252.87
Vapor pressure, MPa*	≤ 1	0.013	0.0059	0.857	–
Critical pressure, MPa	3.0	8.1	6.3	11.357	1.3
Critical temperature, °C	435	239.4	241	1324	−239.95
Kinematic viscosity, cSt*	2.5–3.0	0.74	1.2	0.359	–
Surface tension, N/m*	0.027	0.023	0.022	0.021	–
Bulk modulus of elasticity, N/mm^2 (at 20 °C; 2.0 MPa)	553	823	902	0.910	2.8
Cetane number	38–53	< 5	8	–	5–10
Octane number	15–25	106	109	111	130
Autoignition temperature, °C	250–450	385	362	651	520
Heat of vaporization, kJ/kg**	250	1089	841	1370	449
Stoichiometric ratio (φ)	1:14.6	1:6.5	1:9.1	1:6.045	1:34.2
Minimum ignition energy, mJ (at φ = 1)	0.48	0.14	0.28	680	0.017
Maximum flame temperature, °C**	2054	1890	1920	1800	2254
Flammability limits, vol.%	0.5–7.5	6–36	3–19	15–22	4–75
Flash point, °C	52	12	14	–	–

* at a temperature of 20 °C; ** at a pressure of 0.1 MPa; DF is given for comparison

6.2 Alternative Fuels and the Energy Potential They Can Provide on Board a Vessel

For marine power engineering, the energy potential that is on board in the form of fuel or another energy source is critically important. Therefore, the energy capacity of the potential fuel plays a key role in assessing the prospects for its use in the fleet. Volumetric indicators are especially critical, since they determine the dimensions of the tanks for storing the fuel supply on board, required to ensure a given autonomy of the vessel [2]. The share of loss of autonomy can also used for assessment—the amount of reduction in autonomy with an unchanged volume allocated for storing fuel.

It should note that all the alternative fuels under consideration have a significantly lower energy potential than hydrocarbon fuels of petroleum origin. For comparison, Table 6.1 shows the properties of Marine Diesel Oil.

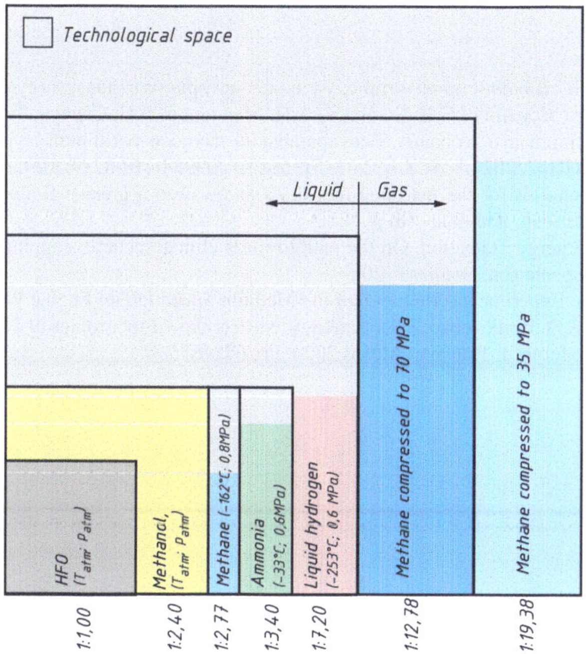

Fig. 6.1 Comparison of the volumes required for on-board storage of various types of alternative fuels compared to traditional petroleum fuels, taking into account the technological space (adapted from [2])

In addition, storing many of these fuels on board a vessel requires process volume and additional equipment to create special conditions to ensure their safe use. Process volume means the required free space in storage tanks, thermal insulation of tanks, space for placing cryogenic and other process equipment.

In Fig. 6.1 provides a comparison of the energy potentials for several main types of fuel considered as alternatives for marine internal combustion engines.

The best in this respect is methanol, which can be stored on board in liquid form at normal temperatures and atmospheric pressure. Its storage does not require a technological volume, and to ensure the specified autonomy of the vessel, the methanol reserves on board will be 2.4 times greater than similar reserves of traditional fuel.

As can be seen from Fig. 6.1, for some alternative fuels, the technological space can be comparable to the volume of the fuel itself. This is especially true for storing gaseous fuels in a compressed state and liquid hydrogen at a temperature of −253 °C. In the first case, durable cylinders are required, in the second—cryogenic tanks with a system for maintaining the required temperatures in them.

Below, the main aspects of the use of alternative fuels in marine internal combustion engines, technical solutions and operating features of such engines will considered in detail.

References

1. Білоусов, Є.В.: Використання метанолу у якості моторного палива для суднових двигунів внутрішнього згоряння / Є.В. Білоусов, А.П. Марченко, В.П. Савчук, Т.П. Білоусова // Двигун внутрішнього згорання. Всеукраїнський науково-технічний журнал. – Харків: вид. НТУ «ХПІ» – 2024. – № 1. – С 43–51. doi:10.20998/0419-8719.2024.1.06
2. Exploring scenarios for the maritime industry's pathway to a greener future. MAN Energy Solutions. Augsburg, Germany. 40 p
3. A Maritime Energy Transition. On the path towards climate-neutral shipping. MAN Energy Solutions, Copenhagen, Denmark. 20 p
4. Results of the Experimental Research of the Medium Speed Diesel Engine Work on Soybean Oil. / I. Shvets,O. Hrabovenko, S. Dotsenko, V. Nesterenko // Proceedings of 24th International Scientific Conference. Transport Means 2020. Pp. 671–675

Chapter 7
Gas Fuels and Their Use in Marine Engines

7.1 History of Creation and Development of Gas Engines

The idea of using gas fuels for internal combustion engines is as old as the engines themselves. In fact, the first engines initially created for the use of gas fuels, and only later did petroleum products take the positions in this sector that they retain to this day.

At the beginning of the twentieth century, piston internal combustion engines confidently displaced steam engines from the industrial sector. Since by this time the bulk of these engines were gas, there was already a developed infrastructure for their use, production, transportation and use of artificial gases.

The inventor of the new efficient compression ignition engine R. Diesel understood that if a new, more efficient engine offered for this infrastructure, it would certainly be a commercial success. Therefore, the inventor attempted to create a gas modification of the already developed engine [1].

According to the original concept of R. Diesel, compressed coal gas was to be into the working cylinder by a special gas injector at the end of the compression stroke. By igniting from contact with hot air and burning, gas fuel would release the heat necessary to perform useful work. However, the low pressure of gas fuel supply to the combustion chamber did not allow obtaining a homogeneous mixture, the formation of which is the main condition for effective combustion. Unlike liquid fuel, which has a greater mass, the drops of which are distributed more or less evenly throughout the combustion chamber by inertia, gas fuel ignited immediately upon exiting the nozzle tip. All subsequent portions of gas fuel entered an inert environment consisting of combustion products, without having direct contact with atmospheric oxygen, which pushed to the peripheral areas of the combustion chamber by the flame front. This circumstance made the combustion process too slow, and afterburning continued on the expansion line until the opening of the exhaust

I. Bilousov, R. Varbanets, *Marine Internal Combustion Engines and Environmental Protection Issues*, Springer Series on Naval Architecture, Marine Engineering, Shipbuilding and Shipping 27,
https://doi.org/10.1007/978-3-032-20940-5_7

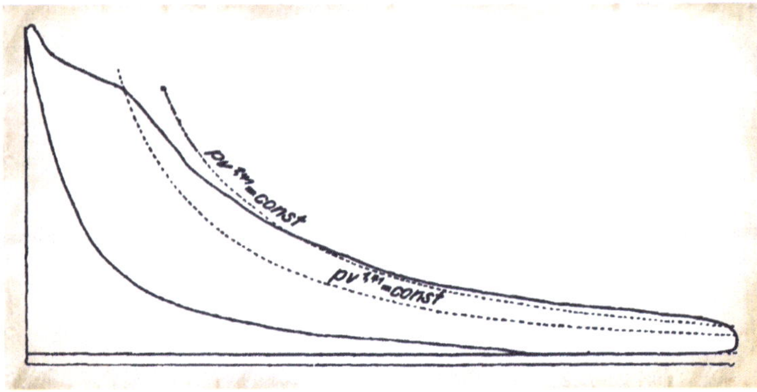

Fig. 7.1 Diagram of the working process of a gas engine tested by R. Diesel (adapted from [1])

valve (Fig. 7.1). As a result, a significant portion of the heat transferred to the cooling liquid, significantly reducing the efficiency of the working process.

As a result, the inventor forced to retreat from the original idea. In various experiments, Diesel tried to feed a very lean gas-air mixture into the working cylinder to obtain a homogeneous charge by the end of the compression stroke. Initially, the contents of the combustion chamber to be ignited by feeding additional gas fuel through a gas injector, and then, the mixture formed in two stages was to burn more efficiently than in the first experiments. However, this version of the engine did not give the expected result. The inventor's next step was to feed the entire portion of gas fuel on the intake stroke, but such a mixture began to detonate even before the piston reached TDC. To avoid this, Diesel tried to feed a small portion of liquid fuel with a very significant advance angle to ignite the gas-air mixture. Despite the fact that high-quality and, accordingly, expensive coal gas was used for the experiments, the best result obtained by Diesel during his experiments was a consumption of 350 l/(hp × h), which was slightly better than other types of gas engines used at that time. However, other types of engines were much simpler and cheaper to produce. Diesel's gas engine could not compete with such engines, and there was no need to. Soon, diesel engines on liquid fuel found application on ships, railway locomotives, cars, etc., where they still occupy a strong position today.

Time has shown that R. Diesel was on the right track. Only the insufficient technological capabilities of that time did not allow him to implement his ideas. In fact, the technologies for using gas fuel in piston engines that exist today, developed by a talented inventor more than a hundred years before they successfully implemented in our time. A new stage associated with the revival of interest in the use of gas fuels in marine diesel engines occurred in the early 1970s, when the world hit by an oil crisis. Thus, the Sulzer Company developed and tested low-pressure gas engines based on RD and RNMD series diesel engines, and by 1986, together with the Japanese company Diesel United, a high-pressure gas supply system for the RTA84 engine designed and tested. However, by that time, the crisis had passed, and these technologies were no longer in demand.

The pause did not last long, after the crisis the volume of transportation increased significantly, and with it the deadweight of the tanker fleet. With the growth of the volume of crude oil transportation by tankers, the problem of disposal of large volumes of degassing products of the transported cargo became more acute. During bunkering and transportation of tankers that transport crude oil, gas fractions (Volatile Organic Compounds—VOC) released from the cargo, which are mainly propane-butane mixtures. These gases mixed with inert gases supplied to the tanks to prevent the possibility of combustion of the cargo. As it saturated with hydrocarbons, the inert gas replaced, and the gas mixtures removed from the tanks sent for processing, during which hydrocarbon gases extracted from them. During the voyage of a tanker with deadweight of 300,000 tons from the Persian Gulf to northwestern Europe with a cargo of light Persian oil, the amount of evaporation can reach 0.6% of the total cargo volume. This amount is comparable with the fuel consumption of the main engine. Naturally, the idea of using VOC as fuel for a ship's power plant was obvious.

Interest in technologies for using natural gas as a motor fuel began to revive with the emergence and growth of sea transportation of gas fuels by specialized vessels—gas carriers.

When transporting liquefied natural gas (LNG) in special cryogenic tanks at temperatures below −160 °C, it constantly evaporates. The amount of evaporated gas per day can be 0.135–0.15% of the total cargo volume. If we take as an example, an LNG tanker with a cargo tank capacity of 74,000 m^3, then about 74 m^3 will evaporate per day. Considering that the expansion coefficient of natural gas when passing from a liquid to a gaseous state is about 600:1, then up to 44,400 m^3 of gas will evaporate per day at atmospheric pressure. The resulting gas can be re-liquefied, but this procedure is quite expensive. Therefore, the most rational is to use the evaporated gas as fuel for the ship's power plant of the tanker.

The next stimulating factor for the revival of interest in gas fuels was the tightening of environmental standards against the background of rising oil prices at the end of the last century and the beginning of this century. The accumulation of experience in the operation of tankers and gas carriers has allowed expanding the scope of use of gas fuel in the engines of other types of ships.

Today, the use of natural gas considered a promising direction for container ships, passenger ships, car carriers, ferries, etc. The use of liquefied petroleum gas considered in relation to technical fleet vessels that operate in coastal waters. As a rule, they are subject to increased environmental requirements. The relatively high density of liquefied petroleum gas, combined with moderate pressure values in cylinders, has determined their wide use for coastal vessels.

The use of gas fuel allows to significantly reducing the amount of harmful emissions in comparison with petroleum-based fuels. In particular, to eliminate sulfur emissions, to radically (by 90%) reduce nitrogen oxide (NO_x) emissions and to significantly (by 20–25%) reduce emissions of particulate matter and carbon dioxide (CO_2). Other advantages of gas fuels include the absence of liquid fractions, which eliminates oil dilution in the piston ring operating area. The almost complete absence of ash content leads to improved lubrication conditions and an increase in the service life of circulating oil. As a result, the service life of engines running on gas fuels can be increased by 1.3–1.5 times.

7.2 Properties of Gas Fuels

The most promising gas fuels include natural gas, which consists mainly of methane (CH_4) and petroleum gases, which are mixtures of mainly propane (C_3H_8) and butane (C_4H_{10}).

Ethane (C_2H_6), which is contained in natural gas in small quantities, usually not considered as an independent fuel, since its isolation is not economically justified.

The butane molecule can have a normal (normal butane—n-C_4H_{10}) and isometric structure (isobutane—i-C_4H_{10}), the properties of which differ. Depending on the ratio of individual components (Table 7.1) included in the composition of petroleum gas, its properties can vary significantly.

Hydrogen (H_2) also belongs to the category of gas fuels, however, given that now its use in the fleet is very limited, we will consider issues related to its use in a separate chapter.

The fuels considered in Table 7.1 considered gas fuels very conditionally. In practice, gas mixtures containing predominantly methane (for example, natural gas)

Table 7.1 Properties of the main gases included in the composition of gas fuels

Fuel	Methane	Propane	Butane	
Chemical formula	CH_4	C_3H_8	n-C_4H_{10}	i -C_4H_{10}
Molar mass, kg/kmol	16.05	44.11	58.14	
Carbon content, % wt.	74.83	81.68	82.63	
Hydrogen content, % wt.	25.17	18.32	17.37	
Density in liquid state, kg/m^3	422.6*	505**	579**	563**
Boiling point, °C***	161.5	42.1	−0.5	−11.7
Lower heat of combustion, MJ/kg	50.0	46.4	45.8	45.6
Vapor pressure, MPa**	–	0.837	0.225	0.260
Critical pressure, MPa	4.59	4.25	3.80	3.65
Critical temperature, °C	−82.6	96.7	152.0	134.7
Cetane number	–	10	10–20	40–55
Octane number	120–130	100–110	91–94	99–101
Methane number	100	90–100	20–30	90–100
Autoignition temperature, °C	537–600	450–470	365–405	410–460
Heat of vaporization, kJ/kg***	513.0	425.0	386.0	344.0
Minimum ignition energy, mJ (at φ =1)	0.30	0.26	0.26	0.25
Stoichiometric ratio with air, φ	1:17.2	1:15.5	1:15.4	
Maximum flame temperature, °C***	1960	1980	1970–1980	
Flammability limits, % vol.	4.4–17.0	2.1–9.6	1.8–8.5	1.8–90
Flash point, °C	−188	−104	-60–-70	-83

* at boiling point −161.5 °C; ** at temperature 20 °C; *** at pressure 0.1 MPa

used only in the gas phase. Propane-butane mixtures can feed into the working cylinder both in a gaseous state and as a liquid, which injected into the combustion chamber with a special injection valve.

Unlike natural gas, propane-butane mixtures can become liquid at ambient temperatures under pressure above 1.6 MPa (the so-called liquefied petroleum gases (LPG)). This makes it possible to store them on board in relatively simple containers.

Most engines using gaseous fuels use a simpler and more efficient external mixture formation, in which a ready-made gas-air mixture participates in the compression process. The main factor complicating the use of gaseous fuels with this method of mixture formation is the possibility of detonation combustion or detonation in gas-air mixtures.

Detonation is a combustion mode in which a shock wave propagates through a gas medium, initiating chemical combustion reactions. In turn, the exothermic reactions occurring behind the wave front maintain its movement due to the heat released. The shock wave and the zone of exothermic chemical reactions following it, called a detonation wave, which propagates through the charge at supersonic speed. A number of physical and chemical processes occurring against the background of an increasing temperature of the gas-air mixture during the compression process cause detonation. Upon reaching a certain temperature, called the autoignition temperature, the reactions acquire a chain character and proceed with a large release of heat. Detonation combustion in the working cylinder accompanied by the occurrence of wave processes, a significant increase in temperature and pressure, which significantly affects the reliability of the engine. As a result, when designing gas engines, it is necessary to provide for design and organizational measures designed to prevent the engine from operating in the zone of modes in which detonation may occur. We will consider these measures in more detail in the relevant chapters.

Their methane number determines the resistance of gas fuels to detonation.

The methane number for gas fuels calculated as the volumetric content (in percent) of hydrogen (H_2) in a mixture with methane (CH_4). Pure methane corresponds to a methane number of 100 (knock-resistant), and hydrogen—0 (unstable). Thus, if the methane number of a fuel is 65, this means that this fuel has the same detonation resistance as a reference mixture consisting of 65% methane and 35% hydrogen. Since gas fuels in most cases are a mixture of gases, their detonation resistance depends on the ratio of their various components and the properties of the gases included in the mixture.

7.3 Technologies for the Use of Gas Fuels in Marine Engines

The specific operating conditions of ships have left their mark on the development of methods for using gas fuels in ship engines. A feature of oil and gas transportation by sea is that cargo transported only in one direction; in the opposite direction, the ship moves in ballast. Thus, the ship's engine must be able to use gas fuel when the ship is moving with cargo and regular liquid motor fuel when the ship is moving without cargo. That is, the engine must be dual-fuel. In addition, depending on the type of cargo, navigation conditions, external conditions and time from the moment of loading, the composition of gases taken from the cargo tanks can vary significantly. In this regard, the engine must be able to operate on both types of fuel simultaneously, using all the available energy of the gas fuel, and in case of its deficiency, compensate for the energy deficit by feeding liquid fuel to the cylinders. The above applies primarily to oil tankers and gas carriers, but is also applicable to ships of other types, for which liquid fuel must remain a reserve. Based on the above, the majority of marine engines today are dual-fuel (DF), i.e. capable of running on gas, liquid fuel, or both fuels at once in various proportions.

There are several fundamentally different approaches to organizing the working process in gas engines [2]:

- converting diesel engines into engines with external mixture formation and spark ignition operating on the Otto cycle;
- using external mixture formation with ignition of the gas-air mixture from a small portion of pilot fuel injected into the working cylinder;
- using internal mixture formation and ignition of the gas-air mixture from a small portion of pilot fuel injected into the combustion chamber.

Each of the above methods can be implemented using both natural and petroleum gases. All of these methods have found their application in marine engines of various types, so let us consider the features of the fuel systems of each type of engine in more detail. Issues related to the storage of gases on board a vessel will not considered, as they are beyond the scope of this book. This topic discussed in sufficient detail in the relevant specialist literature.

7.4 Conversion of Diesel Engines to Spark-Ignition Gas Engines

Conversion of diesel engines into engines with external mixture formation and spark ignition operating on the Otto cycle is the simplest way to convert an engine to gas fuel. Its advantages include [2]:

- simplification of the design (if there is no need to maintain dual-fuel, it is possible to completely abandon the liquid fuel injection system, replacing it with simpler ones: an air-gas mixing system and a spark ignition system);
- operation of the fuel system at low pressures, which reduces the safety requirements for the operation of such systems.

The main disadvantages of using these systems are the loss of the ability to switch to liquid fuel, as well as a decrease in liter capacity associated with a decrease in the weight filling of the cylinders. In addition, at compression ratios typical for diesel

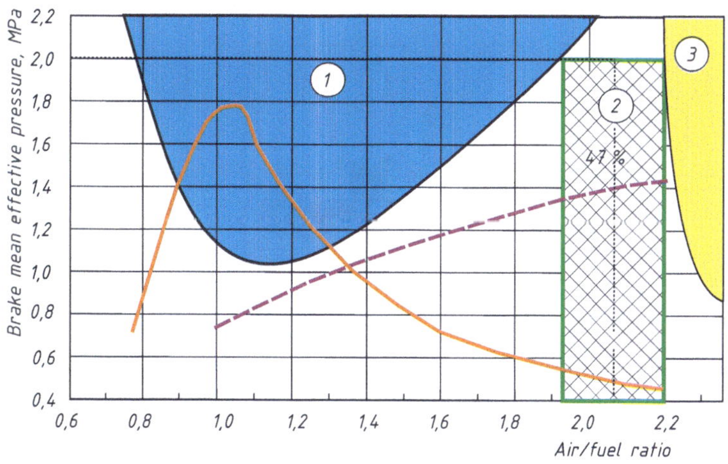

Fig. 7.2 Nature of the working process of a gas engine operating on the Otto cycle, depending on the air-fuel ratio (according to Wärtsilä): (*Horizontal lines*) NO_x does not exceed 1 g/(kW × h); (*Dashed lines*) thermal efficiency of the engine: (1) detonation zone, (2) zone of optimal parameters, (3) zone of combustion impossibility (adapted from [2])

engines, detonation occurs over a wide range of loads, which prevents the use of such engines in installations with direct power transmission to the propeller. Stable operation of engines, which ensures high power and efficiency indicators, observed only in a relatively narrow range (Fig. 7.2), when operating on relatively lean mixtures. It is evident from the figure that the optimum air excess ratio lays in a very narrow range.

When it decreases below 1.9; the engine operates with detonation, and when it increases above 2.2; there is a risk of misfires. This circumstance significantly complicates the qualitative regulation of engines (by changing the amount of gas supplied). In such engines, quantitative regulation by throttling the airstream at the engine inlet using an air damper or by changing the turbocharger capacity is more traditional. Based on the above considerations, engines with external mixture formation and spark ignition used primarily in power plants of oil and gas drilling platforms, where it is possible to ensure their reliable supply of gas fuel throughout the entire period of operation. Less often, they used as part of multi-machine generator sets, ships with electric power transmission to the propeller. In marine engines with relatively large cylinder diameters, the energy of an electric spark is usually insufficient to achieve stable ignition and efficient combustion of lean gas-air mixtures. Therefore, the pre-chamber-torch ignition scheme shown in Fig. 7.3 is widely used in them. The pre-chamber installed in place of the standard fuel injector and is a cavity connected to the main combustion chamber by means of several channels made in the tip of the pre-chamber protruding from the cylinder cover. An electric spark plug and an additional gas supply valve installed in the pre-chamber. During the filling stroke, gas enters the working cavity of the cylinder through a special mixer, which distributes it in the airflow (Fig. 7.3a). The gas supply by the mixer

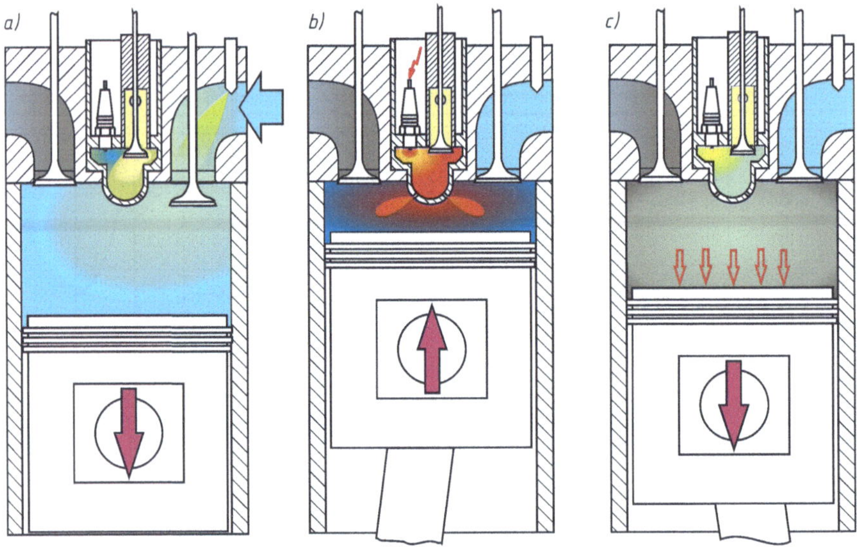

Fig. 7.3 Working process of a gas engine with pre-chamber-torch ignition: (**a**) filling the working cylinder with a lean and pre-chamber with a rich gas-air mixture; (**b**) ignition of the mixture in the pre-chamber by an electric spark and ignition of the mixture in the combustion chamber by plasma jets; (**c**) working stroke (adapted from [2])

selected in such a way as to prepare a lean mixture. In parallel with the gas-air mixture entering the working cylinder, an additional portion of pilot gas fuel enters the pre-chamber cavity through the gas supply valve. As a result, an enriched combustible mixture formed in the pre-chamber. When an electric spark is applied, the mixture in the pre-chamber easily ignites, the pressure increases sharply, after which the burning gases in the form of plasma jets fly out into the main combustion chamber, effectively igniting the lean working mixture, facilitating its complete combustion (Fig. 7.3b). Then the working stroke is performed (Fig. 7.3c). A significant advantage of this method of ignition of the working mixture is that the power of the ignition source in it is significantly greater than in a spark-ignition engine. In addition, the working mixture ignited not at one point near the cold wall, but in the center of the charge. As a result, the efficiency of the working process implemented using this technology can reach 48%.

Figure 7.4 shows a diagram of the fuel system of the Rolls-Royce KV-G3 series pre-chamber-torch ignition engine. The gas supply to the working cylinder is forced and controlled by a special gas-valve, which driven by the engine.

The gas fuel system of the Bergen C26:33 Gas engine from Rolls-Royce made according to a similar scheme (Fig. 7.5).

A gas-valve used to supply gas to the inlet channel of the cylinder cover. The inlet valve lever drives it. This ensures synchronization of the gas fuel process with the air intake into the cylinder, which prevents gas from entering and accumulating

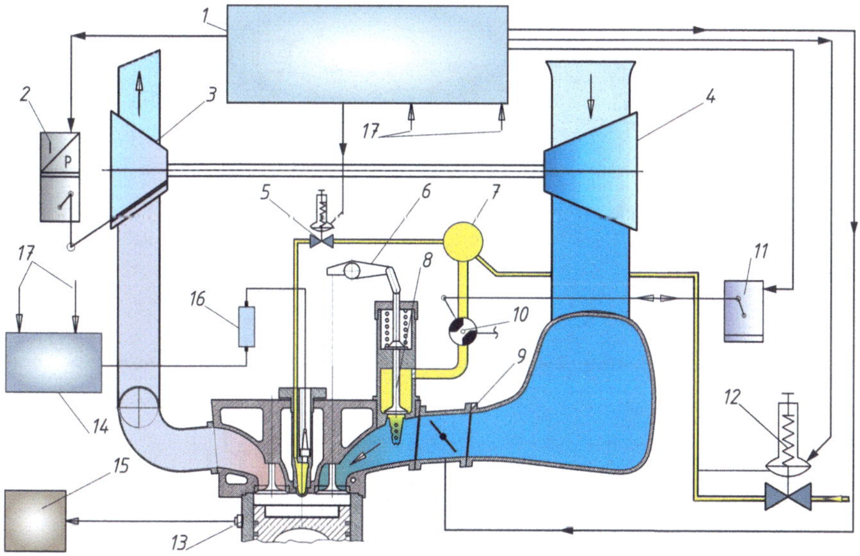

Fig. 7.4 Schematic diagram of the gas supply system of the KV-G3 engine by Rolls-Royce: (1) engine control unit; (2) servo drive for controlling the turbocharger turbine guide vane; (3) turbine; (4) compressor; (5) pressure regulator in the pre-chamber gas supply circuit; (6) gas-valve drive; (7) gas manifold; (8) gas-valve; (9) air damper; (10) spool throttle for regulating the gas supply to the engine; (11) servo drive for regulating the gas supply to the engine; (12) pressure regulator in the gas supply circuit of the working cylinder; (13) knock sensor; (14) spark ignition control unit; (15) knock sensor data processing unit; (16) ignition coil; (17) inputs from sensors for speed, crankshaft position, air inlet temperature, etc. (adapted from [3])

in the inlet receiver. Given the large volume of the inlet receiver, the accumulation of a gas-air mixture in it can cause an explosion with serious consequences.

Gas fuel supplied to the valve through a branch pipeline from the common gas line located along the engine. To regulate the amount of gas fuel entering the engine (the composition of the mixture), a slide valve regulator is installed in front of the gas-valve, made in the form of a hollow cylinder with a slot. As the cylinder rotates, the slot gradually closes the inlet on the spool valve body, throttling the gas flow, thereby reducing the fuel supply to the engine. The spool valve controlled through a mechanical drive system that connected to the engine control unit.

In most modern engine models (e.g. Wärtsilä 34SG), electrically driven valves controlled by an electronic unit (Fig. 7.6) are used to control the gas fuel flow. This allows for more efficient regulation of both the main (intake) and pilot (pre-chamber) portions of gas fuel.

To prevent combustion products from entering the gas supply line to the pre-chamber, an automatic check valve installed in it, which suspended on a spring calibrated to a certain opening pressure (Fig. 7.6).

To ignite the enriched gas-air mixture, an electric spark plug connected to the ignition system installed in the pre-chamber.

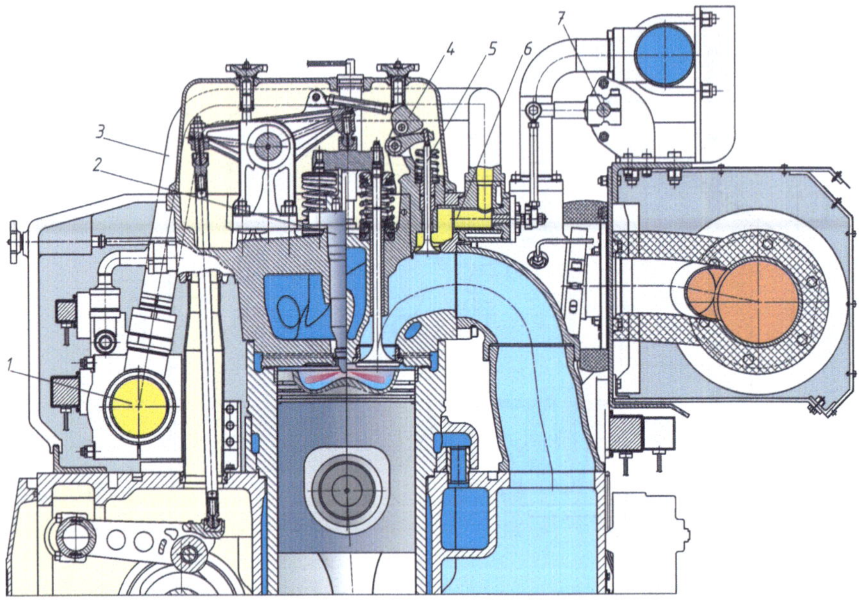

Fig. 7.5 Layout of the main components of the fuel system of the Bergen C26:33 Gas series gas engine, from Rolls-Royce: (1)main gas line; (2) ignition module; (3) gas supply line to an individual cylinder; (4) gas fuel supply valve drive mechanism; (5) gas supply valve; (6) gas supply spool regulator; (7) gas supply spool regulator drive mechanism (adapted from [4])

The ignition system is individual for each cylinder. It consists of an ignition module, ignition coil, high-voltage extension and spark plug (Fig. 7.7). The Teflon insulated high-voltage extension cord has a 5 kOhm resistor installed to reduce interference created by the spark. The Cylinder Control Unit (CCU) performs ignition control. The ignition timing determined by the Main Control Unit (MCU) based on the processing of data received from various sensors. The value of the timing and the duration of the spark delay selected to be optimal for the corresponding engine load.

Modern gas engines are equipped with a combustion control system, which used to detect detonation in one or more cylinders by measuring the vibration level. Based on the analysis of signals from knock sensors installed on individual cylinders, the system can automatically change the spark advance angle or reduce the engine load to 85% of the nominal. In the event of excessive detonation, the system can emergency stop the engine.

Modern gas engines use electronic control systems. For example, the Wärtsilä 34SG engines are equipped with a system called WECS 3000 (Wärtsilä Engine Control System). It consists of several types of units that are located in different places on the engine (Fig. 7.8):

- Main Control Unit (MCU);
- Cylinder Control Unit (CCU);

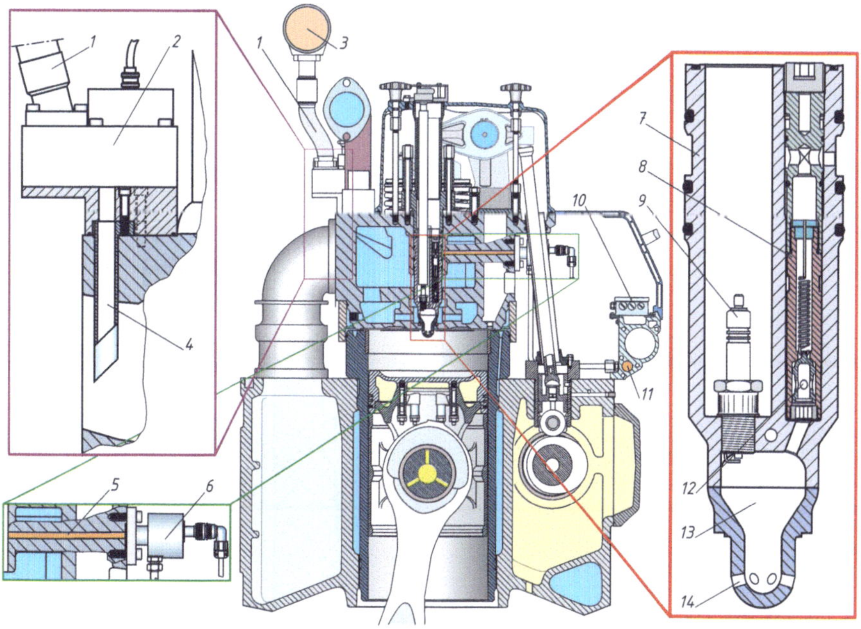

Fig. 7.6 Wärtsilä 34SG prechamber flare-ignited gas engine: (1) gas fuel supply line to the main gas-valve; (2) main gas-valve with electromagnetic drive; (3) main gas line; (4) mixing tube; (5) pilot gas supply channel to the prechamber; (6) gas-valve for controlling gas supply to the prechamber with electromagnetic drive; (7) prechamber housing; (8) housing of the prechamber check valve; (9) spark plug; (10) electronic ignition unit; (11) pilot gas fuel supply line; (12) prechamber check valve; (13) prechamber; (14) prechamber nozzle tip holes (adapted from [5])

- Knock Detection Unit (KDU);
- Distributed Control Unit (DCU);
- Sensor Multiplexer Unit (SMU).

All units can communicate with each other via the CAN bus.

Each engine is equipped with one MCU, which installed in a cabinet on elastic supports, located on the flywheel side of the engine. Other units installed in different places on the engine near the sensors and actuators with which they interact or which they control.

The MCU controls the ignition parameters, gas pressure, time and duration of gas supply. It also controls the engine speed, engine load, turbocharger, engine start and monitors it during operation.

The CCU units control and monitor processes related to the operation of two separate cylinders, such as ignition, gas supply to the pre-chamber and to the intake port in accordance with the reference values of time and duration received from the MCU. The angular position and speed of rotation of the crankshaft calculated by the CCU based on the processing of pulses received from the angle sensor (encoder). The sensor installed on the camshaft of the engine.

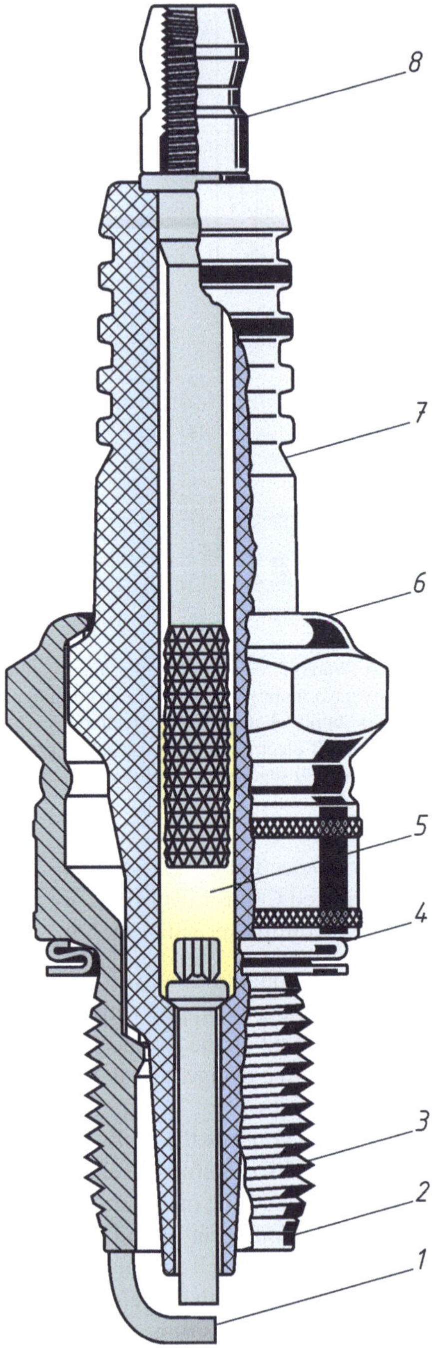

Fig. 7.7 Electric spark plug: (1) side electrode; (2) central electrode; (3) mounting thread; (4) thermal cone; (5) seal; (6) ceramic resistor; (7) hex head housing; (8) electrode; (9) ceramic insulator; (10) contact (adapted from [3])

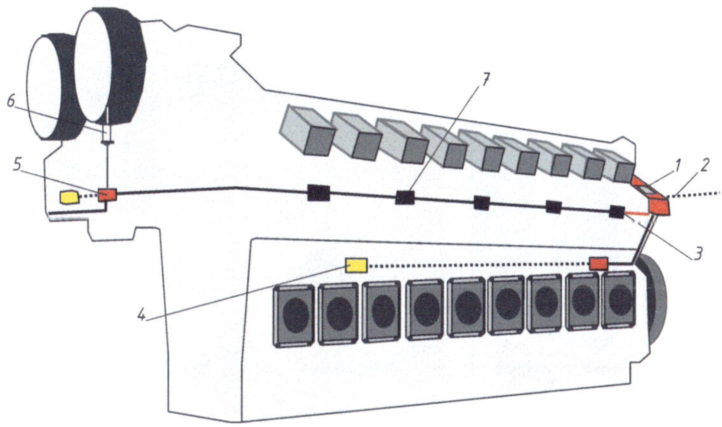

Fig. 7.8 Location of the main components of the WECS 3000 monitoring system of the Wärtsilä 34SG gas engine: (1) Main Control Unit (MCU) (2) Connection socket for external systems; (3) Knock Control Unit (KDU); (4) Sensor Multiplexer Unit (SMU); (5) Distributed Control Unit (DCU); (6) Sensors connected to the SMU or DCU; (7) Cylinder Control Unit (CCU) (adapted from [5])

The KDU unit used to monitor signals from knock sensors installed on each cylinder cover. The KDU filters, amplifies and converts the received signals into digital information, which it sends via the CAN bus to the MCU for further processing. One KDU simultaneously serves up to nine cylinders.

The DCU and SMU units are devices used for measurement. They convert various types of sensor signals into digital information, which sent over the network to the microcontroller for further processing.

For efficient combustion of gas fuel in a gas engine, it is necessary to maintain the fuel-air ratio in a very narrow range (Fig. 7.2). If the gas-air mixture is too lean, misfire will occur, and if the mixture is too rich, detonation may occur. To maintain optimal mixture parameters, it is not enough to regulate the amount of gas supplied to the cylinders, it is also necessary to manage the air supply. For this purpose, gas engines are equipped with turbocharger control devices. This can be control of the turbine guide vane, as shown in Fig. 7.4, or control of the bypass of part of the exhaust gases past the turbine (Fig. 7.9a).

To maintain an optimum air-fuel ratio in the mixture, the Wärtsilä 34SG engines are equipped with an exhaust gas bypass valve (Fig. 7.9b), which bypasses part of the gases past the turbocharger turbine. This regulates the turbine power, which allows maintaining the air pressure in the supercharging receiver at an optimum level throughout the entire range of operating conditions.

The exhaust gas bypass valve is installed on the engine and consists of a drive connected to the throttle valve. The valve allows smooth regulation of the flow of bypassed gases, for which it is equipped with a proportional pneumatic drive. Compressed air for controlling the drive supplied from the general ship system under a pressure of 0.6–0.7 MPa. Air enters the system through a solenoid valve,

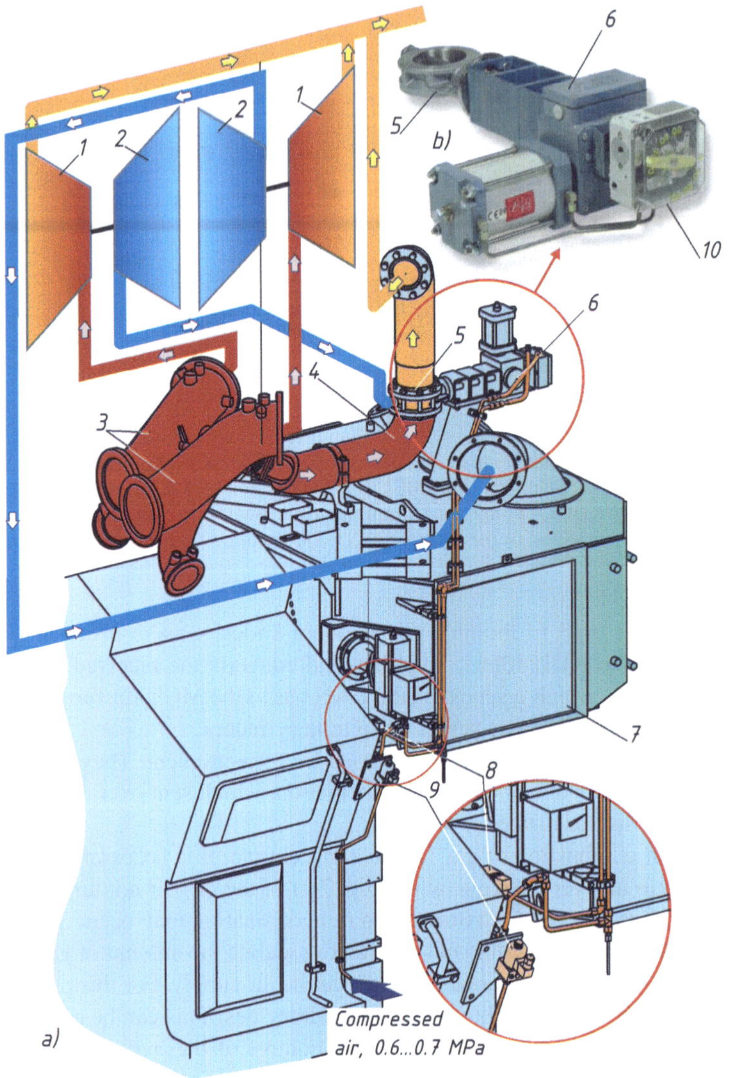

Fig. 7.9 Wärtsilä 34SG gas engine exhaust gas bypass control system: (**a**) connection of main components; (**b**) waste gate with actuator; (1) turbocharger turbine; (2) compressor; (3) exhaust manifolds; (4) bypass; (5) regulating valve; (6) regulating valve actuator; (7) charge air cooler; (8) I/P converter; (9) solenoid valve; (10) waste gate positional (adapted from [5])

after which it enters I/P converter, which allows transforming an electric current signal (I) into proportional pneumatic pressure (p). Then, air under the pressure specified by I/P converter enters the bypass valve drive positional, which rotates the shut-off element (flap) to the required angle.

The engine management system (WECS) monitors the average temperature of the exhaust gases at the outlet of the cylinders. If this temperature is higher than the

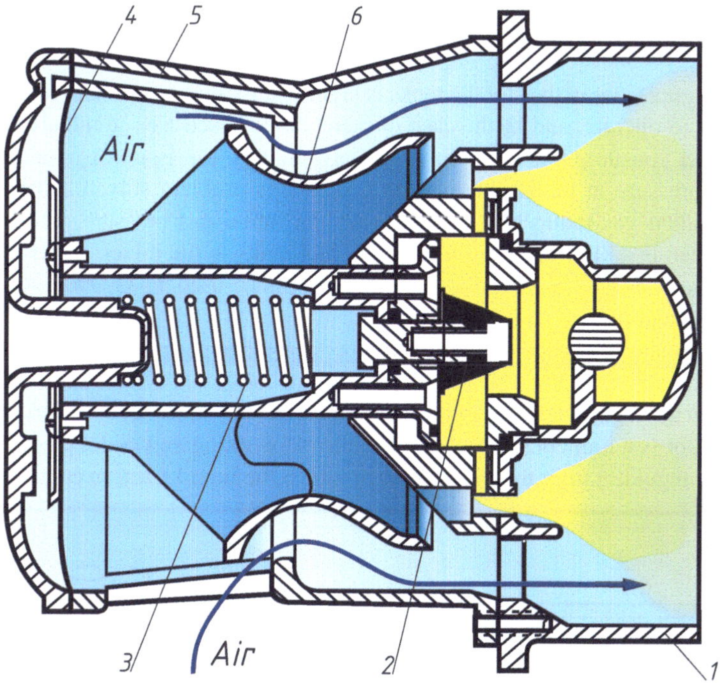

Fig. 7.10 Gas mixer of the G 3500 series engines by Caterpillar: (1)mixer body; (2) gas-valve; (3) gas-valve spring; (4) gas-valve drive diaphragm; (5) vacuum supply channel under the diaphragm; (6) diffuser (adapted from [6])

specified level, the engine management system will begin to close the bypass valve until the required temperature restored. The main purpose of the solenoid valve is to prevent air leakage through I/P converter when the engine stopped.

Gas mixers To prepare the gas-air mixture at the engine inlet, various types of mixing devices are used, an example of the design of one of which shown in Fig. 7.10. Caterpillar uses mixers of this type in gas engines of the G 3500 type. After cleaning, the air enters the engine, passing through a diffuser. Because of the change in flow speed in the diffuser, a vacuum occurs, which transmitted through a special channel to the cavity behind the diaphragm that actuates the gas control valve.

Under the action of vacuum, the diaphragm bends and moves the valve, which has a conical protrusion on the end that changes the flow section of the gas supply channel. Thus, the amount of gas supplied by the mixer is always proportional to the amount of air entering the engine, thereby maintaining an optimal air-fuel ratio. When the engine stops, the lack of vacuum will cause the valve to block the gas supply to the engine with its end surface. In engines with small cylinder diameters, a simplified scheme for feeding gas fuel to the intake receiver used. In this case, one gas mixer installed at the inlet to the receiver, which serves all cylinders simultaneously (in some designs, a group of cylinders).

Gas pressure regulator For normal engine operation, gas must supply to the mixer at a strictly defined pressure.

The gas pressure in the storage tanks is significantly higher than that required for the mixer to operate, and in the case of using compressed gas; it can also change during fuel consumption. Therefore, for gas engines, the issue of maintaining a constant pressure at the engine inlet is acute. This problem solved using special devices called gas reducers. Depending on the initial gas pressure, one to three reducers can install in the section from the fuel tanks to the mixer. An example of the design of a pressure regulator from Caterpillar, used in G 3600 series gas engines, shown in Fig. 7.11.

This regulator designed to work with propane-butane gas mixtures, and can used as a second stage of pressure reduction when using compressed natural gas. The pressure at the regulator inlet is within 2420 kPa, at the outlet—21 kPa. Structurally, the regulator is a unit consisting of two reducers: working and outlet pressure. Gas enters the regulator after the evaporator, in which the liquid fraction converted into

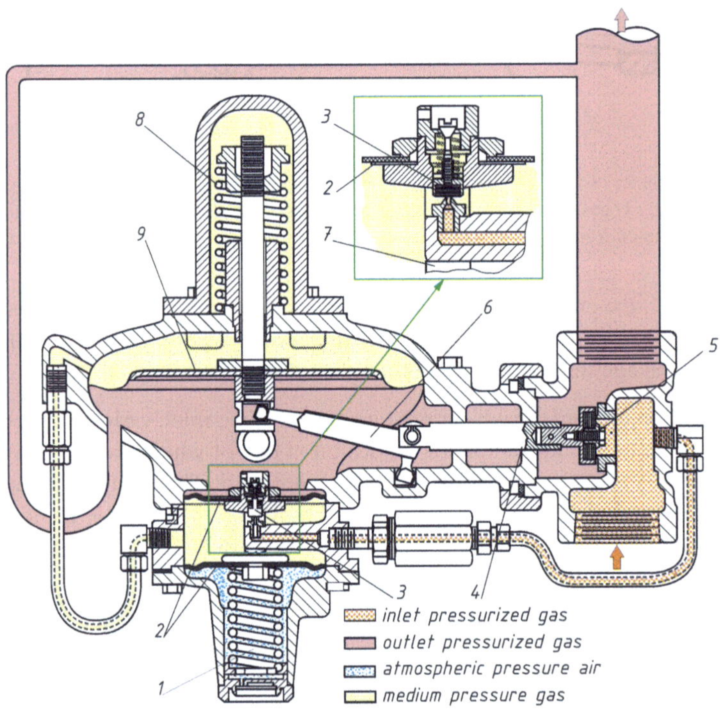

Fig. 7.11 Design and operation diagram of the dual gas reducer of the G 3600 series engine by Caterpillar: (1)working pressure reducer spring; (2) working pressure reducer diaphragms; (3) dual working pressure reducer valve; (4) outlet pressure regulating valve drive rod; (5) outlet pressure regulating valve; (6) outlet pressure regulating valve drive lever; (7) intermediate stop of the working pressure reducer valve drive; (8) outlet pressure reducer spring; (9) outlet pressure reducer diaphragm (adapted from [6])

gas, and after preliminary cleaning. Initially, the gas enters the cavity of the working pressure reducer, which serves to maintain a constant pressure in the working cavities of the regulator. The working pressure reducer consists of two diaphragms, on the upper of which a double valve mounted, together with its body and seat. At the beginning of operation, in the absence of pressure, the spring of the working pressure reducer holds the valve open through an intermediate stop. Therefore, when the cavity of the reducer begins to fill, the increase in pressure leads to the fact that the lower diaphragm is pressed back, overcoming the force of the spring, and releases the upper diaphragm. The gas fills the cavity of the reducer and the cavity above the diaphragm of the outlet reducer.

The outlet pressure reducer serves to reduce and maintain the pressure at the engine inlet. Structurally, it consists of a diaphragm loaded with a spring and a gas-valve driven by the diaphragm through a system of levers.

When the engine is not running, the spring lifts the diaphragm and holds the valve closed through a lever transmission—gas does not flow to the engine. When gas flows from the working pressure reducer into the cavity above the diaphragm, the force acting on it increases and, overcoming the force of the spring, the diaphragm moves downwards, opening the valve. Gas begins to flow to the engine. At the same time, the gas fills the cavity under the diaphragm of the outlet reducer and creates a force that tends to move the diaphragm upwards, thereby closing the valve. As a result, equilibrium established between the valve position, the gas flow through it and the pressure at the regulator outlet.

7.5 Four-Stroke Gas-Diesel Engines with External Mixture Formation and Pilot Ignition

The fuel equipment of four-stroke gas-diesel engines with external mixture formation distinguished by the presence of two, and in some cases, three fuel supply systems for gas and liquid fuel. As a rule, gas-diesel engines created based on existing diesel engine models with additional equipment of the necessary components and systems. In some cases, the cylinder diameters increased to compensate for power losses associated with the transition to the gas cycle. The system for supplying gas fuel together with incoming air is practically no different from the previously considered fuel system of gas engines with spark ignition (Fig. 7.12a). Considering that gas fuels have a low ability to self-ignite, an external flame source is required to ignite the gas-air mixture. In gas-diesel engines, a small portion of liquid fuel injected for these purposes, called an ignition or pilot supply (Fig. 7.12b). Otherwise, the working process is not much different from the previously considered gas engine (Fig. 7.12c).

There are two main approaches to organizing the working process of a gas-diesel engine with external mixture formation:

- supplying a pilot portion using the standard fuel equipment of the engine;
- using special fuel systems for injecting a pilot portion of fuel.

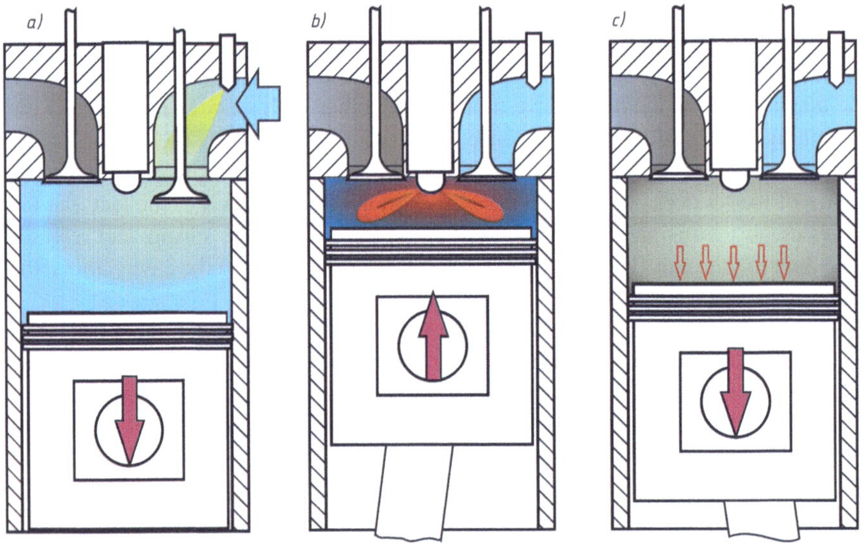

Fig. 7.12 Scheme of the working process of a four-stroke gas-diesel engine with external mixture formation: (**a**) filling the working cylinder with a lean gas-air mixture; (**b**) injection of pilot fuel; (**c**) working stroke (adapted from [2])

7.5.1 Supply of Pilot Fuel Using the Standard Engine Fuel Equipment

With this conversion method, the diesel engine retrofitted with gas system components without significant modification of the standard liquid fuel supply system. When running on gas, the fuel pumps are switched to the minimum stable supply mode, and the engine load is regulated by changing the amount of gas entering the working cylinder (qualitative regulation) or the gas-air mixture (quantitative regulation). The undoubted advantages of the first approach are the need for minimal modification of the engine itself and its fuel system, the ability to switch to operation on liquid fuel at any time. The main disadvantage of this approach is the impossibility of significantly reducing the consumption of liquid fuel for organizing pilot injection. Typically, in engines of this type, the consumption for pilot supply is 15–20% of the consumption on liquid fuel. This explained by the impossibility of obtaining stable injection at low feeds using standard fuel equipment. Standard fuel equipment adjusted to the nominal engine-operating mode; with a decrease in the cyclic fuel supply, the uniformity of supply deteriorates. In addition, a decrease in the cyclic fuel supply leads to a decrease in injection pressure, which worsens the quality of fuel atomization and can lead to misfires.

An example of gas-diesel engines of marine electric generator sets made in accordance with the first approach, shown in Fig. 7.13. MAN Energy Solutions developed such dual-fuel generator sets as part of a comprehensive solution for

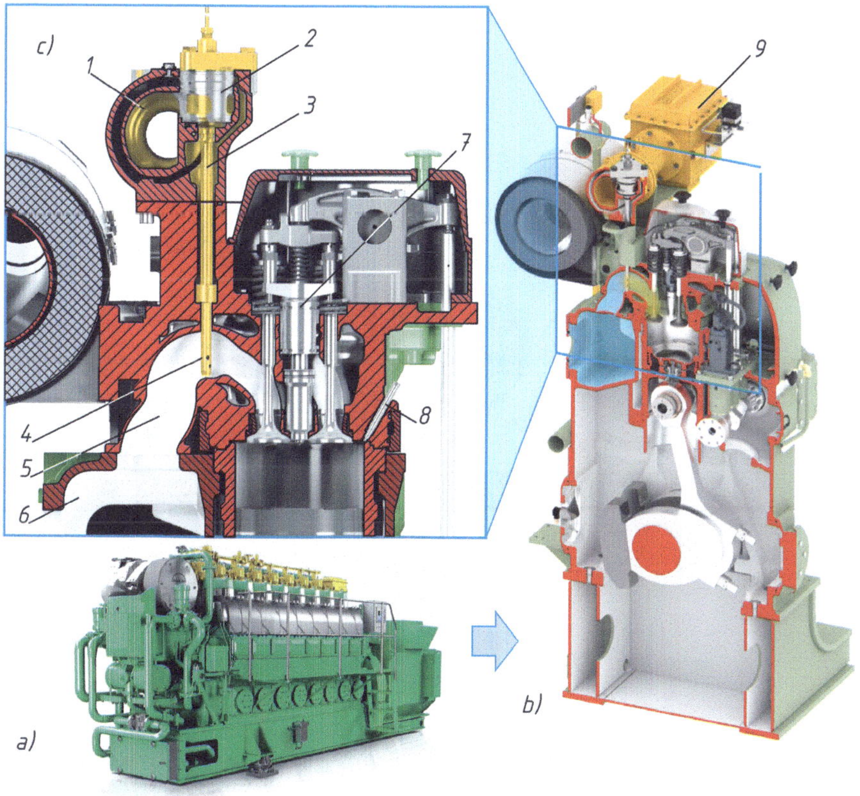

Fig. 7.13 Marine gas diesel engine MAN L28/32DF: (**a**) general view; (**b**) general structure; (**c**) location of main elements of the engine gas fuel supply system: (1) common gas line with external ventilated circuit; (2) gas-valve of the working cylinder; (3) pipe for supplying gas fuel to the mixer; (4) mixer; (5) air duct; (6) air receiver; (7) nozzle for injecting liquid fuel; (8) knock sensor; (9) gas regulator housing (adapted from [7, 8])

ships equipped with low-speed dual-fuel main engines, MAN B&W ME-GI. As a drive, the generators are driven by dual-fuel engines of the MAN L23/30DF and MAN L28/32DF series, created on the basis of the well-proven MAN L23/30 and MAN L28/32 diesel engines, respectively.

In addition to the reliability associated with the use of one of the most reliable engines in this segment as a base, this approach makes it quite easy to upgrade engines already in operation to a gas-diesel version. All components required for this unified with the existing engine design, which significantly simplifies and reduces the cost of such modernization.

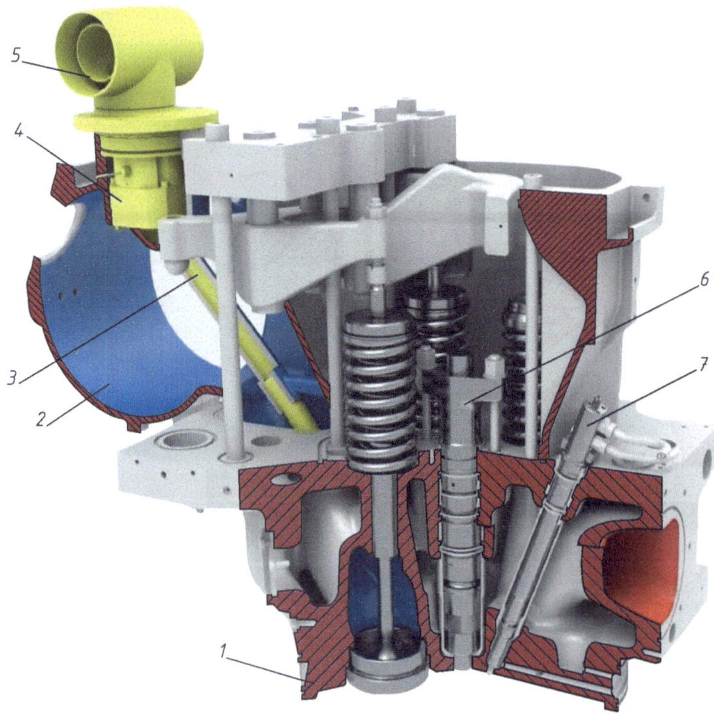

Fig. 7.14 Cylinder cover of MAN 35/44DF gas-diesel engine: (1) cylinder cover; (2) intake receiver; (3) gas supply pipe to intake channel; (4) gas supply control valve; (5) gas line; (6) liquid fuel supply injection; (7) pilot fuel supply injection (adapted from [9, 10])

7.5.2 Use of Special Fuel Systems for Pilot Fuel Injection

This approach boils down to retrofitting the engine with an additional fuel system, specially designed to supply a pilot portion of fuel. An example of this approach is the series of gas engines from MAN, Wärtsilä and others, created because of the existing model range of diesel engines.

In some engines, to compensate for power losses when switching to gas fuel, manufacturers have increased the cylinder diameter. For example, in the Wärtsilä 50DF engine, created because of the 46 series diesel engine, the cylinder diameter increased from 460 to 500 mm, MAN 51/60, created because of MAN 48/60, the cylinder diameter increased from 480 to 510 mm.

All engines designed for continuous operation on gas, but liquid fuel such as MDO can use as a backup. Operation on heavy fuels such as HFO also allowed, but this requires certain preparation and the use of more expensive circulating oil. When running on liquid fuel, the engine uses a standard fuel system. For pilot injection, MAN installs an additional injector (Fig. 7.14), and Wärtsilä installs a dual injector with two nozzle tips for main and pilot fuel injection (Fig. 7.15).

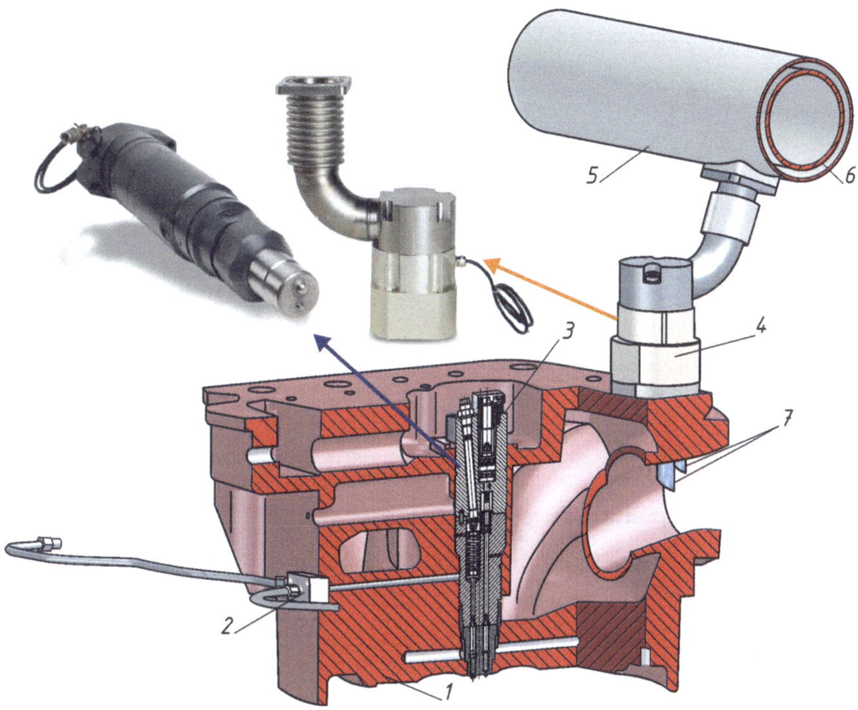

Fig. 7.15 Cylinder cover of the Wärtsilä 20DF gas diesel engine: (1) cylinder cover; (2) fuel supply to the pilot injection system; (3) twin injector; (4) gas supply control valve; (5) outer casing of the gas line; (6) gas line; (7) mixer (adapted from [10])

To prepare the gas-air mixture, the engines are equipped with a gas supply system to the mixers installed on each cylinder cover. Thus, the gas-air mixture is prepared directly at the cylinder inlet (Fig. 7.12). This solution avoids the risk of explosion in the intake receiver, which has a large volume and often designed as a cavity in the crankcase. Therefore, a gas explosion in the receiver can lead to the destruction of the entire engine.

To supply the pilot portion, modern gas-diesel engines use an accumulator fuel injection system.

Accumulator systems ensure stable supply and quality of fuel atomization regardless of the engine-operating mode. In addition, the parameters of all fuel system elements specially optimized to ensure stable supply of small cyclic portions. As a result, the amount of pilot fuel was reduced to 1–5%.

The diagram of the fuel system for the injection of reserve and pilot fuel of the Wärtsilä 50DF gas-diesel engine, shown in Fig. 7.16.

In the systems of supply of pilot liquid fuel under high pressure of the accumulator type, the radial axial fuel pump of L'Orange GmbH (Fig. 7.17) is widely used, which is driven by an electric motor. The electric motor controlled, as well as all other processes, by the electronic control unit of the engine.

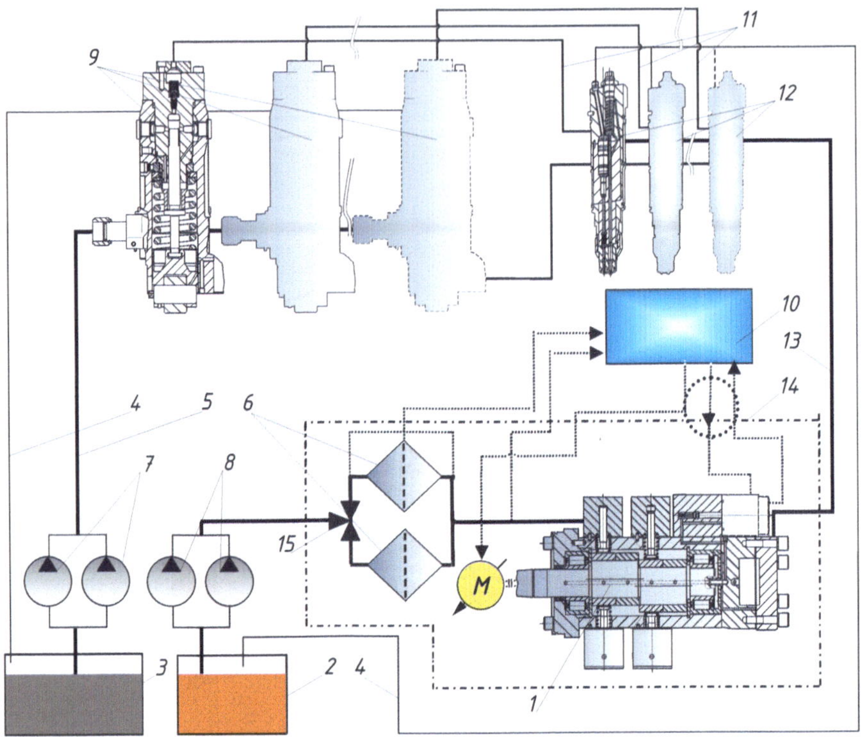

Fig. 7.16 Schematic diagram of the fuel supply system of Wärtsilä gas-diesel engines with an additional circuit for pilot fuel injection: (1) electrically driven pilot injection system high-pressure pump; (2) light fuel tank for pilot injection; (3) main supply heavy fuel tank; (4) return line; (5) heavy fuel pressure line; (6) fuel filters; (7) heavy fuel feed pump; (8) light fuel feed pump; (9) standard heavy fuel high-pressure pumps; (10) electronic control unit; (11) high-pressure lines; (12) injectors with twin nozzles; (13) high-pressure line of pilot fuel; (14) fuel pump unit of the pilot injection system (adapted from [11])

After the pump, the fuel enters the common accumulator for all injectors under a pressure of 90 MPa, from where it supplied to the injectors with two injector tips via separate pipelines. The injector design shown in Fig. 7.18. In fact, two injectors are located in one housing, one of which has a design traditional for engines with a volumetric injection system, and the other is equipped with an electromagnetic valve for controlling the supply, typical for injectors of accumulator systems. The nozzle housing also has two needle valves: a small one for injecting pilot fuel and a large one for injecting the main fuel.

Apart from the use of a special injector, the engine fuel system designed to supply liquid fuel is no different from the basic system used in Wärtsilä engines, series 46. An electronic engine control unit controls pilot injection.

The control unit also controls the transition from gas to diesel fuel. If necessary, the engine can switch from gas to diesel fuel without stopping at full load. The reverse transition can carry out at a load of up to 80% of the nominal value.

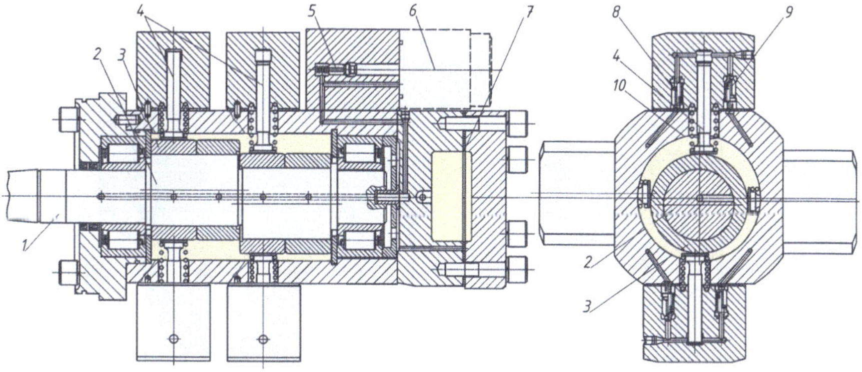

Fig. 7.17 Radial-plunger pump from L'Orange GmbH for Common Rail pilot fuel supply systems: (1) pump drive shaft; (2) drive eccentric; (3) drive roller; (4) plunger; (5) throttling feed control valve; (6) electromagnetic drive of the feed control valve; (7) accumulator; (8) filling valve; (9) discharge valve; (10) return spring (adapted from [11, 12])

An electronic unit controls the gas fuel supply. An example of a diagram of the gas fuel system in gas-diesel engines for marine use shown in Fig. 7.19.

The gas pressure in the fuel system depends on the engine load and the calorific value of the gas, but in all cases, it does not exceed 0.4 MPa.

To prevent gas leaks into the engine room, all gas lines placed in protective casings, the interior of which constantly ventilated. To prevent mechanical damage to the supply control valves due to thermal expansion, they connected to the main gas line using sealed bellows. Flame arresters and methane content sensors in the ventilation gases installed at the outlet of the ventilation system. If methane detected in the latter, the automatic protection system turns on the alarm, and if the concentration reaches 60% of the lower flammability limit of the gas-air mixture, an emergency engine shutdown occurs.

In case of damage to the gas system, the engine automatically switched to liquid fuel, and the gas is displaced using nitrogen supplied from the inert gas system. The remaining gas fuel removed through the emergency relief valve into a special combustion device.

To prevent gas fuel from entering the crankcase through leaks in the piston rings, engines must started only on liquid fuel. When maneuvering the vessel, the engine must also switch to liquid fuel. The crankcase valves must adjusted taking into account the possibility of an explosion in the crankcase of the gas-air mixture formed because of gas leakage through the piston rings. The crankcase ventilation system must ensure effective removal of crankcase gases. It is mandatory to equip it with a methane sensor and an oil mist sensor. Most gas-diesel engines have a separate option for installing crankshaft bearing heating sensors and an inert gas crankcase ventilation system.

The exhaust tract of a gas engine must not connect to the exhaust tracts of other engines.

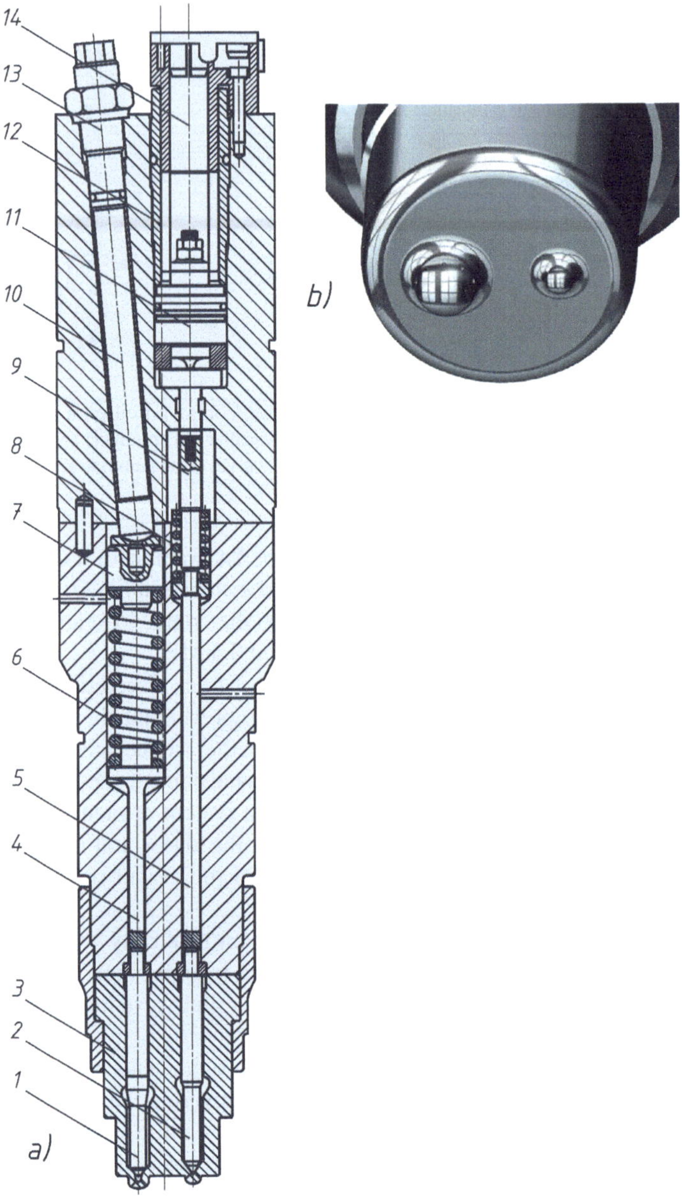

Fig. 7.18 Wärtsilä 50DF gas-diesel engine injector with twin nozzles for reserve and pilot liquid fuels (**a**) general view of nozzle tip (**b**): (1) reserve fuel supply needle valve; (2) pilot fuel supply needle valve; (3) nozzle body; (4) reserve fuel needle valve rod; (5) pilot fuel needle valve rod; (6) main fuel needle valve spring; (7) spring stop; (8) pilot fuel needle valve spring; (9) pilot fuel needle valve locking piston; (10) spring adjusting rod for setting reserve fuel valve opening pressure; (11) pilot fuel nozzle control solenoid; (12) — mounting cup; (13) adjusting bolt; (14) installation stop with electrical connector (adapted from [13])

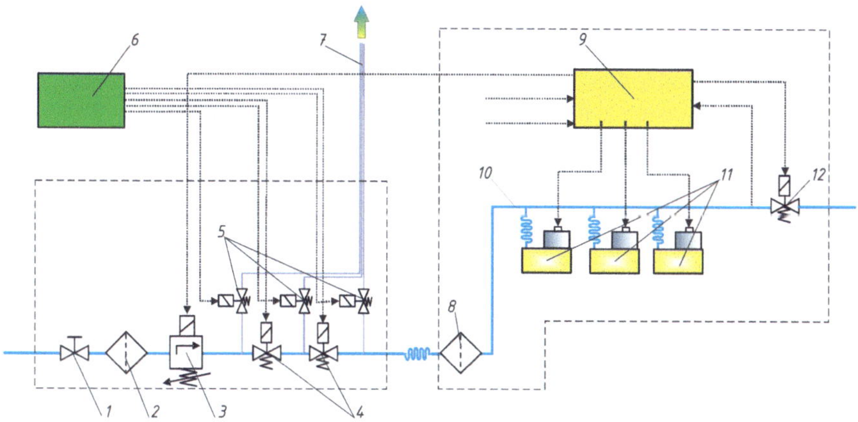

Fig. 7.19 Schematic diagram of the gas supply system for Wärtsilä gas-diesel engines: (1)main gas-valve; (2, 8) filter; (3) gas pressure regulator; (4) electrically controlled safety valves; (5) electrically controlled safety valves of the ventilation system; (6) electronic control unit of the gas installation; (7) ventilation pipes; (9) electronic control unit of the engine; (10) gas line; (11) valves for controlling gas supply to the engine; (12) emergency gas discharge valve (adapted from [11])

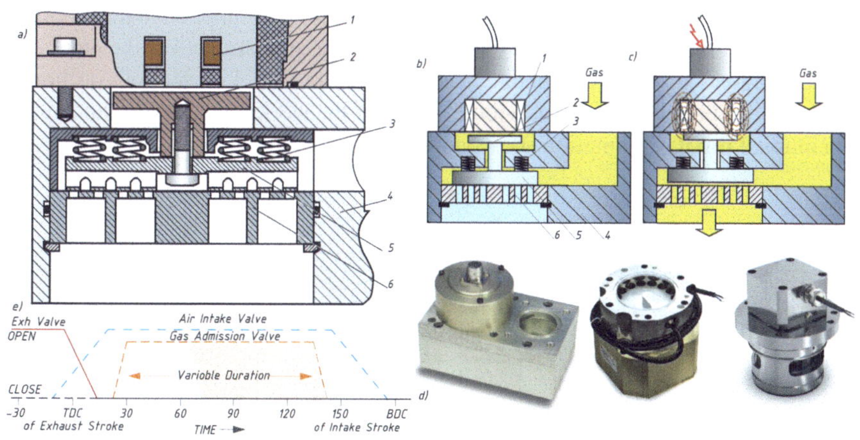

Fig. 7.20 Gas supply metering valve (**a**). Operating sequence of the gas-valve in the closed (**b**) and open (**c**) state. Different types of valves from Woodward SOGAV (**d**). Gas-valve opening phases (**e**): (1) solenoid coil; (2) solenoid anchor; (3) return springs; (4) valve body; (5) valve plate; (6) valve seat (adapted from [2, 13])

The engine controlled by dosing the gas fuel using electrically controlled solenoid valves installed at the inlet of each engine cylinder (Fig. 7.20). The valve opens some time after the start of the intake process and closes before it ends. This ensures that all the gas fuel enters the working cylinder. The amount of fuel supplied by the gas-valves depends on the valve opening time, which determined by the electronic control unit. This procedure performed for each cylinder individually.

The valve assembly consists of a housing in which a valve is placed, which has the shape of a flat cylindrical plate, and a seat with a system of axial holes for the passage of gas fuel. The valve is connected to the solenoid anchor (Fig. 7.20b), which, when voltage is applied to the coil, attracts the anchor and opens the valve (Fig. 7.20c). The valve closes under the action of the return spring and the gas pressure in the supravalve cavity. When operating on gas, minor deviations from the specified mode can lead to detonation. Therefore, a detonation sensor installed on each cylinder, which connected to the engine control unit.

Solenoid Operated Gas Admission Valve (SOGAV) series electrically controlled valves from Woodward are widely used in marine gas and gas-diesel engines with external mixture formation. The general appearance of some gas-valves shown in Fig. 7.20d. SOGAV valves provide accurate measurement of the mass flow of gas at the inlet to the working cylinder. This allows gas and dual-fuel engines to operate stably, without detonation, on a lean mixture with high efficiency and reduced emissions of harmful substances.

7.6 Four-Stroke Gas-Diesel Engines with Internal Mixture Formation

In engines made according to this scheme, gas and pilot fuel supplied to the working cylinder under high pressure at the end of the compression stroke. Unlike other gas engines, engines of this type operate according to the Diesel cycle, both on liquid and gas fuel. The sequence of individual stages of the working process in an engine with internal mixture formation shown in Fig. 7.21.

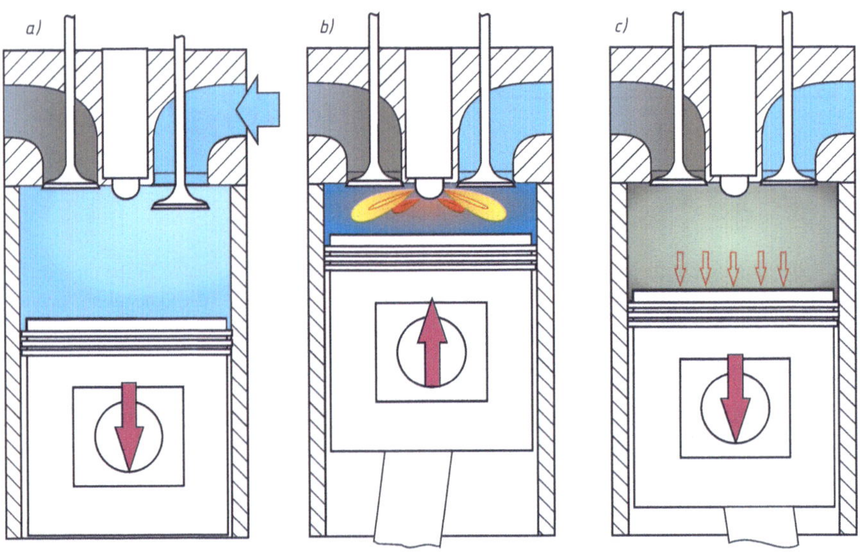

Fig. 7.21 Scheme of the working process of a four-stroke gas-diesel engine with internal mixture formation: (a) filling the working cylinder; (**b**) injection of gas and pilot fuel; (**c**) working stroke (adapted from [2])

To supply two fuels, a special injection valve with several nozzle tips or separate injection valves are used, which, as a rule, first inject the pilot fuel, and after its self-ignition, with a small delay under pressure, the gas fuel injected. Figure 7.22 shows an example of an engine made according to this scheme, developed by Wärtsilä because of the Vasa 32 diesel engine, which received the designation—Vasa 32GD.

To supply liquid fuel, the standard fuel system of the diesel engine is used, and it ensures the supply of liquid fuel in the entire range of operating modes for the dual-fuel cycle: from supplying a pilot portion to a complete transition to liquid fuel. Gas fuel supplied to the injector from the common gas line under a pressure of 35 MPa through special high-speed control modules installed on each cylinder cover. The

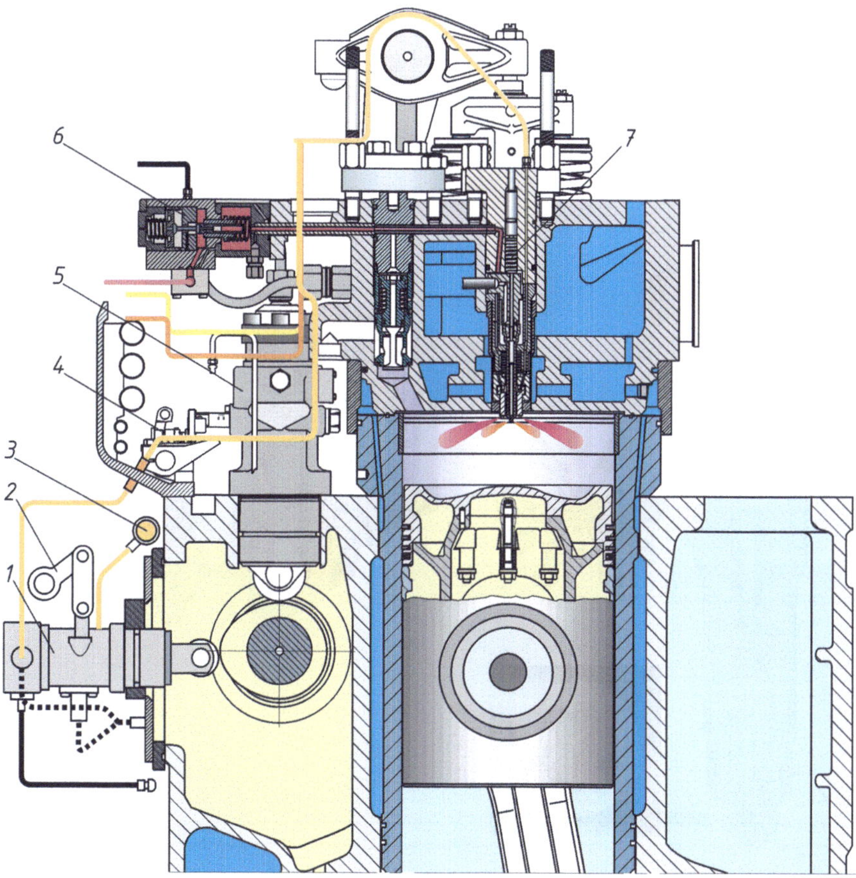

Fig. 7.22 Location of the main components of the fuel system of the gas-diesel engine with internal mixture formation Vasa 32GD: (1) high-pressure pump used as a generator of control pulses; (2) drive from the control mechanism of the pulse supply pump; (3) control fuel supply line; (4) drive from the control mechanism of the liquid fuel supply pump; (5) high-pressure fuel pump for supplying liquid fuel; (6) high-speed control module for the supply of gas fuel; (7) four-valve injector for injection of gas and liquid fuel (adapted from [10, 14])

engine controlled by adjusting the duration of gas injection. In fact, the engine gas system was an analogue of the Common Rail system, in which the compressor maintains a constant fuel pressure inside the gas circuit.

A hydraulic drive used to drive the high-speed gas fuel modules, in which additional fuel pumps installed on the engine crankcase covers used as synchronous pulse generators (Fig. 7.22). Control of the supply of these pumps allows for the formation of pulses of the required duration, which, acting on the valves of the high-speed module, vary the magnitude and time of their opening depending on the specified engine-operating mode. In this way, the amount of gas fuel supplied to the working cylinder through the central injector together with liquid fuel dosed (Fig. 7.23).

Heavy (HFO) and diesel (MDO) fuels can used as liquid fuels, and in some cases, for engines adapted for oil platforms, crude oil. Natural gas used as gas fuel, but there are models adapted for associated gases of some industries. That is, this scheme characterized by extremely high fuel flexibility. Supplying gas fuel at the end of compression virtually eliminates the possibility of detonation, and therefore the sensitivity of the engine to the methane number of the gas fuel.

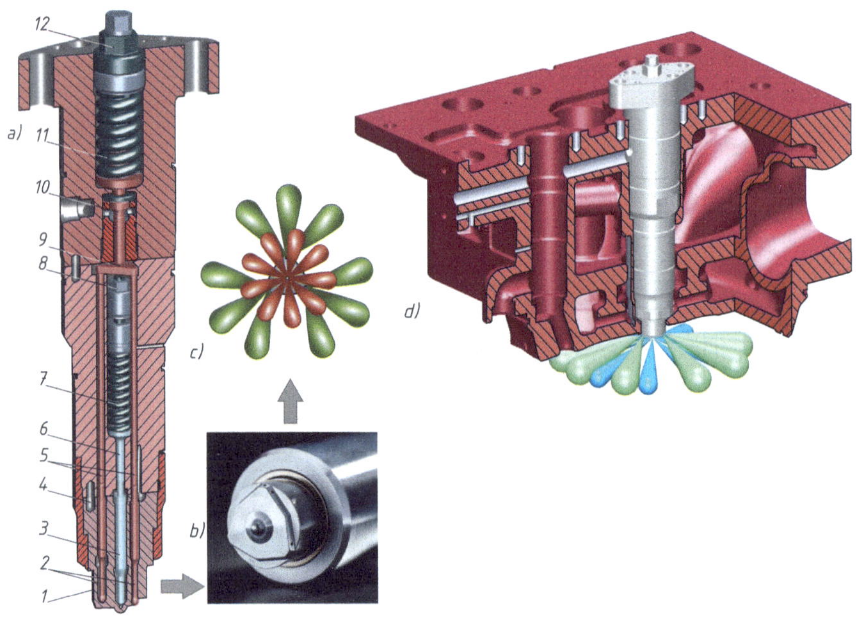

Fig. 7.23 Four-valve injector of Vasa 32GD gas-diesel engine (**a**), general view of nozzle head of injector (**b**), distribution of spray plumes of gas and liquid fuel (**c**), location of injector in cylinder cover (**d**); (1) injector body; (2) needle valves of gas fuel nozzles; (3) needle valve of liquid fuel nozzle; (4) adjusting pin; (5) stems of gas fuel needle valves; (6) stem of liquid fuel needle valve; (7) spring of liquid fuel needle valve; (8) liquid fuel valve adjustment screw; (9) crossmember of gas fuel needle valve stems; (10) crossmember drive stem; (11) spring of gas fuel needle valves; (12) gas fuel spring tension adjustment screw (adapted from [15])

This technology has also been adapted for the use of liquefied petroleum gas (LPG), although the conditions for supplying and burning petroleum gases differ significantly from the use of natural gas. Natural gas injected into the cylinder in the supercritical outflow mode in the gas phase, while LPG injected in a liquid state. In addition, LPG behaves differently from diesel fuel due to its higher rate of evaporation when introduced into the cylinder. However, full-scale engine tests have proven that the optimum choice of injection pressure and speed produces a heat release curve that is virtually identical to that obtained when the engine is running on liquid fuel.

The injector combines four nozzles in one body, three for gaseous fuel injection and one for liquid fuel injection (Fig. 7.23a). Gaseous fuel injected under high pressure through three nozzle tips located at the vertices of a triangle at the end of the injector nozzle body (Fig. 7.23b). Each nozzle has three nozzle openings, so that nine gas jets formed during gas injection. The nozzle tip for liquid fuel injection is located in the center of the nozzle and has nine holes that deliver liquid spray plumes between the gas ones (Fig. 7.23c, d). Pilot fuel injected somewhat earlier than gas fuel, creating open flame centers that effectively initiate the process of ignition of gas fuel supplied to the combustion chamber with some delay. The share of pilot fuel, depending on the operating mode, can fluctuate within 5–13%.

Joint regulation of the supply of gas and liquid fuels allows the engine to operate reliably on both fuels simultaneously in different proportions. As can be seen from Fig. 7.24, the joint use of two fuels is available in the range from 35% to 87.5% of the nominal load. This allows burning gaseous and liquid fuels simultaneously, and

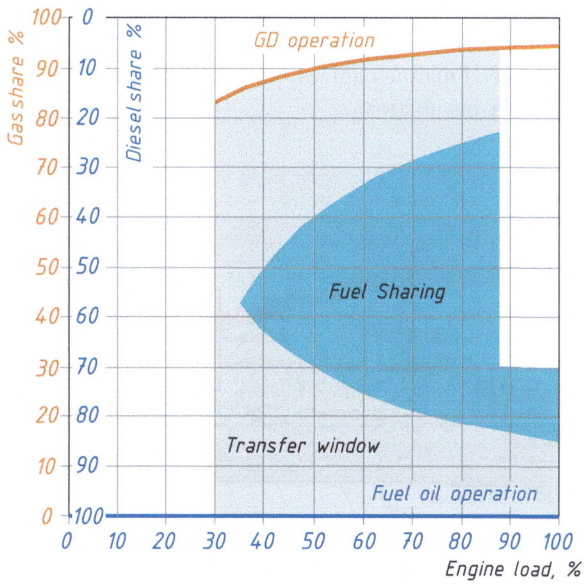

Fig. 7.24 Operating modes of the Wärtsilä 32GD on gas or liquid fuels, as well as on two fuels simultaneously (adapted from [16])

the ratio between the types of fuel can flexibly adjusted both automatically and by the operator's instructions. This is especially convenient, for example, if it is necessary to use all available gas fuel, supplementing the missing energy potential with liquid fuel. High flexibility in choosing the ratios of the two fuels makes the engines insensitive to the calorific value of the gas fuel used, since the lack of energy obtained from the gas fuel can always be compensated for by increasing the supply of liquid fuel and vice versa. This is extremely important when using gaseous fuels from sources with unstable composition and properties of gases.

Having gained positive experience from the successful operation of the Vasa 32 gas-diesel engines with the cylinder size S/D = 35/32, Wärtsilä extended it to a newer version of the diesel engine with the same cylinder diameter Wärtsilä 32 GD (S/D = 40/32) and with a larger cylinder diameter—Wärtsilä 46 GD (S/D = 58/46). These engines already use an electronic control system for the supply of gas and pilot fuel (Fig. 7.25).

The gas injection system consists of gas control valves for each cylinder and a main control unit that controls the duration of the opening time for each valve individually. The control unit receives an input signal from the controller, which used as a reference, and then the signal converted into a pulse that determines the opening time for the high-speed valve.

The main disadvantages of the internal mixture formation technology are the complexity of the engine design, the need to use high pressures of gas fuel, which reduces the safety level. In addition, the need to maintain high pressure increases the energy costs of compressing gas fuel.

In general, the development in recent years of technologies based on external mixture formation, discussed earlier, made it possible to obtain commensurate economic and environmental benefits at lower costs for their implementation. In this regard, for medium-speed marine engines, this technology rarely used, which cannot say about stationary installations.

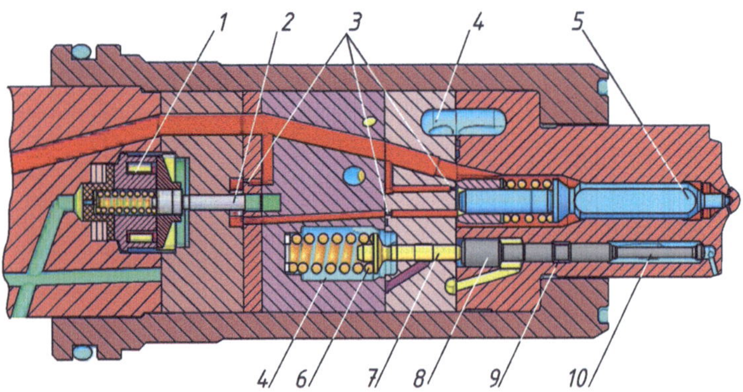

Fig. 7.25 Four-valve injector of the Wärtsilä 32GD gas-diesel engine; (1)solenoid, (2) pilot valve; (3) throttle channels; (4) gas accumulation cavity; (5) liquid fuel needle valve; (6) gas-valve spring; (7) rod; (8) gas-valve drive hydraulic piston; (9) sealing oil cavity; (10) gas fuel needle valve (adapted from [15])

7.7 Use of Gaseous Fuels in Two-Stroke Low-Speed Engines

In two-stroke engines, it is quite difficult to organize external mixture formation, since before entering the working cylinder; the air fills the sub-piston cavity, which has a large volume. The presence of a large amount of gas-air mixture increases the risk of explosion and the severity of its consequences. Therefore, in modern two-stroke low-speed engines, internal mixture formation used, in which gas fuel supplied to the working cylinder after the purge ports have closed.

There are two main approaches to internal mixture formation [14]:

- gas is supplied to the working cylinder immediately after the purge ports have closed at the end of the gas exchange process at the initial stage of the compression stroke [17, 18];
- gas is supplied to the combustion chamber immediately after the self-ignition of the pilot fuel at the end of the compression stroke [19].

The advantages of the first approach are good mixing of the gas-air mixture during the compression process and the ability to supply gas fuel to the working cylinder under relatively low pressure (Fig. 7.26). Due to the latter feature, such systems called *low-pressure systems*.

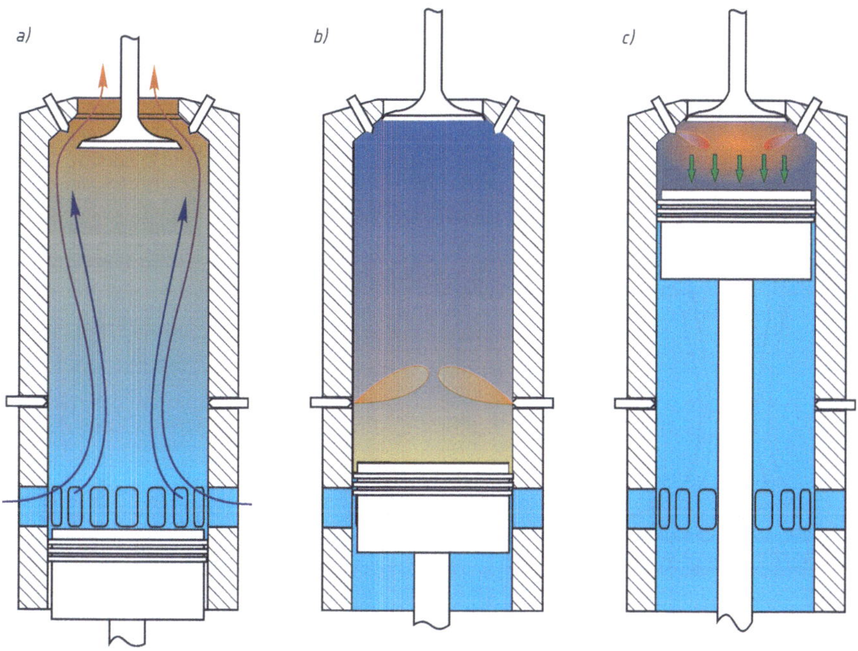

Fig. 7.26 The working process of a two-stroke dual-fuel engine with gas supply to the working cylinder immediately after closing the purge ports at the beginning of the compression stroke: (**a**) cylinder purge; (**b**) gas supply; (**c**) ignition, combustion and power stroke (adapted from [2])

The disadvantages of such a mixture formation scheme are:

- the possibility of detonation combustion, which results in increased requirements for the detonation resistance of the fuel (methane number);
- the probability of the gas-air mixture entering the under-piston space through leaks in the piston rings during the compression stroke;
- incomplete combustion of fuel in the stagnant zones of the combustion chamber.

Currently, low-pressure systems are used in two-stroke low-speed engines of the "X" series manufactured by Winterthur Gas and Diesel Ltd., which received an additional index—DF, and engines of the "ME" series from MAN B&W, which are designated by the GA index. These gas-fueled engines operate on a cycle close to the theoretical Otto cycle. At the same time, they retain the ability to operate on liquid fuel used as a reserve, on a cycle close to the Diesel cycle. The advantage of feeding gas fuel directly into the combustion chamber under high pressure is that it eliminates the possibility of detonation, since only the mixture of air and exhaust gases is involved in the compression process. Gas fuel is fed into the combustion chamber after the self-ignition of the bulk portion of fuel has occurred, which facilitates its effective ignition and combustion. The disadvantage of this method is that gas fuel must feed into the engine working space under high pressure, which complicates the fuel system and everything related to its operation (Fig. 7.27).

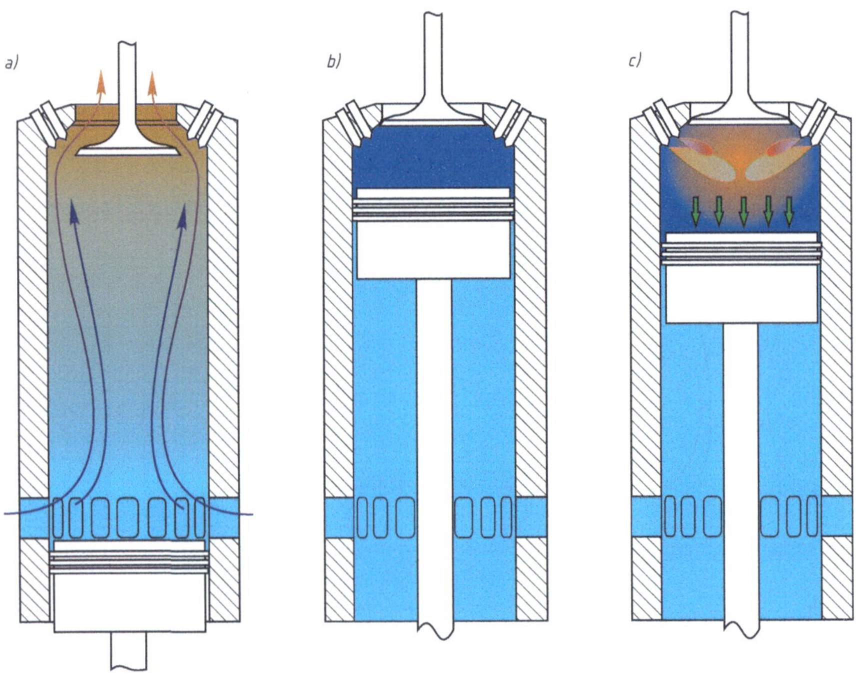

Fig. 7.27 Working process of a two-stroke gas-diesel engine with gas supply at the end of the compression stroke: (**a**) cylinder scavenging; (**b**) gas supply, ignition and combustion; (**c**) power stroke (adapted from [2])

Such systems called *high-pressure systems* or direct *gas injection systems* (GD).

To compress natural gas, it is necessary to use multi-stage compressors, which increases energy costs, reducing the overall efficiency of the plant.

Currently, high-pressure systems used in low-speed gas-diesel engines from MAN, which used as the main engines not only on gas carriers or oil tankers, but also on other types of ships. These engines, developed because of diesel engines of the ME series, received an additional index—GI.

The Japanese company Mitsubishi is also developing in this direction, which, based on diesel engines of the UEC series, has developed its own version of a low-speed gas-diesel engine, which received the index UEC-LSGi.

7.7.1 Two-Stroke Gas-Diesel Engines with Low Pressure Gas Supply

After an analysis of possible concepts for creating a dual-fuel low-speed engine conducted by Winterthur Gas and Diesel Ltd.—WinGD (inherited the developments of Wärtsilä) in early 2011, it was decided to develop a technology for feeding gas fuel to the engine under low pressure. The first studies carried out at a test base in the Italian city of Trieste. For this purpose, one cylinder with a diameter of 50 cm of a six-cylinder laboratory diesel engine converted to operate on gas fuel. By 2013, this engine completely converted to operate on gas fuel as the main fuel and on liquid fuel as a backup. At the same time, on gas fuel, the engine operated on a cycle close to the theoretical Otto cycle.

MAN has developed own version of low-pressure dual-fuel engines. They produced under the GA (Gas Admission) index. For the first time, an engine running on gas fuel using low-pressure technology demonstrated at the company's research center in Copenhagen on March 18, 2021.

The low-pressure gas fuel supply technology, designed for burning a lean fuel-air mixture, based on proven solutions previously developed on four-stroke medium-speed engines.

During the compression process, the gas well mixed with air and ignited using a pilot portion of fuel. At the same time, the size of the pilot fuel portion does not exceed 1% of the cyclic supply of liquid fuel at full load and depends little on the engine-operating mode on gas fuel. The use of natural gas reduces NO_x formation by 90% compared to the use of liquid fuel, which explained by a more uniform temperature distribution throughout the combustion chamber and a smaller number of high-temperature zones in which NO_x formed. The use of this technology makes it possible to meet the requirements of the International Maritime Organization (IMO Tier-III) for NO_x emissions without any subsequent treatment of exhaust gases after the engine.

The gas supply system does not require high-pressure compressors, which significantly reduces equipment costs, on-board energy consumption and maintenance during operation. This allows for a significant simplification of the fuel system, the use of cheaper and more reliable screw or centrifugal compressors. The presence of low pressure in the gas lines reduces the risk of gas leaks, and therefore makes them safer. In addition, the technology unified with the element base that previously used for four-stroke engines, which significantly reduces the cost of its implementation. The presence of an existing supply network of system components that have already undergone long-term testing on four-stroke dual-fuel engines significantly simplifies the operation of new, low-speed engine models.

7.7.1.1 Systems for Supplying Gas Fuel to the Engine

The main component of the engine is the module for feeding gas fuel to the working cylinder (Safe Gas Admission Valve—SGAV). To ensure efficient mixture formation, each cylinder liner of the engine is equipped with two such modules located diametrically opposite (Fig. 7.28).

The height at which the gas modules installed on the cylinder liner should be sufficient to ensure the supply of the required amount of gas fuel at a given gas fuel pressure before the piston blocks the gas nozzle holes. In practice, this is approximately one third of the piston stroke delayed from BDC. Since the gas fuel is supplied at the beginning of the compression process, a gas pressure of 1.0–1.2 MPa is sufficient to obtain a homogeneous gas-air mixture, even when the engine is operating at full load. In practice, the system is designed for a maximum pressure of

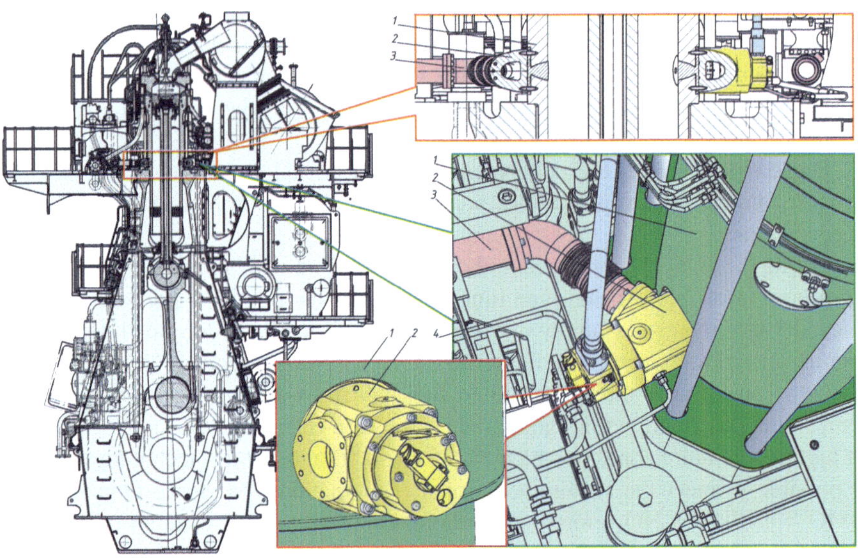

Fig. 7.28 Installing the gas module on the cylinder liner of the W-X62DF engine: (1) cylinder liner; (2) gas fuel supply module; (3) gas fuel supply line to the supply module; (4) supply line for control oil of the gas module hydraulic drive system (adapted from [20])

1.6 MPa, which ensures the required cyclic supply of gas fuel even when using gas fuels with a calorific value of up to 28 MJ/m^3. At the same time, the same duration of gas supply maintained and the same maximum output power provided.

The design of the RT-flex50DF engine safe gas admission valve, shown in Fig. 7.29.

The layout of the modules developed to ensure maximum reliability of their operation and ease of maintenance. A cartridge design used, which allows for the replacement of all moving parts without removing the valve body or gas pipeline system. This ensures short periods of downtime for repair or replacement of valves and reduces the risk of gas leaks caused by maintenance.

To open the gas admission valve, a hydraulic drive used that takes oil from the existing control circuit used to drive the exhaust valves, which eliminates the need for an additional hydraulic system. To prevent leakage of gas fuel into the hydraulic drive system, sealing oil used, which supplied to the valve guide bushing under pressure exceeding the gas pressure.

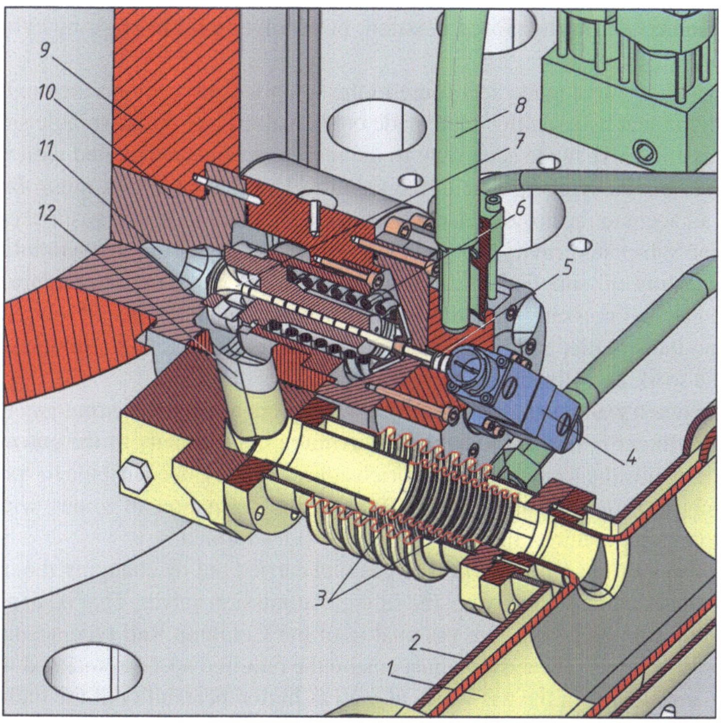

Fig. 7.29 Installing the safe gas admission valve on the RT-flex50DF engine cylinder liner: (1) gas supply line; (2) outer casing of the gas supply line; (3) outer and inner bellows; (4) electromagnetic control valve; (5) hydraulic piston of the gas admission valve drive; (6) valve closing spring; (7) gas admission valve; (8) hydraulic line for the valve drive; (9) cylinder liner; (10) gas-valve housing; (11) channel for supplying gas fuel to the working cylinder; (12) channel for supplying gas fuel to the gas admission valve (adapted from [21])

This oil also lubricates the valve guide. The valve opened using a hydraulic piston, and closed and held by a cylindrical spring. A bistable electromagnetic valve controls oil flows in the gas-valve drive system, which used in RT-flex engines to control the fuel supply. We have considered the design and operation of bistable valves in more detail in the fifth section of the main book [2]. The signal to the control valve comes from the microprocessor engine control module. To ensure stable tightness, a valve design with a spherical seat is used. In addition, the valve is equipped with a displacement sensor, through which feedback carried out with the control unit. If the valve does not return to the closed position, the control unit automatically stops the gas supply and switches the engine to liquid fuel.

In the GA series engines from MAN, the gas module is equipped with two valves, one of which opens, providing supply to the other valve, which directly supplies gas fuel to the working cylinder. The first valve called a window valve (WV), since it sets the time window within which gas can supply. The second, the gas admission valve (GAV), directly connects the gas module cavity to the working cylinder (Fig. 7.30). To increase safety, each of the valves controlled by separate control units, which supply control oil to the drive hydraulic cylinders from the engine control hydraulic line. In the closed position, both valves held in place by cylindrical springs.

The window valve made according to the so-called suspension scheme. It interacts with the seat by a sealing edge made on the valve plate on the spindle side. The gas admission valve made according to the reverse scheme and sealed with the seat edge made on the side opposite to the spindle. This design allows minimizing the volume adjacent to the nozzle channel, which is the main source of gas fuel entering the engine sub-piston cavity. To prevent gas fuel from entering the hydraulic drive system, sealing oil supplied to the gas-valve spindle guide under pressure higher than the gas fuel pressure. In addition to the sealing function, this oil lubricates the guide and the spindle, and its part leaking from the gap together with the gas fuel enters the working cylinder, where it burns.

The geometry and arrangement of the nozzle channels for feeding gas into the working cylinder optimized to achieve maximum homogeneity of the gas-air mixture. This avoids the formation of fuel-rich zones, which can contribute to spontaneous combustion of the mixture and lead to the formation of zones with high combustion temperatures, in which nitrogen oxides formed.

The dosing of the cyclic portion of gas fuel carried out by changing the opening time of the gas admission valve in the safe gas admission valves. That is, in essence, the gas fuel supply system is a gas analog of the Common Rail type accumulator fuel supply systems. The system must ensure the required speed associated with the opening and closing of the gas valve, as well as high repeatability of the fuel supply processes from cycle to cycle in the entire range of operating modes.

When choosing the start and end time of gas fuel supply in low-pressure two-stroke engines, the following considerations is used:

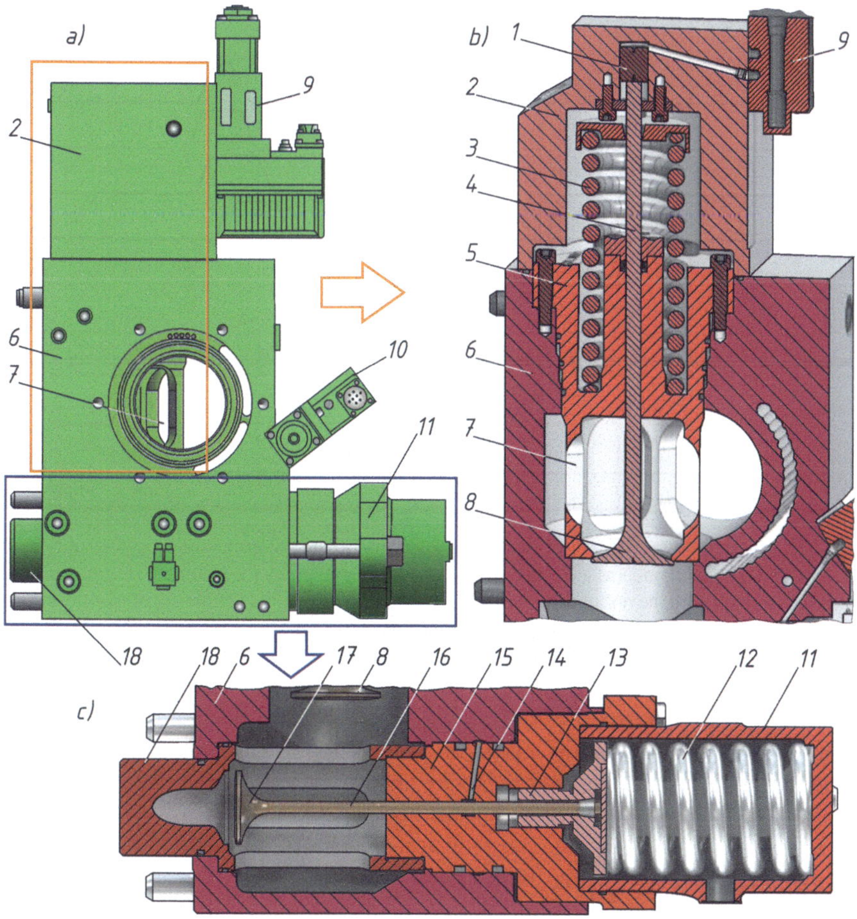

Fig. 7.30 Safe gas admission valve of GA series engines from MAN: (**a**) general view of the module; (**b**) window valve; (**c**) gas admission valve; (1) window valve drive piston; (2) window valve drive housing; (3) spring; (4) window valve spindle; (5) window valve housing; (6) safe gas admission valve housing; (7) windows; (8) window valve plate; (9) window valve control unit; (10) gas admission valve control unit; (11) gas admission valve spring housing; (12) spring; (13) gas admission valve drive hydraulic piston; (14) sealing oil supply channel; (15) gas admission valve housing; (16) gas admission valve spindle; (17) gas admission valve plate; (18) nozzle insert with gas admission valve seat (adapted from [22])

– the supply of gas fuel to the working cylinder must be started as late as possible to reduce the risk of its contact with hot residual gases in the upper part of the cylinder, which can lead to spontaneous combustion and premature combustion. The same is true for preventing methane from entering the exhaust gases during cylinder scavenging (primary methane slip). In fact, scavenging should end with the exhaust valve closing before the gas-air mixture reaches the top of the cylinder;

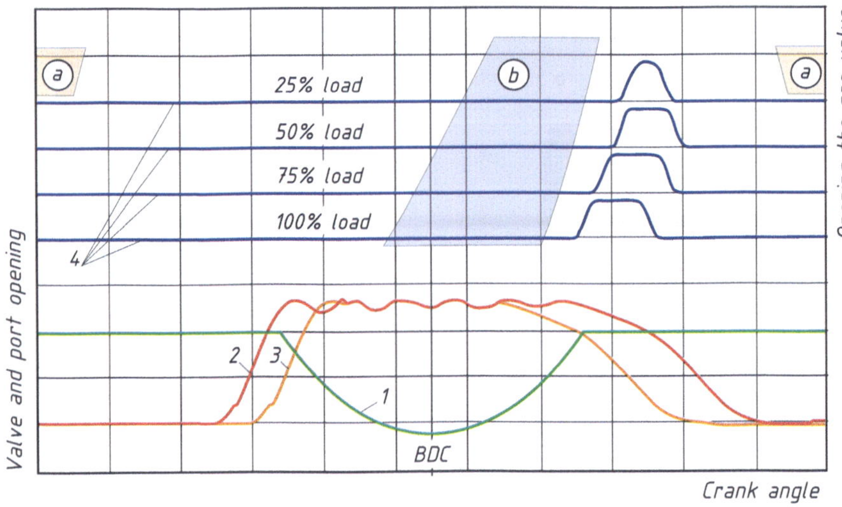

Fig. 7.31 Gas admission valve opening curves in relation to the gas exchange processes of a low-pressure two-stroke engine: (**a**) gas valve nozzle overlap boundary by the piston; (**b**) region of limitations associated with the possibility of direct gas fuel breakthrough; (1) time-section of the purge ports; (2) time-section of the exhaust valve at 100% load; (3) time-section of the exhaust valve at 25% load; (4) gas admission valve opening curves at different loads (adapted from [23])

- the supply must be stopped before the piston approaches the nozzle holes of the gas module. This will allow a layer of clean air to form above the piston surface and retained during the compression process, which will prevent the gas-air mixture from entering closed areas, such as the gap between the piston and the cylinder liner. This will potentially reduce secondary methane slip, improving the environmental performance of the engine.

As follows from the above considerations, the gas-fuel supply system must form a stratified charge in which a homogeneous gas-air mixture in the middle part must be limited at the top and bottom by a layer of clean air. Typical gas-valve opening curves, combined with the gas exchange process in a two-stroke low-speed low-pressure engine, shown in Fig. 7.31.

As the engine load increases, the boost air pressure increases, which leads to a proportional increase in the backpressure at the outlet of the nozzle openings of the gas module. As a result, the outflow rate of gas fuel decreases, which can lead to a decrease in the cyclic portion entering the working cylinder during the time allocated for opening the gas admission valve. In order to compensate for the increase in backpressure, in low-pressure gas engines, the automatic control system increases the gas fuel pressure before the gas module in proportion to the increase in boost pressure (Fig. 7.32). Thus, the pressure under which the gas fuel enters the engine working space is equal to the difference in pressure before the gas module and the air charge pressure in the working cylinder. Regulation of the gas fuel pressure before the gas module makes it possible to reduce the time required for gas supply

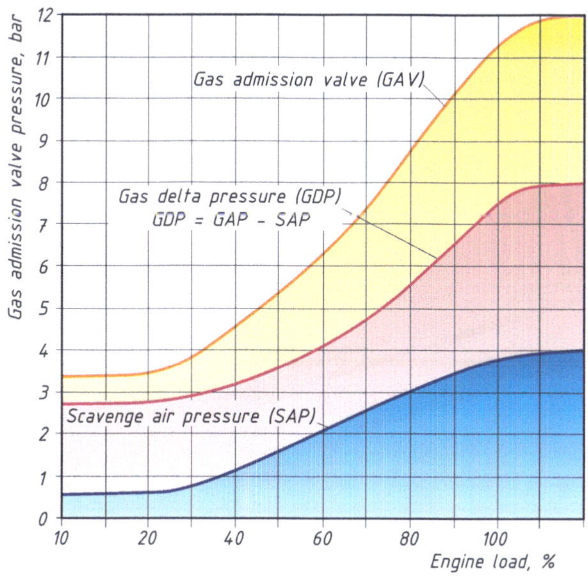

Fig. 7.32 Gas fuel pressure before the gas admission valve, at the outlet of the nozzle opening of the gas module and backpressure in the working cylinder depending on the engine load (adapted from [21])

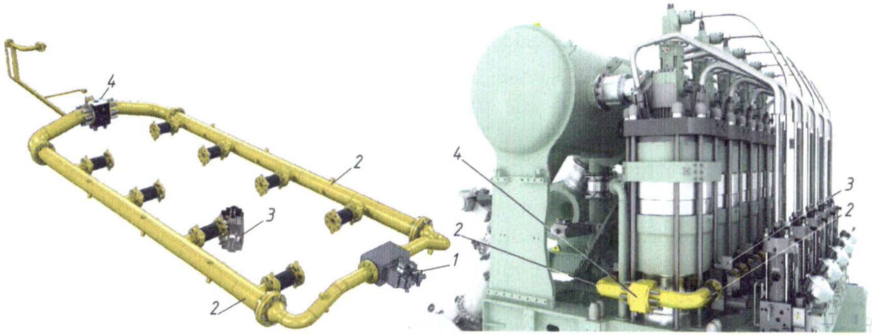

Fig. 7.33 Gas piping of the GA series engine from MAN: (1) gas regulating unit; (2) main pipelines; (3) safe gas admission valve; (4) nitrogen purge block (adapted from [24, 25])

in order to ensure optimal time intervals and angles of opening and closing of the gas admission valve, synchronizing them with the position of the piston in the working cylinder (Fig. 7.31).

The gas fuel supplied to the engine through a Gas Valve Train (GVT) and a Gas Regulating Unit (GRU), which regulates the pressure in the engine gas system. Fuel supplied to the safe gas admission valves through double-walled bellows from the gas manifolds located on both sides of the engine (Fig. 7.33). All gas lines, the space around the valves, flanges, etc. are double-walled. The space between the walls used

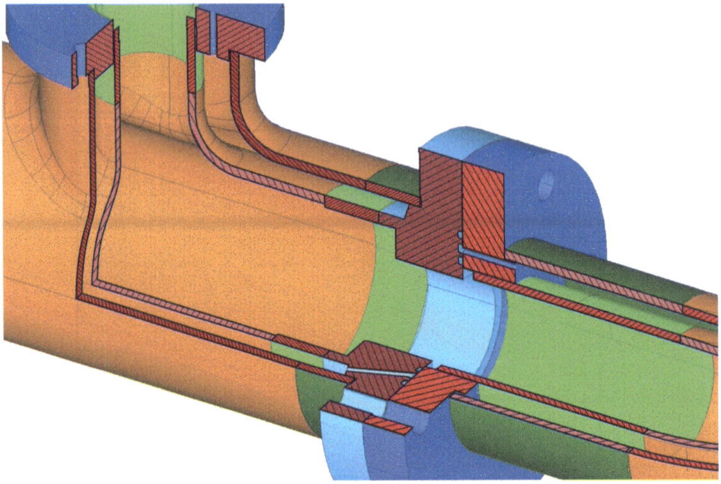

Fig. 7.34 Fragment of a double-wall gas main with a branch to the gas modules (adapted from [22])

to collect leaked gas (Fig. 7.34), it constantly ventilated with a capacity of at least 30 air exchanges per hour. Gas sensors installed at the outlet of the ventilation system; if they detect methane content above the established norm, they transmit a signal to the control unit to emergency stop the engine or switch it to liquid fuel. For safety purposes, all gas pipelines of the engine piping are made of stainless steel. The pressure in the ventilated space maintained below the pressure in the engine room. The exhaust fan motor is located outside the engine room. Air for the ventilation system taken from the gas-safe zone, and after purging the ventilated space, it is removed outside the engine room. The gas pipelines on the engine designed for a pressure that is 50% higher than the normal operating pressure. The gas pipelines must protect from vibrations, as well as from falling heavy objects. In engines running on gas fuel according to the Otto cycle, unlike the diesel cycle, it is impossible to influence heat generation by controlling the fuel injection process, since all the fuel enters the working cylinder and mixes with air before ignition. Instead of quantitative, these engines use qualitative regulation by changing the mixture composition, i.e. the fuel-air ratio or air-fuel ratio. The higher the air-fuel ratio, the leaner the fuel-air mixture and vice versa, a decrease in air-fuel ratio leads to enrichment of the mixture. Correct regulation of this ratio together with optimally selected ignition timing allows you to achieve the optimal combustion rate, which leads to maximum efficiency of the working process with minimal NO_x emissions. The correct choice of air-fuel ratio determines the engine's ability to operate at high values of mean effective pressure without premature ignition of the mixture, detonation or combustion misfires.

Figure 7.35 shows the limitations imposed on the engine operating process depending on the mixture composition. There is a certain enrichment limit, which may vary depending on the amount of residual gases, the presence of lubricating oil

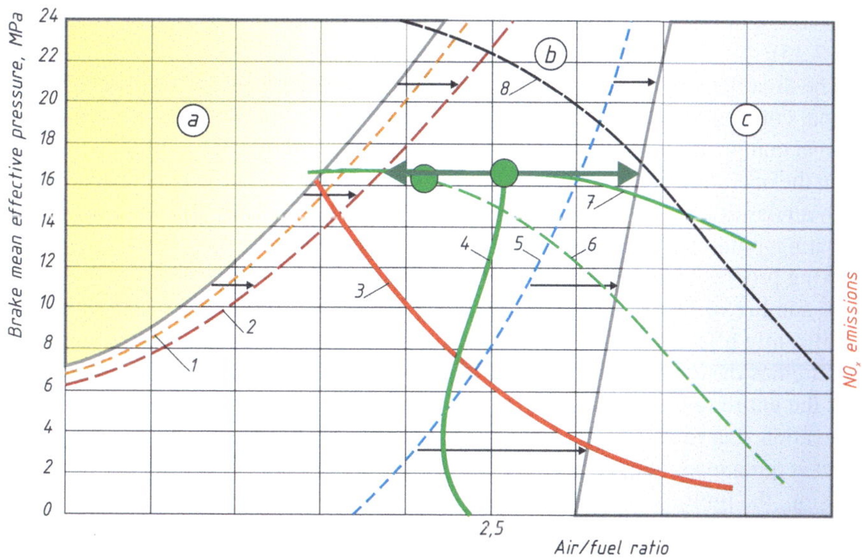

Fig. 7.35 Limitations imposed on the operating process of a two-stroke low-speed engine operating on the Otto cycle depending on the mixture composition: (**a**) detonation combustion zone; (**b**) normal combustion zone (working zone); (**c**) zone of unstable combustion and misfires; (1) curve of the effect of the presence of residual gases on the boundary of the working zone; (2) curve of the effect of the presence of lubricating oil vapors on the boundary of the working zone; (3) NO_x emission curve; (4) curve of the optimal mixture composition over the entire range of operating modes; (5) boundary of the working zone when using direct injection of pilot fuel into the combustion chamber; (6) mixture leanness limit when using direct injection of pilot fuel into the combustion chamber; (7) mixture leanness limit when using prechamber-torch ignition; (8) turbocharger characteristic (adapted from [23])

vapors, and other factors. If this limit is exceeded, the combustion rate will increase significantly, which will lead to an increase in the dynamics of the operating process, an increase in thermal and mechanical loads on engine parts, and an increase in NO_x emissions. With further enrichment of the mixture, especially with an increase in the engine load, spontaneous combustion of the mixture may occur even before the start of pilot injection, as well as detonation combustion (zones a, Fig. 7.35). In practice, the average indicator pressure (engine operating-mode) and the required detonation margin determine the enrichment limit of the mixture. Leaning the combustible mixture gradually slows down combustion, which leads to a decrease in the efficiency of the operating process and a decrease in NO_x emissions (curve 3, Fig. 7.35). Further leaning of the mixture leads to increased combustion instability, up to complete misfire (zones c, Fig. 7.35). This leads to increased emissions of unburned fuel ("methane slip"), which significantly worsens the environmental performance of the engine.

The limit of maximum leaning of the gas-air mixture depends on the efficiency of the ignition source (curve 6, Fig. 7.35). It can significantly expanded by using a powerful ignition system, for example, a pre-chamber-torch system (curve 7, Fig. 7.35).

The area between the boundaries of the enriched and lean mixture zones is the normal combustion zone or the working zone (zones b, Fig. 7.35). In this zone, the optimal composition of the combustible mixture ensured over the entire load range, when the operating point has the required reserve, being at a sufficient distance from the boundaries of both zones (curve 4, Fig. 7.35). In this case, the position of the operating point provides an optimal compromise between the efficiency of the working process and NO_x emissions.

A number of factors, primarily the composition and properties of the gas fuel, significantly affect the size of the working zone. Modern low-speed low-pressure gas engines designed to use combustible gases with a methane number of 65–80 over the entire range of operating modes. The oil mist formed because of cylinder lubrication (curve 1, Fig. 7.35) has some effect on combustion. In practice, this effect is quite insignificant, even if some drops of lubricating oil ignite during compression, the ignition energy of individual drops is too small to ignite the lean gas-air mixture and affect combustion.

In practice, the composition of the gas-air mixture is controlled by changing the purge air pressure (curve 8, Fig. 7.35), which in turn is regulated by the turbocharger turbine bypass valve (Fig. 7.36). When operating on the Otto cycle, in the medium and low load ranges, to maintain the optimal mixture composition, the bypass valve flap opens, bypassing part of the exhaust gases past the turbine, as a result of which the air supply by the compressor decreases. When operating on the diesel cycle, the bypass

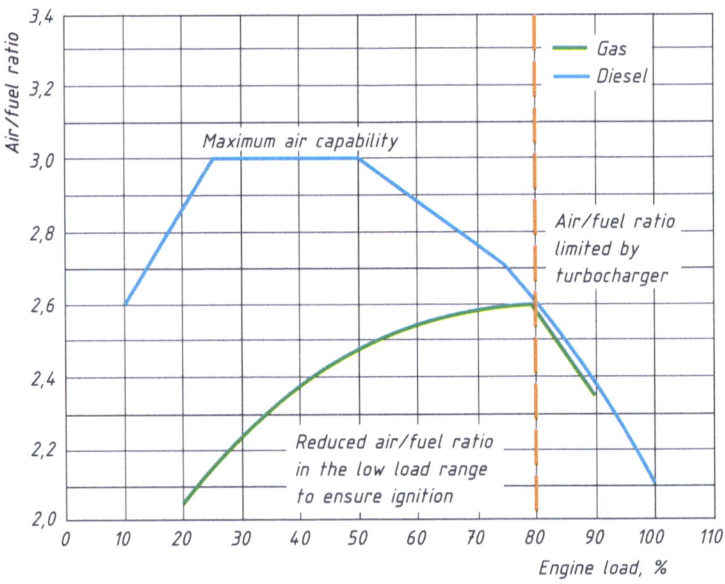

Fig. 7.36 Fuel-air ratio when operating on the diesel cycle and its regulation when operating on the Otto cycle with a gas-air mixture (adapted from [23])

valve is always closed. Flexibility in regulating the mixture composition also provided by the control system of the exhaust valve of the working cylinder, changing the phases of its opening and closing on command from the electronic control unit.

The required purge air pressure depends on the amount of gas fuel supplied to the working cylinders, which determined by the engine-operating mode and actual operating conditions, such as ambient temperature, pressure, etc. The values of purge air pressure determined by the algorithms of the engine management system, which take into account the current parameters of the working process and external factors affecting its operation. To do this, the engine management system constantly monitors the pressure in the working cylinders and misfires. In addition, the engines are equipped with knock sensors to ensure additional safety.

As can see in Fig. 7.35, the working zone in which the mixture composition is optimal narrows significantly with increasing engine load. For this reason, in order to provide the necessary reserve for the stability of the working process, the average effective pressure of engines operating on the Otto cycle is limited at 1.7−1.85 MPa. This is approximately 80% of the MCR of the same engine when operating on the diesel cycle.

7.7.1.2 Pilot Fuel Supply Systems

To ensure dual-fuel operation of the engine, it is equipped with two or three independent fuel supply systems, each of which controlled by an electronic microprocessor module using a separate program depending on the fuel used and the engine-operating mode.

The standard fuel system of the base engine used to supply reserve liquid fuel. In GA series engines from MAN, this is a system with hydraulic drive of fuel pumps. For engines manufactured by WinGD, an accumulator-type system used, which is typical for all RT-flex and X series engines [2]. At the same time, the engines retain the ability to operate on heavy grades of fuel in the entire range of load-speed modes.

We have analyzed fuel systems for liquid fuel injection in engines of the series under consideration in detail in Chap. 5 of the main book. For this reason, we will not dwell on them in detail, focusing on a more detailed analysis of the pilot fuel systems. In the RT-flex-DF and X-DF series engines from WinGD, pilot fuel is supplied by a separate low-capacity accumulator system (Fig. 7.37a).

This solution allows for a stable supply of small portions of fuel with the ability flexibly regulate the operating mode of the pilot ignition system. A radial plunger fuel pump from L'Orange GmbH (Fig. 7.17) with an electric drive used to supply fuel to the accumulator under high pressure. The main elements of the pilot ignition system mounted on the engine as a separate module, which significantly simplifies their monitoring and maintenance (Fig. 7.37b)

The pressure in the accumulator of the pilot fuel injection system maintained at 120 MPa. When the engine is running on liquid fuel, to prevent coking of the pilot injector nozzle, they continue to operate in the minimum stable supply mode. To prevent fuel from getting on engine components in the event of damage, all high-pressure pipelines placed in a protective casing. Light diesel fuel used as a pilot fuel,

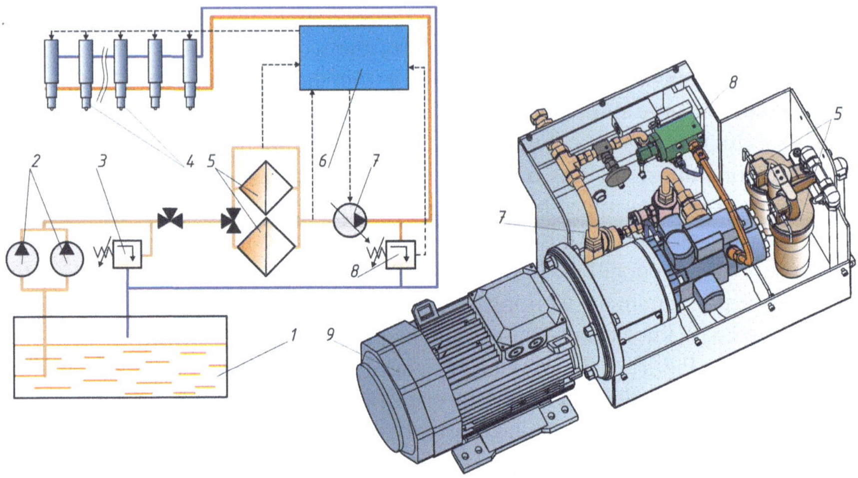

Fig. 7.37 Schematic diagram of the common-rail pilot fuel injection system of the RT-flexDF gas-diesel engine (**a**) and the general arrangement of the main components in the supply unit (**b**): (1) pilot fuel tank; (2) booster pumps; (3) low-pressure line bypass valve; (4) pilot fuel injection nozzles; (5) fine filters; (6) electronic control unit; (7) high-pressure fuel pump; (8) high-pressure line bypass valve; (9) electric motor; (*Solid lines*) hydraulic lines; (*Dashed lines*) control lines (adapted from [20])

but the possibility of using heavy fuel oil (HFO) for pilot injection, which the engine uses as a reserve fuel, considered.

To improve the conditions for ignition of the lean gas-air mixture, two ignition modules installed on each cylinder, consisting of a swirl chamber with a pilot fuel injection installed in it (Fig. 7.38).

The prechamber is made of heat-resistant steel in the form of two separate liners and cooled along the outer surface with water supplied from the cylinder cover cooling circuit. The internal cavity of the swirl chamber connected to the combustion chamber by a tangential channel. During the compression process, air flows from the working space into the swirl chamber, forming a vortex, into which the injection supplies a portion of pilot fuel by the end of the compression stroke.

The presence of a vortex movement of the charge contributes to good mixing of air with fuel and effective self-ignition. Fuel combustion leads to an increase in pressure in the vortex chamber (Fig. 7.39). Under the action of this pressure, plasma jets ejected through an opening in the end of the tangential channel into the main combustion chamber. These jets penetrate the entire volume of the gas-air mixture in the main combustion chamber, igniting it.

This method of activating the combustion process allows for the efficient burning of even very lean gas-air mixtures, significantly expanding the range of the engine's operating zone in terms of mixture composition (Fig. 7.35, curve 7). Engines equipped with a pre-chamber-torch ignition system can operate at higher air-fuel ratio than engines with direct injection of pilot fuel into the main combustion chamber (Fig. 7.35, curve 6).

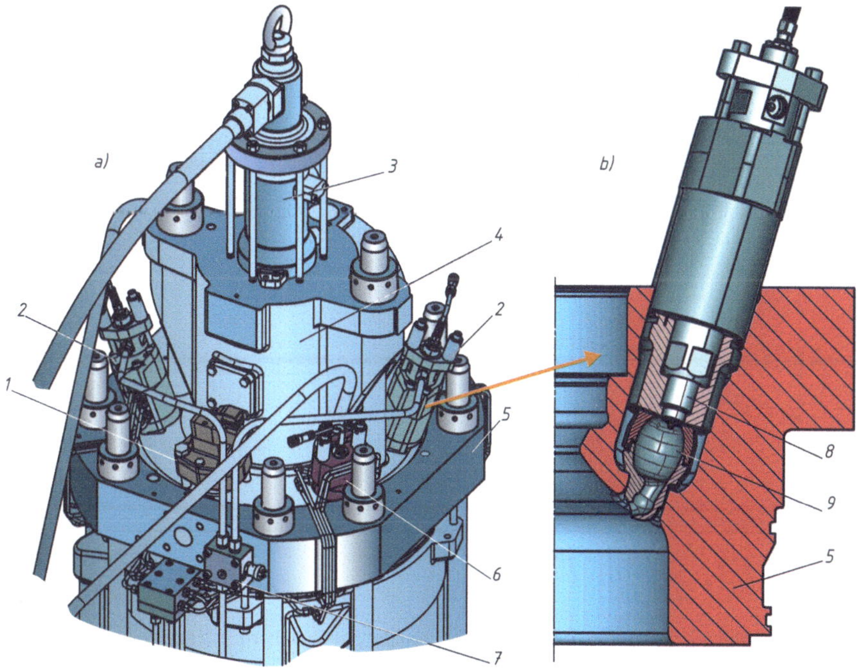

Fig. 7.38 Equipment of the cylinder cover of the RT-flex-DF gas-diesel engine (**a**) and the pre-chamber-torch module of the pilot ignition of the gas-air mixture (**b**): (1) air-starting valve; (2) ignition module; (3) hydraulic cylinder of the exhaust valve drive; (4) exhaust valve housing; (5) cylinder cover; (6) injection valve for supplying reserve liquid fuel; (7) pilot fuel supply manifold; (8) pilot fuel supply injection valve; (9) swirl-chamber (adapted from [2])

Optimized configurations of the swirl-chamber and the nozzle tip of the pilot fuel injection nozzle allow achieving combustion of the gas-air mixture to be virtually insensitive to the pilot fuel injection parameters and quantity. Pilot fuel injection serves exclusively to activate the combustion process of the gas-air mixture, and any excess amount of it cannot accelerate the combustion process or make it more complete. This concept ensures good distribution of ignition energy in the combustion chamber with a minimum amount of pilot fuel of 0.5–1.0% of the total energy (Fig. 7.40) consumed at full engine load.

Uniform distribution of plasma jets over the combustion chamber volume promotes uniform combustion of the gas-air mixture without local hot spots. This reduces the emission of nitrogen oxides in the combustion products, since the latter are formed mainly in local high-temperature zones. As a result, high efficiency of the working process achieved, and emissions of unburned hydrocarbons ("methane slip") reduced. The level of "methane slip" in low-speed engines is significantly lower than in four-stroke gas engines operating on a lean mixture. This is due to lower rotation speeds, longer combustion time and a larger combustion space with a favorable volume-to-surface ratio. Figure 7.41 shows the dependence of "methane

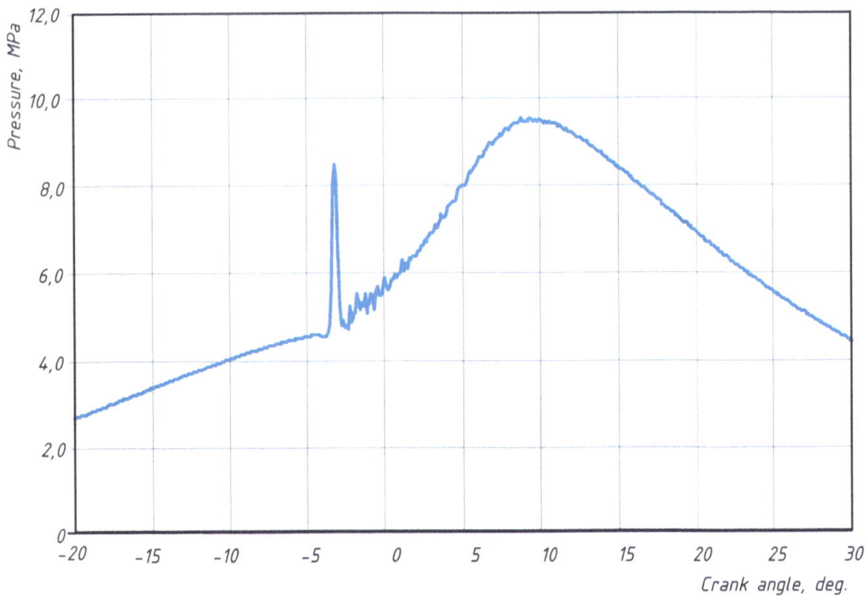

Fig. 7.39 Change in pressure in the swirl-chamber of the pre-chamber-torch ignition system of the RT-flex-DF series engines (adapted from [23])

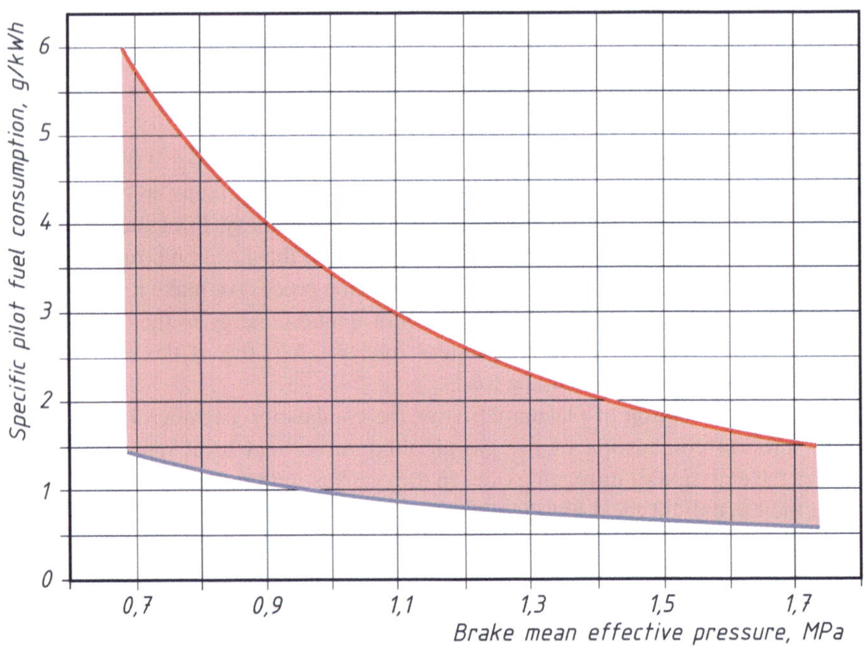

Fig. 7.40 Dependence of the specific pilot fuel consumption of the X-DF series engines in gas mode over the entire range of operating modes (reduced to the ISO standard) (adapted from [23, 24])

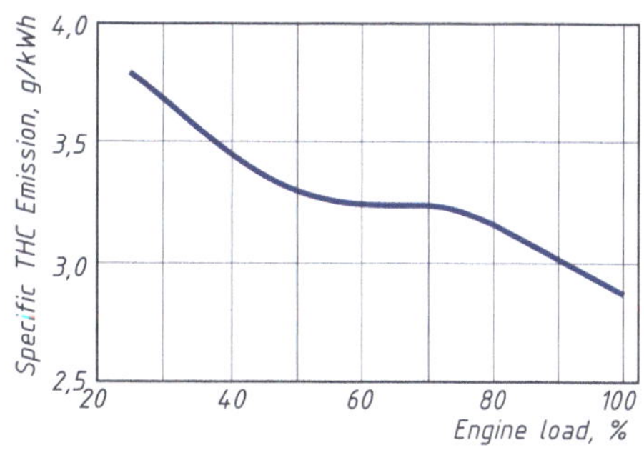

Fig. 7.41 Total Hydro Carbon (THC) emissions measured for the 6X72DF engine from WinGD ("methane slip") (adapted from [23])

slip" on the engine-operating mode. The actual mechanism by which methane gets into the exhaust gases is its incomplete combustion in the near-wall regions and dead volumes (secondary slip mechanism).

In the GA series engines from MAN, the pilot fuel supplied by two hydraulically driven modular pump-injectors made using Fuel Booster Injection Valve (FBIV) technology installed on the cylinder cover. These modules are an adapted version designed for small feeds, which is why they called Micro-Boosted Injection Valve (MBIV). We have considered in detail the issues related to the design and use of FBIV modules in Chap. 5 of the main book [2].

The electronic control unit of the working cylinder CCU (Cylinder Control Unit) controls the ignition modules. It generates and transmits an electrical signal to the electro-hydraulic injection control valve FIVA (Fuel Injection Valve Actuation).

7.7.1.3 Features of Operation of Low-Pressure Gas Engines

The operation of the engine on gas fuel allows obtaining at the output flange a power level of 80% of the MCR on liquid fuel (Fig. 7.42). If it is necessary to obtain higher power, the engine converted to liquid fuel. Starting and stopping the engine for safety purposes also carried out on liquid fuel. It should note that the obtained power on gas fuel corresponds to the main operating mode of marine engines.

The main factor limiting the power is the occurrence of detonation in the working space of the cylinder or misfires associated with the exit of the working process from the working zone (Fig. 7.35). The need to maintain the ability of the engine to operate on liquid fuel requires maintaining a sufficiently high compression ratio. As a result, conditions for the occurrence of detonation combustion can formed in the working cylinder, the cause of which is uncontrolled premature spontaneous ignition of the gas-air mixture. Due to the many centers of spontaneous ignition in the cylinder, wave processes occur, accompanied by high-frequency noise and vibrations (Fig. 7.43).

In this case: the cost of work on compressing the mixture in which heat is already being released increases; mechanical loads on engine parts increase; vibration leads to accelerated destruction of engine structural elements. In order to avoid detonation, it is necessary to maintain the composition of the lean gas-air mixture in a

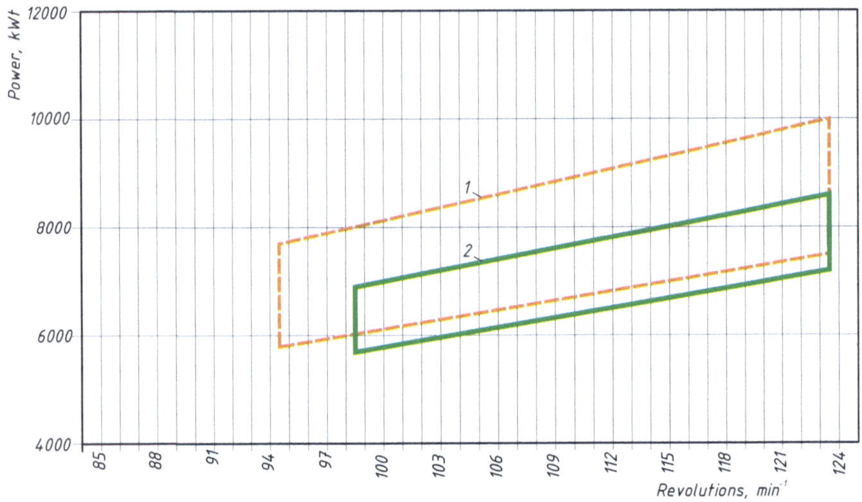

Fig. 7.42 Fields of recommended operating modes of RT-flex50B engines on liquid fuel (1) and RT-flex50DF on gas (2) (adapted from [26])

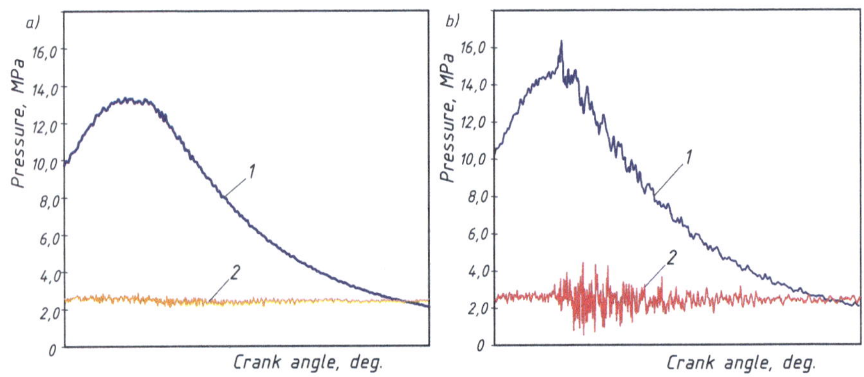

Fig. 7.43 Normal (**a**) and detonation (**b**) combustion of the gas-air mixture in the working cylinder of the low-speed gas-diesel engine RT-flex50 DF (adapted from [27])

narrow range (Fig. 7.35), which entails a decrease in the energy potential of the charge and, consequently, a decrease in the work it produces. The methane number determines the resistance of gaseous fuel to detonation. Therefore, all low-pressure engines are sensitive to the methane number of the gas fuel used.

The rate of physical and chemical processes leading to detonation depends on the composition of the gas-air mixture. For the engine to operate at nominal power, the methane number of the gas should not be less than 80; when the power is reduced, the methane number can also be reduced to 80–55 units (Fig. 7.44a). This makes the engine's operating process extremely sensitive to the air temperature at the inlet (Fig. 7.44b). The higher the temperature at the beginning of the compression

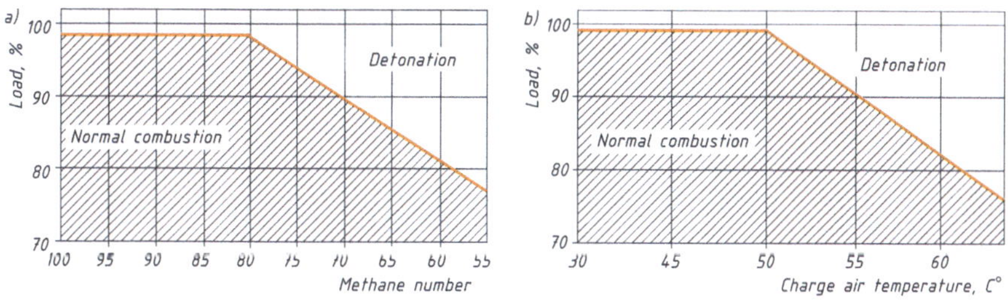

Fig. 7.44 Effect of methane number (**a**) and charge air temperature (**b**) on the occurrence of detonation combustion in the working cylinder of a low-speed engine with gas supply during the compression stroke (adapted from [28])

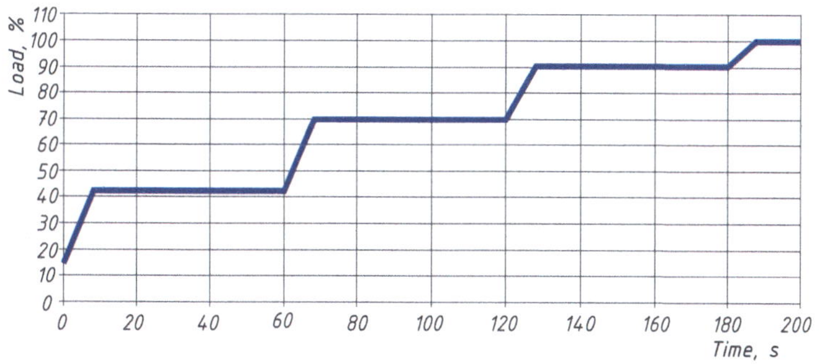

Fig. 7.45 Bringing a gas-diesel engine with a low-pressure fuel system to maximum continuous power (adapted from [26])

process, the faster the physical and chemical processes leading to detonation combustion occur.

The need to maintain the composition of the gas-air mixture in a narrow range (Fig. 7.45) does not allow a rapid increase in engine power by significantly increasing the gas supply to the working cylinder, since this may lead to misfires. A decrease in load should also not accompanied by a sharp decrease in gas supply, since this will lead to detonation combustion. Thus, engines with gas fuel supply on the compression stroke characterized by a slow response to load changes. The transition from mode to mode should carry out gradually with delays in intermediate modes to stabilize the composition of the gas-air mixture (Fig. 7.45). This feature complicates the use of the engine during maneuvers and requires its conversion to liquid fuel. At the same time, it should be noted that environmental requirements are most stringent in coastal areas, where maneuvering is often necessary, which to some extent neutralizes the advantages of using gas fuel.

In order to prevent engine operation with detonation, special sensors installed on each cylinder, recording wave processes. In case of detonation, the sensors transmit a signal to the electronic control unit, which reduces the load on the engine or

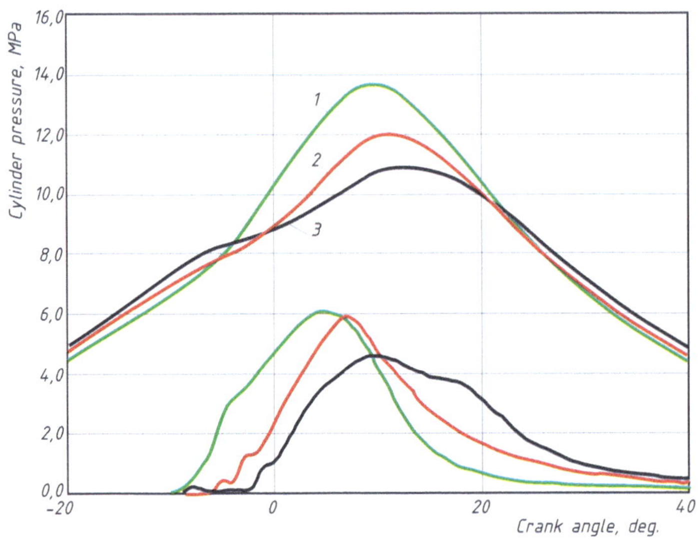

Fig. 7.46 Comparison of cylinder pressures and heat release rates during operation according to: (1) gas cycle; (2) mixed cycle (40% liquid and 60% gas fuels) (3) diesel cycle. Engine load 75% of MCR (adapted from [26])

switches it to liquid fuel. Switching from one type of fuel to another can carried out without stopping the engine at powers up to 80% of the MCR.

A promising direction considered joint parallel operation of liquid and gaseous fuel supply systems. In this case, part of the energy in the cylinder obtained by burning gas fuel, which supplied in the amount necessary to ensure detonation-free compression, and the missing part of the energy, obtained by injecting liquid fuel into the combustion chamber (Fig. 7.46).

The latest modifications of the X-DF series engines have a combustion control function (Dynamic Combustion Control—DCC). If high power is required, external environmental conditions change, or with a low methane number of gas fuel, the pressure in the cylinders can exceed the normal operating range. In this case, a small amount of liquid fuel (3–15% of the energy consumed) is injected by the main fuel injectors, which allows reducing the amount of gas fuel supplied to maintain an optimal air-fuel ratio.

In addition to knock sensors, the engine is equipped with pressure sensors in the working cylinder, the main task of which is to monitor misfires. If they occur, unburned gas-air mixture can accumulate in the exhaust receiver, which can lead to an explosion and damage to the engine. Misfires or inefficient combustion of fuel can cause by malfunctions of the pilot ignition system. There is also a control of the presence of gas fuel in the sub-piston space, which can lead to serious consequences. To carry out such control, special sensors installed in the sub-piston cavity. The reason for gas entering the cavity under the piston may be loose closing or damage to the gas admission valve. When the piston passes through the filling holes, gas can enter the sub-piston cavity through leaks in the valves. In addition, some of the gas-air mixture will constantly leak into the sub-piston cavity through leaks in the piston rings. Incomplete combustion of gas fuel due to the above reasons results in a small amount of methane entering the exhaust gases. In general, methane emissions into the atmosphere by the ship's power plant do not exceed 0.3%, and the total amount of greenhouse gases emitted is 20–30% lower than that of engines running on liquid fuel. Currently, Wärtsilä has launched production of a new generation of low-speed two-stroke engines designed to run on gas fuel, based on previously developed X-series diesel engines. These engines include W-X62DF, W-X72DF, W-X82 and W-X92. In general, the developed technology for converting engines to gas fuel while maintaining the ability to run on liquid fuel is aimed at both new engines and upgrading those already in operation. At the same time, according to the company's estimates, the cost of conversion will be 20–25% of the original cost of the engine.

7.8 Two-Stroke Gas-Diesel Engines with Gas Fuel Injection at the End of the Compression Stroke

The ME-GI series engine is the result of many years of work by MAN B&W specialists, which has been going on since the early 1990s. The first commercial engine of the ME-GI series was first unveiled on November 9, 2012, at the Hyundai Heavy Industries Co., Ltd. plant in South Korea.

Two-stroke low-speed diesel engines of the ME series are used as base models for gas-diesel engines with gas fuel injection at the end of the compression stroke.

Engines modernized for the gas-diesel process received an additional index GI (Gas Injector).

The standard fuel system of the engines used to ensure their operation on liquid fuel and for the injection of a pilot portion when operating on gas fuel. This simplifies the design (there is no separate pilot fuel injection system).

Initially, high-pressure engines were intended for LNG tankers, which under normal sailing conditions the amount of evaporated gas fuel only covers 80–90% of the engine's needs. For this reason, the use of a standard liquid fuel injection system to supply a pilot portion of 5–8% of the total cyclic supply was considered entirely justified. Subsequently, with the tightening of environmental standards, engines of this type became an attractive alternative for other types of vessels, because of which design changes made in subsequent versions of the fuel system that allowed reducing the pilot fuel consumption to 0.5–1.0%. It should note that for LNG tankers, under certain sailing conditions, the evaporation share might decrease. In this case, it can cover only 40–50% of the required amount of gas fuel needed by the engine. Therefore, the fuel system must be able automatically replace the missing gas fuel with liquid fuel in any ratio. In addition, the calorific value of the gas entering the engine may change. At the beginning, gas vapors contain a large amount of

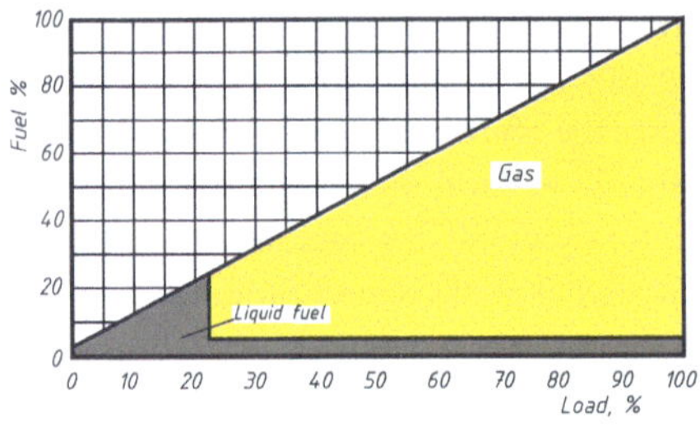

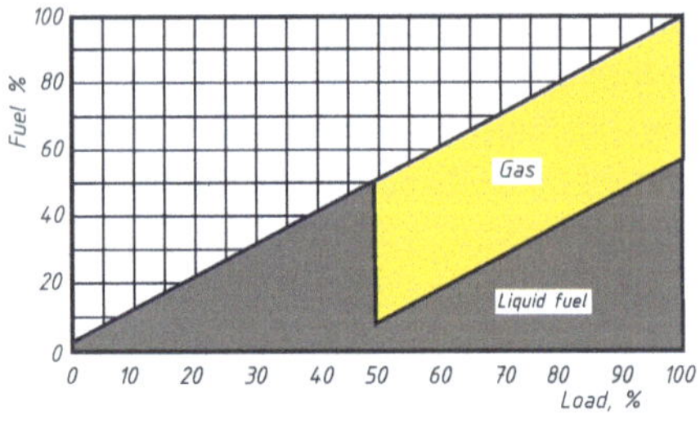

Fig. 7.47 Operating modes of low-speed gas-diesel engines: (**a**) with a constant amount of pilot fuel; (**b**) using all available gas (adapted from [29])

nitrogen, which, having a lower boiling point (−195.75 °C), evaporates first. Therefore, for low-speed gas-diesel engines, two modes are considered:

- with a constant pilot fuel supply (Fig. 7.47a), when the engine operates on liquid fuel (MDO, MGO, HFO) in the starting mode and at low loads. Starting from 25% load, a constant pilot supply is established, and the required power regulated by changing the amount of gas supplied to the cylinder;
- when using all available gas fuel (Fig. 7.47b), when the engine operates on liquid fuel at low and medium loads. At higher loads, all gas fuel enters the cylinders, and the required power regulated by supplying liquid fuel.

The transition from one type of fuel to another, as well as the transition from mode to mode, carried out automatically without reducing the power on the shaft in the entire range of operating-modes.

Equipment for supplying gas fuel under high pressure includes compressors, heat exchangers, a system for supplying gas fuel to the working cylinders, gas supply control modules and gas injectors.

7.8.1 Supply of Gas Fuel to Engines Under High Pressure

The high-pressure gas fuel system can integrate with the ship's system for reliquefying the evaporated gas fuel. This solution is typical for ships carrying liquefied natural gas or ethane. For other types of ships, it implemented as an autonomous system that serves the main and auxiliary engines. The system includes tanks for storing liquefied fuel with their servicing subsystems, a pump-evaporation unit and a gas valve train (GVT). Since the consideration of the issues of storing and processing gas fuels on board a ship is beyond the scope of this book, for a more detailed study of these issues, we recommend that you refer to the relevant literature. Below, we will consider only the main purpose and some functional capabilities of this equipment necessary for understanding its interaction with ship engines.

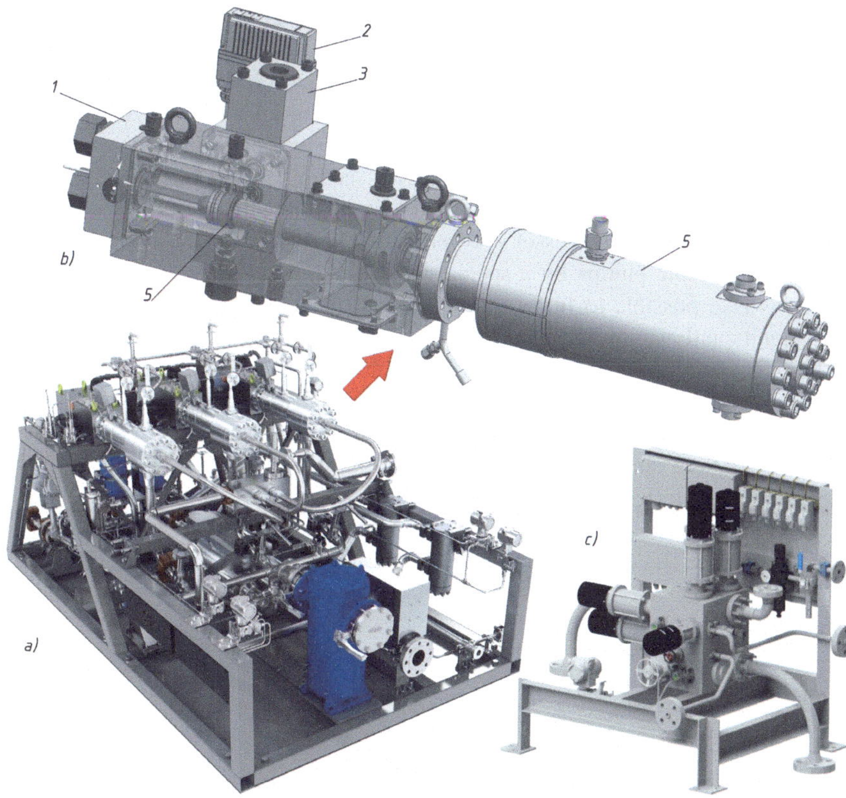

Fig. 7.48 General view of the pump-evaporation unit (PVU) (**a**), cryogenic pump section (**b**) and gas valve train (GVT) (**c**); (1) hydraulic distribution block; (2) explosion-proof valve; (3) electronic exhaust valve actuation; (4) plunger section; (5) discharge section (adapted from [30])

The Pump Vaporiser Unit (PVU) is a pumping unit for feeding natural gas or ethane into MAN B&W ME-GI/GIE two-stroke engines. It is a key element of the Fuel Gas Supply System (FGSS).

The pumping unit (Fig. 7.48a) is equipped with three hydraulically driven cryogenic pumps, an evaporator, filters and a control system with safety functions.

A low-pressure centrifugal pump or a booster pump installed in the fuel storage tank supplies the gas fuel to the PVU at a temperature of approximately −163 °C.

The gas then fed to high-pressure piston-type cryogenic pumps with three cylinders, driven by linear hydraulic pistons (Fig. 7.48b).

After compression, the gas fuel enters a compact heat exchanger, where it heated to 45 °C. An aqueous solution of ethylene glycol used to supply heat to the fuel heater. A high-pressure filter traps fine particles present in the gas fuel before the gas is sent to the gas valve train and then to the engines. The gas pressure regulated by changing the flow of hydraulic fluid supplied to the cryogenic pumps.

Each cryogenic pump section driven separately, which provides the necessary redundancy.

Separate control of the three pump sections provides 100% redundancy. Each section of the cryogenic block is capable of delivering up to 50% of the gas fuel consumed by the engines at full load. Thus, the pump-evaporation unit is capable of delivering the required capacity with only two sections in operation. This allows one pump section to taken out of service and repair without deteriorating the performance characteristics of the vessel's power plant. Oil supplied to the hydraulic drive system of cryogenic pumps by a special hydraulic pump station with an electric drive. It is also possible to integrate the drive with the engine hydraulic system, which significantly simplifies the design of the unit.

The pump-evaporation unit delivers gas fuel to the ME-GI main engine with a pressure of 15.0–26.5 MPa depending on the engine load (Fig. 7.49).

The parameters of the gas fuel that the unit must provide at the engine inlet given in Table 7.2. At the same time, the pump-evaporation unit supplies gas fuel to dual-fuel engines of electric generator units under a pressure of 0.6–1.2 MPa. The

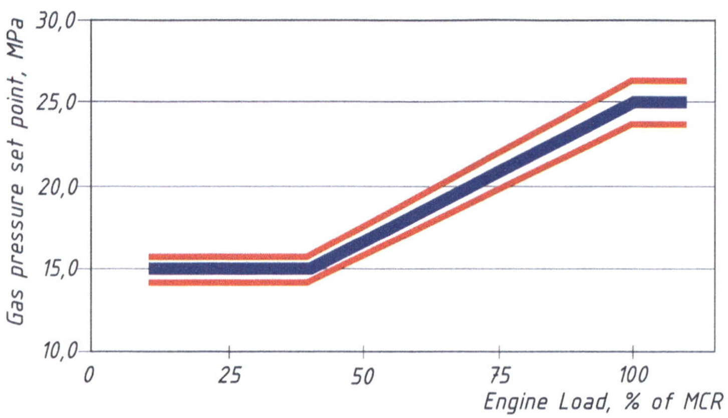

Fig. 7.49 Gas fuel pressure in front of the ME-GI engine depending on its operating mode (adapted from [30])

Table 7.2 Requirements for gas fuel supplied to ME-GI series engines

Parameter	Meaning
Pressure:	
Nominal, MPa	25.0
Maximum value, MPa	30.0
Pulsation limitation, MPa	± 0.2
Set value tolerance, %	±5
Temperature:	
Below ambient temperature or max 30 °C, whichever is lower	
Quality:	
No condensation, no oil drips, water or mist	

control of the pump-evaporation unit integrated with the engine control system. From the engine control unit, the requirements for gas pressure and its mass flow are transmitted to the PVU control system and then to the pump module. This allows obtaining stable parameters of gas fuel in the entire range of engine operating modes. Currently, customers offered a standard series of pump-evaporation units for various capacities of the main and auxiliary engines.

Gas valve mechanism (Fig. 7.48c) controls the gas supply to the engine when it is running on gas fuel. It installed between the pump-evaporator unit and the engine to provide blocking functions when it is necessary to stop the gas supply and purging functions when the engine switched to liquid fuel. After stopping the gas supply, the entire internal space from the valve to the engine purged with inert nitrogen.

Gas mains. After the gas valve mechanism, the gas fuel enters the engine gas line, which also functions as a pressure accumulator. All gas pipelines on the engine made of welded parts and only at the points of connection of the tubes that drain the gas fuel to the safe gas admission valves, flange connections are used, which are necessary for servicing the elements of the gas system. The pipeline design designed to compensate for thermal expansion when the engine is heated. All gas system pipes designed for a pressure exceeding the working pressure by 50%, and during factory tests, they tested with a pressure of 150% higher than the working pressure. All gas pipes placed in protective shells capable of withstanding the pressure that may occur when the main line ruptures. The internal space between the shell and the pipeline connected to the forced ventilation system, which provides approximately 30-fold air change per hour. The cavities adjacent to the main elements of the fuel system, in which a gas leak may occur, also ventilated. A gas presence sensor connected to the emergency alarm system installed at the outlet of the ventilation system. The pipelines arranged so that they not damaged in the event of heavy objects falling during engine repair and maintenance. Otherwise, the safety measures are similar to those typical for the previously considered installations with four-stroke gas-diesel engines. A fragment of one of the design options for the gas main shown in Fig. 7.50.

To improve the safety of engine operation, the power plant is equipped with an inert gas system (N_2), which allows purging both the entire gas fuel supply system and its individual elements under a pressure of 0.4–0.8 MPa. Such purging is mandatory when switching to liquid fuel or when any section of the gas supply system damaged. The high-pressure gas-diesel engine feed system is shown schematically in Fig. 7.51a.

7.8.2 Injection of Gas Fuel into the Combustion Chamber of the Engine

Gas supplied to the combustion chamber immediately after the pilot portion of liquid fuel has been to the cylinder and ignited (Fig. 7.51b). In this way, a high degree of fuel burnout achieved and the risk of gas entering the sub-piston space through leaks in the piston rings prevented.

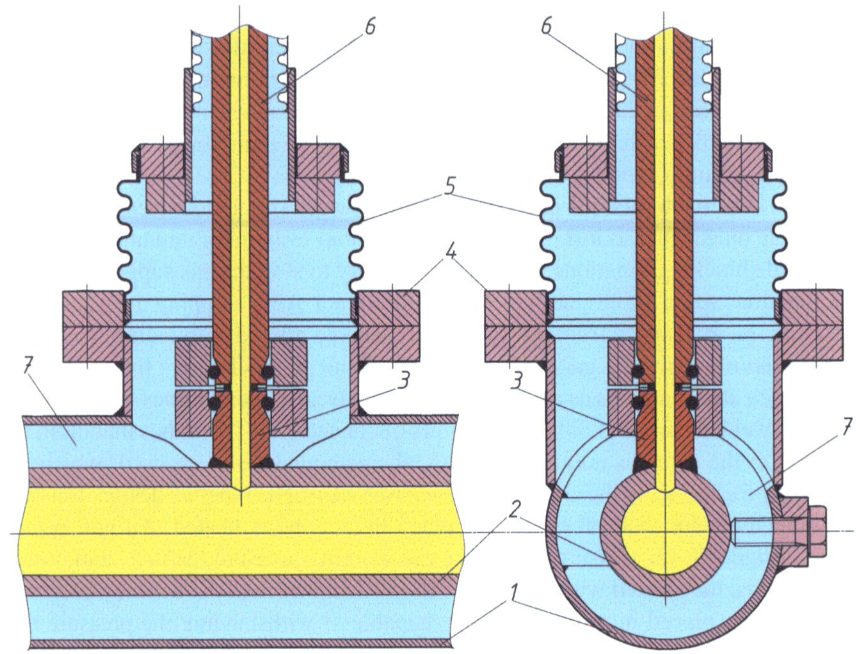

Fig. 7.50 Fragment of the gas line of the ME-GI series gas-diesel engine: (1) protective casing; (2) gas line; (3) connecting nipple; (4) connecting flange; (5) protective corrugated casing; (6) gas outlet tubes to the feed control unit; (7) ventilated space (adapted from [2])

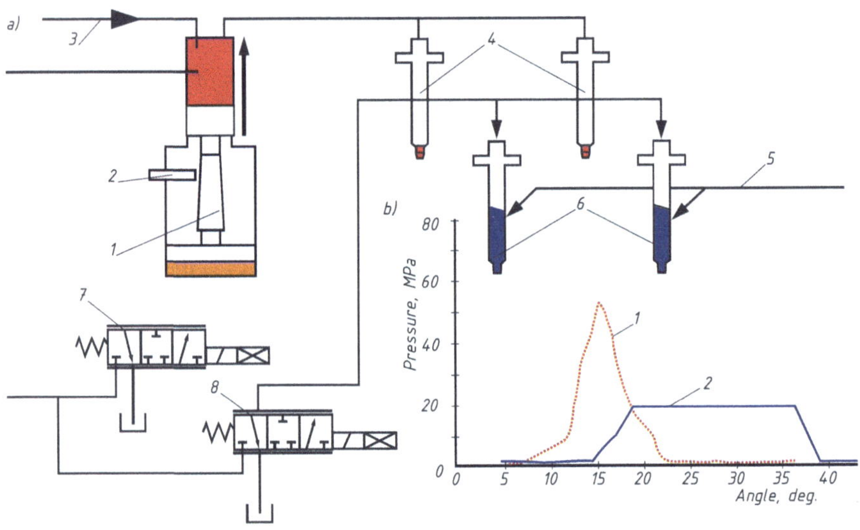

Fig. 7.51 Diagram of the fuel system of the ME-GI series gas-diesel engine (**a**) and the laws for supplying pilot (1) and gas fuel (2) (**b**): (1) high-pressure fuel pump with hydraulic drive; (2) liquid fuel supply value sensor; (3) liquid fuel supply; (4) liquid fuel injection nozzles; (5) gas supply; (6) gas injectors; (7) fuel injection valve actuation (FIVA); (8) electronic liquid gas injection (ELGI); (9) supply of control oil under pressure of 30 MPa (adapted from [31])

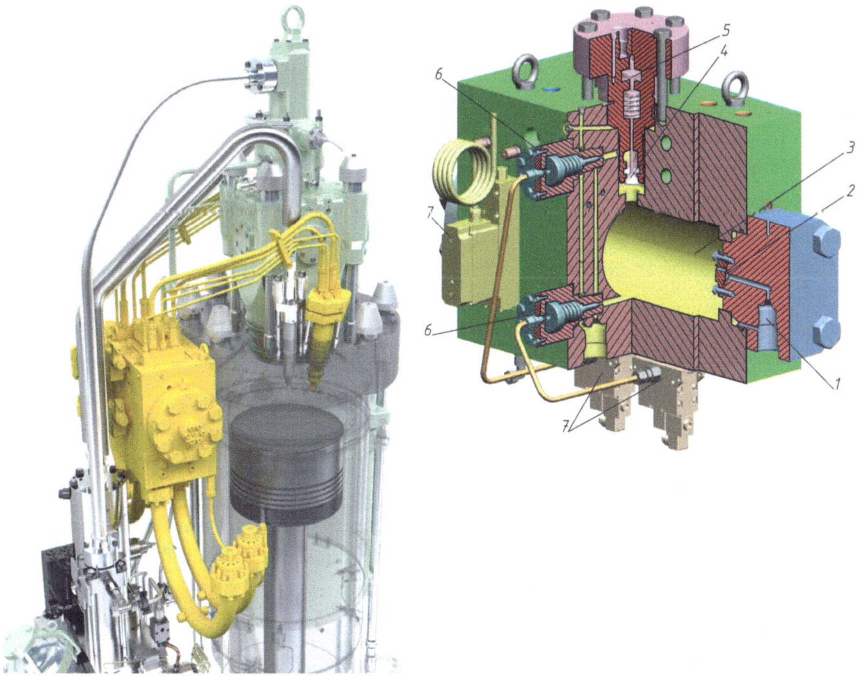

Fig. 7.52 Installation of the gas supply control module on the engine and its general structure: (1) gas fuel inlet unit; (2) check valve; (3) pressure accumulator cavity; (4) window valve; (5) window valve drive piston; (6) inert gas system purge valve; (7) window valve control unit; (8) inert gas system purge valve control unit (adapted from [31])

In first-generation gas-diesel engines, the main elements of gas fuel control arranged in one module, which includes (Fig. 7.52):

- gas accumulator;
- main shut-off valve or window valve;
- valve for purging the system with inert gas;
- valve for controlling the hydraulic drive of the injectors.

The main shut-off valve, as in low-pressure systems, called a window valve (WV). Its opening provides a time synchronization window (angular interval) during which fuel injection is possible. It allows gas fuel to flow from the accumulator through openings in the control unit and cylinder cover to the fuel gas injection valves (FGIV), which are the second regulating devices on the fuel path from the pressure accumulator to the combustion chamber. The window valve prevents gas fuel injection outside the permissible synchronization window, thereby limiting the flow of gas fuel into the working cylinder outside the specified angular (time) interval. The window valve sets the maximum possible interval, thereby limiting the maximum supply of gas fuel. The gas fuel injection valves control the precise timing and amount of gas injected. They hydraulically driven and controlled by electric valves that receive electrical signals from the engine control unit. The housing of the feed control module attached to the cylinder cover, which has internal drillings for supplying gas to the gas injectors installed in the cylinder cover. The gas injectors installed in proximity to the liquid fuel injection injectors.

The location of the gas feed control module and its general structure shown in Fig. 7.52. The gas module diagram and its operating procedure shown in Fig. 7.53.

The gas fuel from the supply line enters the pressure accumulator, which made as a cavity in the housing of the control module, through a check valve. The capacity of the accumulator is equivalent to approximately 20 cyclic gas feeds at nominal load.

The presence of the accumulator in the control unit serves to reduce the pressure drop during fuel injection. Stable pressure is necessary so that the control system can correctly determine the injection time, which sets the cyclic feed value.

In the absence of a control signal from the control unit, the window valve is closed and gas does not enter the gas injectors (Fig. 7.53a). When an electrical signal received from the engine control unit to the window valve control unit, its spool moves and supplies control oil to the window valve hydraulic drive mechanism. The valve opens, and gas flows to the injectors, the needle valves of which remain closed

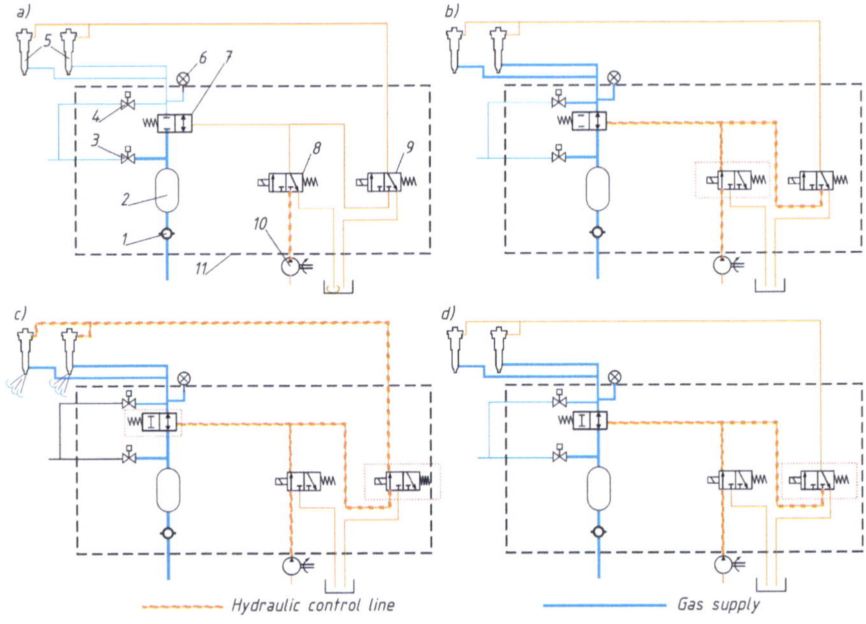

Fig. 7.53 Diagram of the gas supply control module and its operating procedure: (**a**) no gas supply to the gas injectors; (**b**) gas supply to the gas injectors; (**c**) supply of control oil to the injectors, fuel injection; (**d**) closing of the injector valves, injection stop. (1) gas line; (2) accumulator; (3) purge valves; (4) emergency relief valve; (5) gas injectors; (6) gas pressure sensor; (7) window valve; (8) window valve control unit; (9) injector hydraulic drive control unit; (10) control oil pump; (11) supply control module (adapted from [29, 32])

at this point (Fig. 7.53b). Filling the channels between the module and the injectors, the gas acts on the pressure sensor. Information about the actual pressure received from the sensor is sent to the engine control unit and is used to calculate the required injector opening time to ensure a given cyclic supply value.

Based on the information received, the control unit generates a signal sent to the injector hydraulic drive control unit. Under the action of the signal, the unit's spool moves and supplies control oil to the injector drive (Fig. 7.53c). Opening, the injectors supply gas to the engine combustion chamber.

After removing the control signal from the injector hydraulic drive control unit, the spool moves and switches the oil from the hydraulic drive circuit to drain, because of which the injector valves close and the fuel supply stops (Fig. 7.53d).

Removing the signal from the window valve control unit leads to its closure, and the system returns to its original state (Fig. 7.53a).

In the considered scheme of gas supply to the working cylinder, a two-stage connection is used—first through the window valve, and then through the injector valves. This solution allows increasing the level of operational safety, and, if necessary, to clean damaged cavities by blowing them with inert gases through special valves.

The design of the gas injector of the first generation engines and its installation on the cylinder cover, are shown in Fig. 7.54.

Two (three) gas injectors are installed on the cylinder cover in special wells located in close proximity to the liquid fuel injectors. Gas supplied to the injector

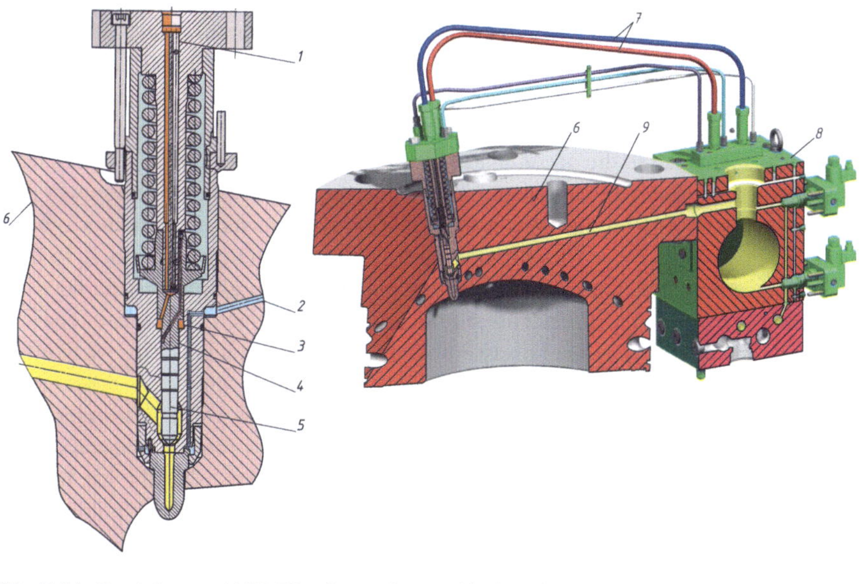

Fig. 7.54 Gas injector of ME-GI series engines and its location on the cylinder cover: (1) control oil supply channel; (2) ventilation channel; (3) cavity of hydraulic control of needle valve; (4) needle valve; (5) gas cavity; (6) cylinder cover; (7) control oil supply; (8) gas fuel supply module; (9) channel for supplying gas fuel to the injector (adapted from [33–35])

through holes in the cylinder cover. Then, through holes in its injector body, the gas enters the cavity above the shut-off cone of the needle valve. To prevent gas leaks, sealing rings installed between the injector body and the cylinder cover. The cavities located outside the seals connected to the general ventilation system. In the closed state, the needle valve held by the force of the spring acting on the plate, made with it as a single unit. In the lower part of the needle valve, there is a flange, precisely fitted to the body, acting as a hydraulic piston. Through the channels in the guide and in the body of the needle valve, oil from the feed control module enters the annular cavity under the flange, forcing the needle valve to open. To prevent gas leakage between the needle valve and the body of the nozzle, special sealing oil constantly supplied to the gap between them under a pressure 0.2–0.3 MPa higher than the gas pressure in front of the nozzle. The sealing oil enters through the channel system into the annular groove on the body of the needle valve.

The next generation of high-pressure gas-diesel engines, designated ME-GI Mk. 2, includes a number of component upgrades that improve reliability, increase efficiency and reduce operating costs.

In particular, the auxiliary piping system has simplified. Channels located in the cylinder cover and the supply control unit housing have replaced some of the external oil supply tubes for injector control. This has increased reliability by eliminating the possibility of damage to these pipes, for example, due to heavy objects falling on them.

A single main pipe used to supply gas fuel to the supply control modules. Instead of check valves, calibrated holes used in the pressure accumulator housings. The design of the gas accumulator itself been optimized to achieve a more compact size.

The gas fuel module has undergone significant modernization (Fig. 7.55).

In the new design, all control valves are located in a separate block, which connected to the accumulator block using seals and guide pins between the flanges of the blocks. The valve block attached directly to the cylinder cover, while all gas supply lines attached to the accumulator block. Thus, if it is necessary to dismantle the cylinder cover, there is no need to dismantle the gas module or disconnect the gas lines, it is enough to disconnect the valve block from the accumulator block. After that, you can remove the cylinder cover together with the valve block. In this case, the accumulator block remains on the engine, which eliminates the need to disconnect it from the gas lines. This significantly reduces the time for maintenance and repair and allows you to not disassemble the gas joints unnecessarily, since after such disassembly it is necessary to replace all the seals and a complex procedure for checking them after reassembly. The design of the window valve has undergone significant changes, namely, to fix it in the closed state; they refused to use a cylindrical spring. Now the window valve held closed by the pressure of the gas fuel in the accumulator cavity (Fig. 7.55d). The valve opened by acting on its spindle with a hydraulic piston, as was the case in the previous version of the module. To close the window valve, a special device called (Window Valve Force Close—WVFC) installed opposite it. It performs a triple function:

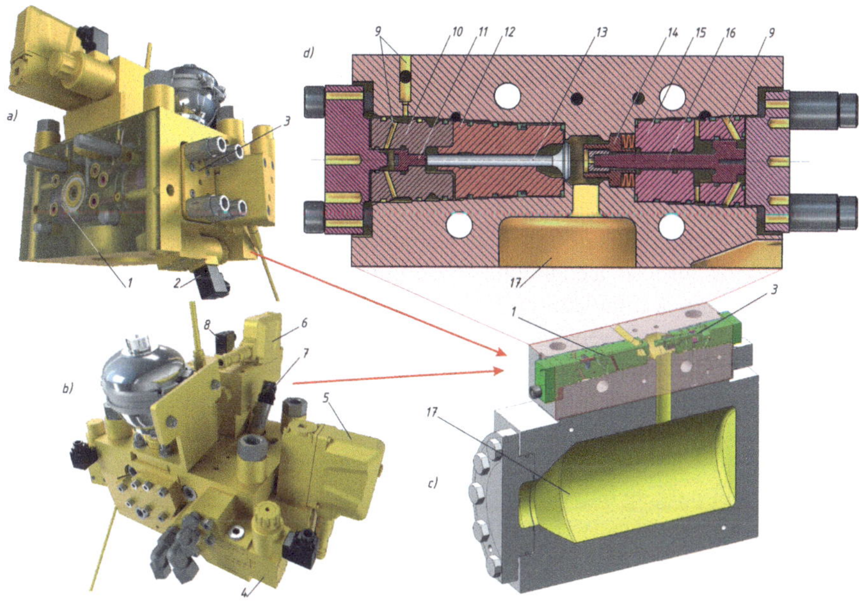

Fig. 7.55 Gas supply control module of ME-GI Mk 2 engines, general view of the valve unit (***a***, **b**), pressure accumulator (**c**) and valve unit device (**d**): (1) window valve (WV); (2) electronic window control valve (ELWC); (3) window valve forced close (WVFC); (4) sealing oil block (SOB); (5) electronic window valve (ELWI); (6) electronic gas injection block (ELGIB); (7) pressure sensor; (8) electronic gas injection valve (ELGI); (9) control oil supply channels; (10) window valve drive hydraulic cylinder; (11) window valve drive piston; (12) window valve housing; (13) window valve; (14) damper; (15) hydraulic cylinder of the window valve forced closing mechanism drive; (16) drive piston; (17) accumulator cavity (adapted from [35, 36])

- acts as a damper to soften the impact of the window valve when it opens, for example, when operating without sufficient gas pressure in the system;
- acts as a drive creating force to close the window valve;
- holds the window valve closed when the gas fuel supply is turned off.

The last option allows you to disconnect individual cylinders from the gas fuel supply, which the engine control unit switches to liquid fuel. Thus, even if any components of the gas system of an individual cylinder fail, the engine can continue to operate at full power.

The window valve and its closing device have separate hydraulic drives, which controlled by separate electro-hydraulic valves receiving a signal from separate electronic control units.

For greater operational safety, the control and safety functions in the engine control system divided into different hardware units. The most important information about the process of gas supply to the engine monitored using two pressure sensors. One sensor transmits information to the control unit, and the other sensor to the safety unit. Both sensors must be operational so that the engine can continue to operate on gas. If any of the sensors fails, the gas fuel supply is disconnected. The control unit controls gas fuel injection by the injector, and the window valve controlled by the safety unit. An external protective shell to prevent gas leakage into the engine rooms in the event of a rupture or leak in the internal gas pipeline protects the gas pipelines. All main pipelines connected to the feed control unit by means of flanged joints. The gas fuel fed through the internal pipe of the double-walled main, and ventilation is through the external space between the gas pipe and the shell.

In order to simplify the design, new gas modules are not equipped with individual purge valves. Instead, a common purge unit for all modules is installed, which is attached to the gas accumulator of the last cylinder (Fig. 7.56a).

The gas fuel unit is a separate unit at the end of the gas supply pipeline, which also connected to the inert gas, ventilation and leakage control systems. The gas fuel unit performs a combined function, namely; a check valve, a regulating valve and a safety valve. It prevents gas fuel from entering the inert gas system while allowing inert gas to enter the gas fuel line during system purging.

The gas fuel-purging unit contains several components necessary to perform the described functions, namely (Fig. 7.56a, b):

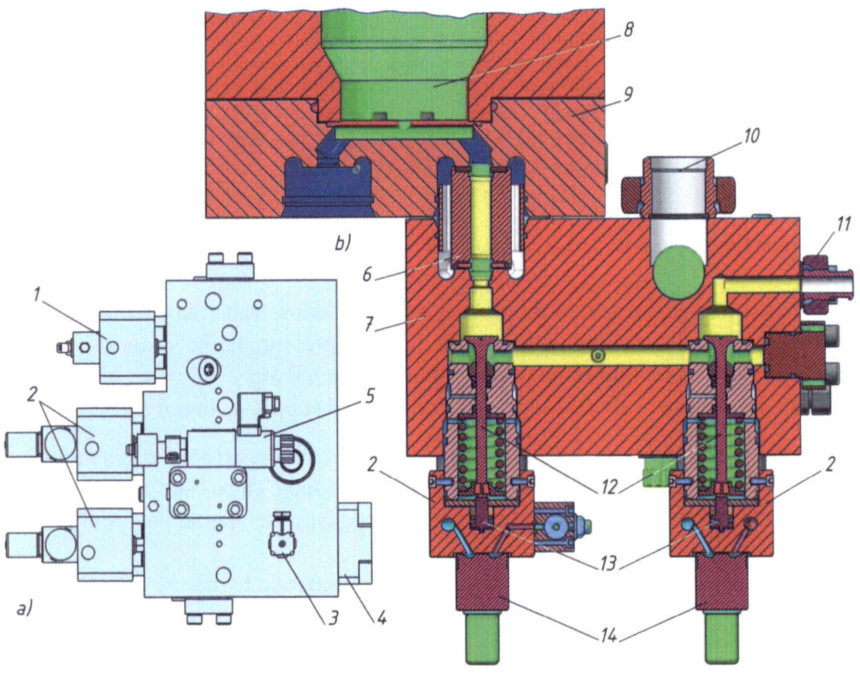

Fig. 7.56 Gas fuel purge unit: (**a**) general view; (**b**) section along the plane of the blocking valves; (1) check valve; (2) blocking valves; (3) pressure sensor; (4) inert gas filter (N_2); (5) pressure relief valve; (6) transition element; (7) purge valve unit body; (8) pressure accumulator cavity; (9) accumulator cover; (10) ventilation pipe; (11) inert gas supply; (12) shut-off element of the blocking valve; (13) hydraulic pistons of the blocking valve drive; (14) electrohydraulic valve for controlling the blocking valve drive (adapted from [37])

- an inert gas filter located at the inert gas inlet of the unit, this filter protects the gas fuel purging unit from dirt and impurities;
- an inert gas pressure sensor located at the inlet of the unit. The sensor measures whether the inert gas pressure is sufficient to perform purging;
- a check valve installed at the gas fuel inlet of the unit. This valve prevents gas fuel from entering the inert gas system. The check valve is equipped with a manual control handle. During normal operation, this handle must always remain in the closed position;
- two blocking valves installed in series and located between the check valve and the inert gas inlet. The block valves control the engine purge by opening to allow inert gas to pass through during purge and act as safety valves if there is a leak in the check valve;
- a bleed valve installed between the two block valves. This valve is open when the engine is running on gas fuel. It removes gas fuel into the ventilation system from the space after the block valve. In the event of leaks in either the check valve or the block valve, this results in the gas fuel supply to the engine being cut.

In the upgraded series of engines, changes also made to the design of the gas injector, although the principles of its operation remained the same (Fig. 7.57).

Gas fuel supplied to the gas injector valve through holes in the cylinder cover. To prevent gas leaks, sealing rings made of heat- and gas-resistant material installed between the cylinder cover and the injector body, as well as between the injector body and the gas valve body. In the event of a gas fuel leak through the sealing rings, the gas will be directed through a special leak collection channel in the injector body into the space between the inner and outer protective tubes of the double-walled gas pipeline system, where the leak will detected by methane sensors.

In new versions of ME-GI engines, the supply of control oil and oil for sealing the gas valve stem carried out through drillings made in the cylinder cover. This required changing the design of the injector body, providing additional annular grooves on it, through which oil from the channels gets into the injector cavity. This also required the installation of additional seals separating these cavities. The method of fastening the injector to the cylinder cover has also changed. In new versions, the gas injector fastened with a flange using six bolts, which allows for a more even distribution of loads in the injector elements.

Operation of gas injectors of the first modifications showed that the low inertia of gas fuel promotes its combustion in the immediate vicinity of the nozzle tip of the injector. This leads to significant heating of the tip and its rapid failure. In subsequent versions of gas injectors, the part of the nozzle tip protruding into the combustion chamber reduced to a minimum. Tool steel S40Cr5Mo1V has proven itself as a material for the manufacture of nozzle tips. The use of new design solutions and materials made it possible to increase the service life of the nozzles to 6000 hours, and the time between scheduled inspections to 2000 hours. The evolution of the design of nozzle tips of gas injectors of gas-diesel engines of the ME-GI series shown in Fig. 7.57b. Gas injectors are installed in the cylinder cover in close

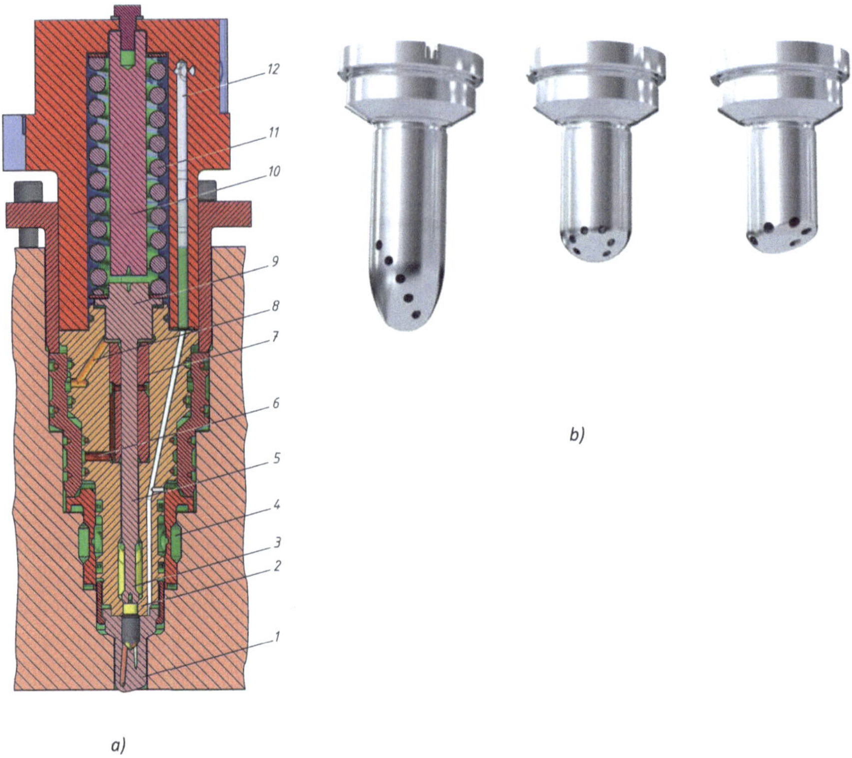

Fig. 7.57 Gas injector of ME-GI Mk 2 series engines (**a**) and nozzle tip variants (**b**): (1) nozzle tip; (2) gas valve body; (3) gas valve shut-off cone; (4) gas fuel supply cavity; (5) gas valve stem; (6) sealing oil supply channel; (7) gas valve stem guide sleeve; (8) control oil supply channel; (9) gas valve hydraulic drive piston; (10) gas valve stop; (11) spring; (12) gas fuel leakage control channel (adapted from [37, 38])

proximity to the liquid fuel injectors, so that the gas fuel torches formed by them intersect the trajectory of the pilot fuel torches (Fig. 7.58).

7.8.3 Pilot Ignition Systems for Gas Fuel

The automatic engine control system uses algorithms in which pilot fuel injected into the combustion chamber first. Its self-ignition and combustion leads to an increase in pressure, albeit insignificant. A sensor installed in the cylinder cover records the increase in pressure. It transmits a signal to the control unit, which indicates the formation of an open flame in the combustion chamber. This signal starts the process of opening the gas valve of the injector and the supply of gas fuel to the combustion chamber. This eliminates combustion misfires, which can lead to both increased methane slippage and consequences that are more serious. When in

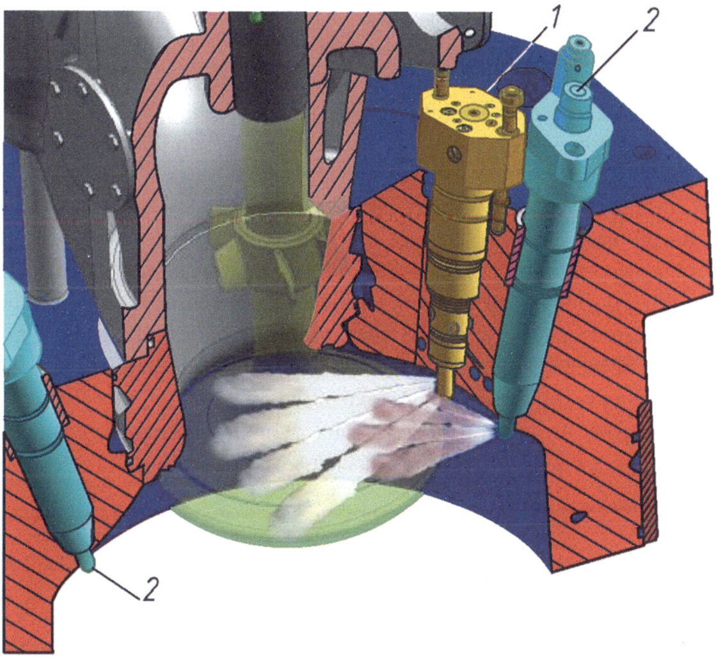

Fig. 7.58 Installation of injectors in the cylinder cover of the ME-GI series gas-diesel engines: (1) gas injector; (2) liquid fuel injector (adapted from [39])

contact with an open flame, the gas-air mixture formed in the torch quickly ignites and burns without reaching the walls of the combustion chamber. This reduces the likelihood of incomplete combustion in the near-wall areas and methane getting into dead zones, such as gaps between the piston head, cylinder liner and piston rings. This significantly reduces the amount of hydrocarbons getting into the exhaust gases. The expansion of the scope of use of low-speed gas-diesel engines on vessels of other types, except for gas carriers, has contributed to the search for solutions aimed at reducing the pilot fuel supply. This allows for a reduction in operating costs, on the one hand, and an improvement in the environmental performance of the engine, on the other. For dual-fuel engines, the pilot fuel largely determines the emission of harmful components in the exhaust gases such as sulfur oxides and particulate matter. Therefore, reducing the pilot supply leads to a reduction in the emissions of these components.

The main problem that specialists face when trying to reduce the pilot supply is a significant deterioration in the quality of atomization when reducing the volume of fuel passing through the nozzle openings. A significant flow section of the fuel channels, designed to pass large portions of liquid fuel, leads to a significant decrease in the exit velocity of the jet when the supply is reduced. This does not contribute to the occurrence of longitudinal and transverse disturbances in the jet, which cause its further disintegration. In addition, fuel systems designed to supply fuel portions in a certain range associated with the main operating modes of the engine, operate extremely unstable outside these ranges. This is especially true for supply systems with a mechanical drive of high-pressure fuel pumps.

For this reason, most manufacturers of dual-fuel engines use separate low-capacity accumulator systems for pilot fuel injection in their designs. However, this approach significantly complicates the design of the engine and its subsequent operation.

When developing a new version of the ME-GI Mk. 2 gas-diesel engines, the concept of injecting pilot fuel through the same injectors that provide the main supply of liquid fuel during operation on the diesel cycle retained. However, to improve the quality of atomization at low feed rates, a new version of the injectors with an adjustable stop of the nozzle spool valve was designed (Fig. 7.59a).

When operating on liquid fuel, the stroke of the nozzle tip spool valve is 2.8 mm. In doing so, it opens two rows of nozzle holes, which together ensure the efficient passage of large volumes of cyclic portions (Fig. 7.59b).

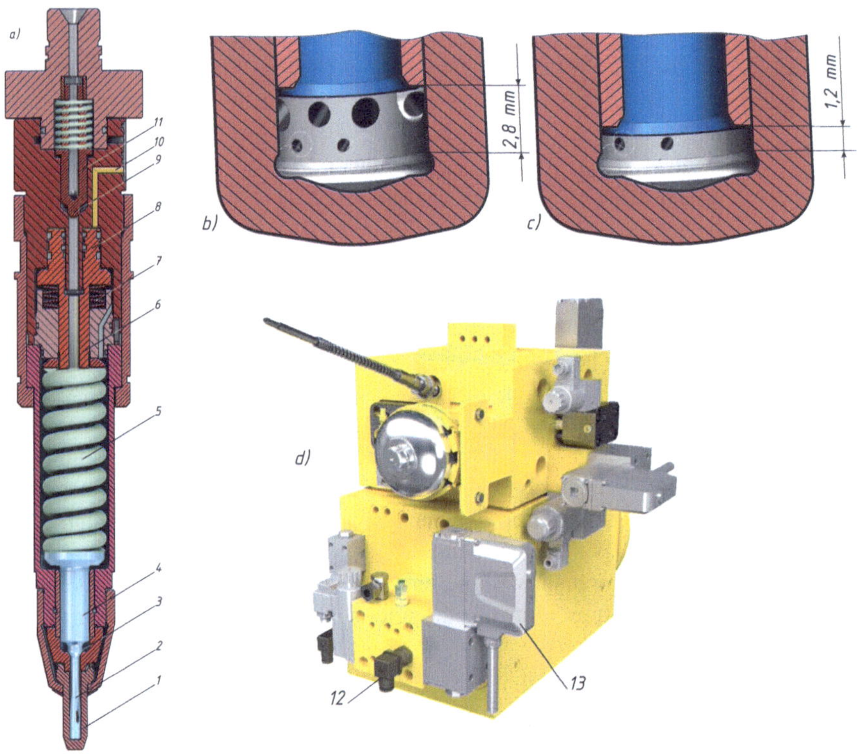

Fig. 7.59 Liquid fuel injection valve (**a**) with two-level stop of the needle and spool valves of the nozzle (**b, c**), location of the pilot injection controls on the gas module (*d*): (1) nozzle tip; (2) nozzle tip spool valve; (3) needle valve; (4) needle valve guide; (5) needle valve spring; (6) needle valve stop; (7) disc spring; (8) hydraulic piston of the stop drive; (9) check valve; (10) channel for supplying control oil to the hydraulic piston; (11) check valve guide; (12) control oil pressure sensor; (13) electromagnetic control valve (adapted from [37])

When switching to gas fuel, a special hydraulic piston installed in the injector body moves the stop of the needle valve of the nozzle and the spool valve of the nozzle tip, limiting their stroke to 1.2 mm (Fig. 7.59c).

As a result, only two nozzle channels are open for fuel to pass, the cross-section of which is 7.5% of the total cross-section of the nozzle channels of the nozzle. Through these openings, the pilot portion enters the combustion chamber, ensuring the required quality of atomization.

Because of using the described two-stage stop adjustment system, it was possible to reduce the pilot fuel consumption to 0.5–1.5% of the full cyclic feed.

To drive the hydraulic piston that ensures a change in the stroke of the needle and spool valves, control oil is supplied to the injector under a pressure of 22.5–30.0 MPa. To control the supply of control oil to the fuel injector, to switch it to pilot injection mode, a special electrohydraulic control valve and a pressure sensor in the control line installed on the gas module (Fig. 7.59d).

For particularly large container ships, MAN developed a gas-diesel version of particularly compact engines, which were designated, MAN 90ME-GI and 95ME-GI.

These engines used the technology of a top-controlled exhaust valve (TCEV) and a hydraulically driven pump-injector for liquid fuel injection (Fuel Booster Injection Valve—FBIV) (Fig. 7.60). We examined these technologies in more detail in the corresponding chapters of the main book [2].

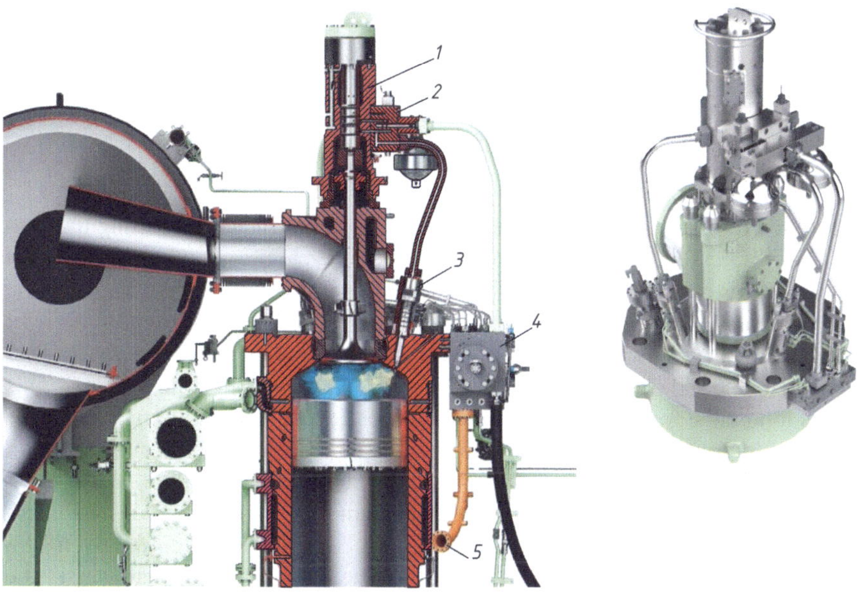

Fig. 7.60 General arrangement of the working cylinder cover with upper control of the exhaust valve and fuel pump-injectors with hydraulic drive of MAN 90ME-GI and 95ME-GI engines: (1) upper control module; (2) control oil supply unit; (3) pump-injector; (4) gas fuel supply control module; (5) double-walled gas fuel supply pipe (adapted from [40])

It should note that the use of these design features has significantly improved the mass-dimensional parameters of the engines. All controls and actuators mounted on the cover of the working cylinder. The gas fuel supply control module, as in other versions of GI engines, is attached to the cylinder cover on the side.

The distribution block, located on the valve body, supplies high-pressure hydraulic oil through two control spools to the exhaust valve drive and the fuel pump-injectors drive. Since the pressure required for liquid fuel injection created directly by the pump-injector, there are no high-pressure pipelines in the system. This, in turn, improved the dynamic characteristics of the fuel injection system.

Three pump-injectors for liquid fuel supply and three gas injectors installed on the cylinder cover of the MAN 90ME-GI and 95ME-GI engines. The pump-injectors provide pilot fuel portions when the engine is operating on the gas-diesel cycle and full cyclic feeds when the engine is operating on liquid fuel. Thus, the engines do not need an additional fuel system for pilot fuel injection, which significantly simplifies its design and maintenance. High-pressure gas fuel distributed through a system of double-walled pipes to each cylinder, entering the distribution block, the design of which is similar to the one we reviewed earlier. Gas injectors provide injection of gas fuel, which comes to them from the distribution module through channels made in the cylinder cover. Gas injectors controlled hydraulically, as in other versions of high-pressure engines. All the innovations considered in the modernized version of the ME-GI Mk 2 series gas-diesel engines made it possible to add another significant option—to switch one of the cylinders to work on liquid fuel, while the remaining cylinders operate on the gas-diesel cycle.

7.8.4 *Features of Operation and Environmental Performance of High-Pressure Engines*

Unlike low-pressure engines, where the engine operates on gas fuel using the Otto cycle, high-pressure engines operate on the diesel cycle using both gas and liquid fuels. For this reason, the nature of the working process differs slightly. From the data provided (Fig. 7.61), it is evident that the forms of heat release rates of diesel and gas-diesel cycles are not only very similar in terms of the maximum rate of heat release and the end of combustion, but also in terms of the form of heat release. This makes it possible for individual cylinders to operate together on different fuels.

It should noted that the operation of an engine on gas fuel on a cycle close to diesel also has a negative side, namely the impossibility of meeting the standards for nitrogen oxide emissions even in accordance with Tier II without additional measures. High-pressure engines are essentially diesel engines and, accordingly, have a level of NO_x emissions similar to a diesel engine, adjusted for differences in the chemical composition of liquid fuel and natural gas, as well as their combustion temperatures (Fig. 7.62). Taking these features into account, the NO_x emission levels when operating on a gas-diesel cycle are somewhat lower, but not so much as to

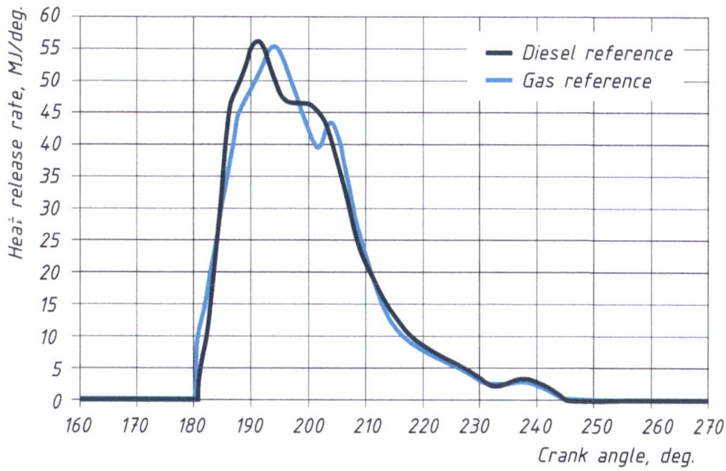

Fig. 7.61 Heat generation rates in the working cylinder of the S70ME-C-GI engine during operation on the diesel and gas-diesel cycle at 100% load (adapted from [41])

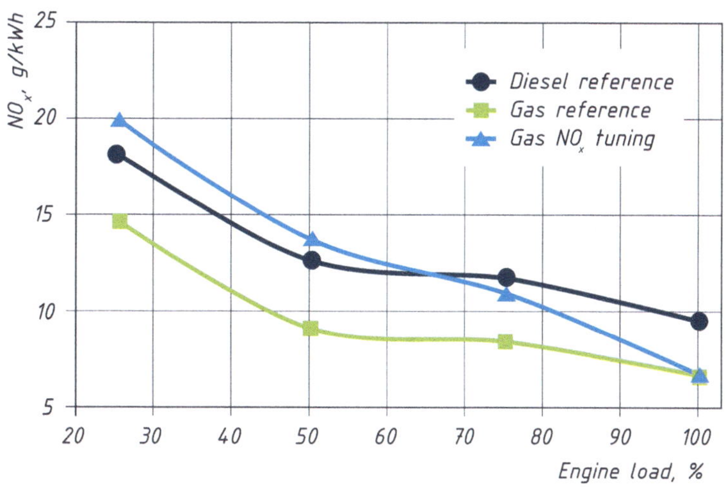

Fig. 7.62 Change in NO_x emissions in the range of operating modes when operating on diesel, gas-diesel and optimized cycles. (adapted from [41])

position this engine as meeting the growing requirements for reducing emissions of this group of harmful substances.

Understanding this, the manufacturer, as an option, considered the possibility of introducing some changes to the operating process (Fig. 7.63), which allowed increasing the fuel efficiency of the engine (Fig. 7.64). At the same time, the level of NO_x emissions does not exceed the similar indicator for the diesel mode in the

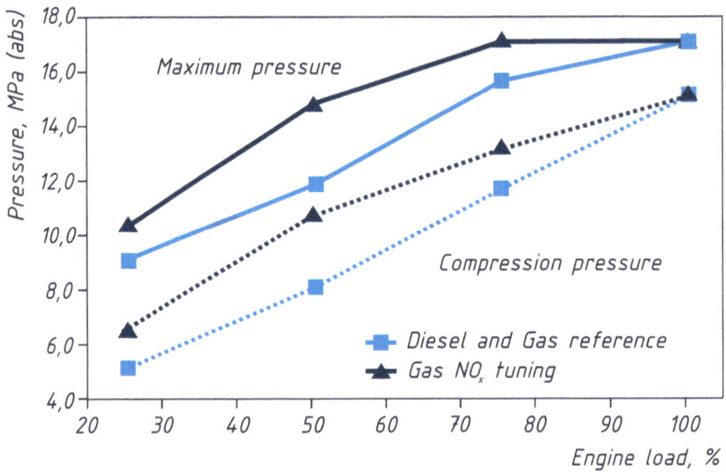

Fig. 7.63 Change in the working process of an engine operating on a gas-diesel cycle to optimize fuel consumption while maintaining the NO_x level not exceeding the same indicator when operating on a diesel cycle (adapted from [41])

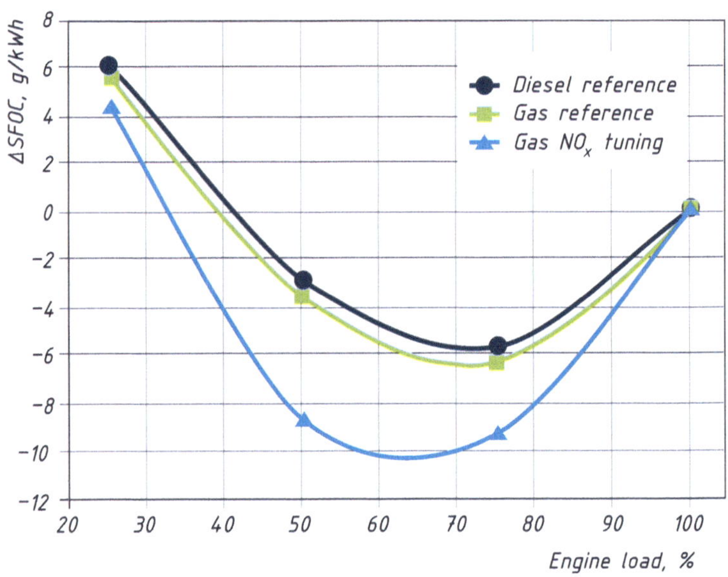

Fig. 7.64 Change in specific fuel oil consumption of an engine operating on a diesel, gas-diesel and optimized cycle (adapted from [41])

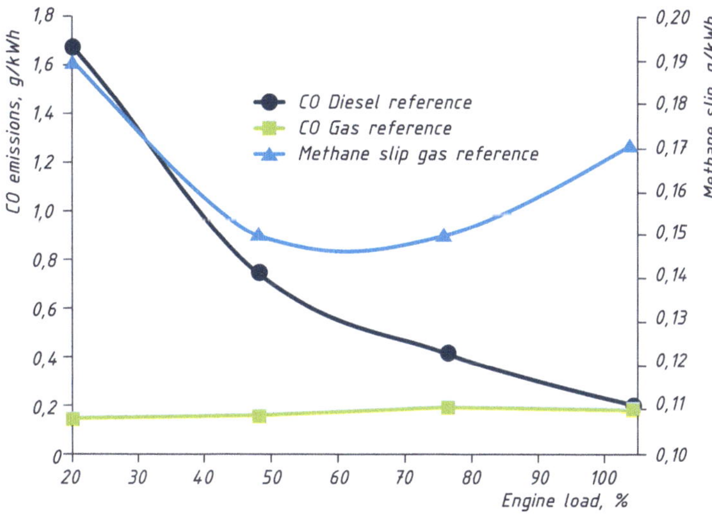

Fig. 7.65 CO and methane emissions during engine operation on diesel and gas-diesel cycles (adapted from [42])

main range of operating loads (Fig. 7.62) As a result; it was possible to increase the fuel efficiency of the engine when operating on the gas-diesel cycle up to 3%.

Almost homogeneous mixture formation (except for a small proportion of pilot fuel) leads to high completeness of fuel burnout, which minimizes the content of incomplete combustion products in the exhaust gases such as carbon monoxide (CO) and hydrocarbons (CH) (Fig. 7.65).

As can be seen from Fig. 7.65, the CO content remains virtually unchanged throughout the entire range of engine operating modes, while when operating on liquid fuel, the combustion quality noticeably decreases with decreasing load.

A significant advantage of high-pressure technology is the low level of methane leakage (Fig. 7.65), since almost all mechanisms of its entry into the engine exhaust gases eliminated.

Contact of gas fuel immediately after exiting the nozzle channel of the injection with the centers of open flame from the combustion of pilot fuel effectively initiates the combustion process. Due to this, combustion occurs within the spray plume of gas fuel, practically without reaching the cold walls of the combustion chamber, where reactions can slow down significantly. In addition, the entry of gas fuel into the dead zones associated with the combustion chamber is limited, since they filled with clean air that got there during the compression process.

Despite the high NO_x emissions during combustion of gas fuels in high-pressure gas-diesel engines, it should noted that this technology is well combined with the previously discussed NO_x emission reduction technologies, such as Exhaust Gas Recirculation (EGR) or Selective Catalytic Cleaning (SCR). Figure 7.66 shows that exhaust gas recirculation is even more effective for the gas-diesel process than for

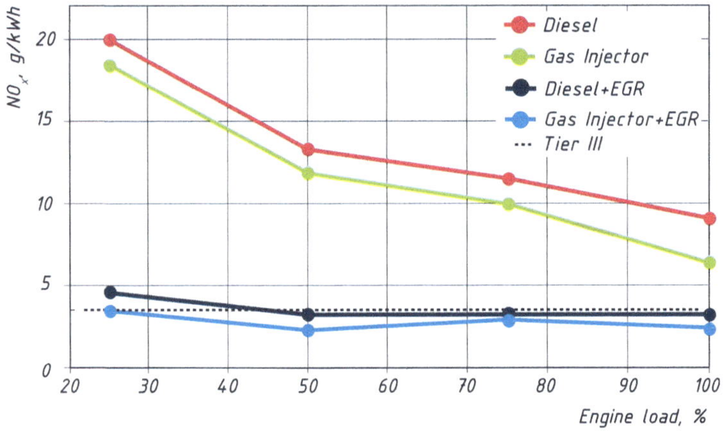

Fig. 7.66 NO_x emissions during engine operation on diesel and gas-diesel cycles with and without EGR technology (adapted from [40])

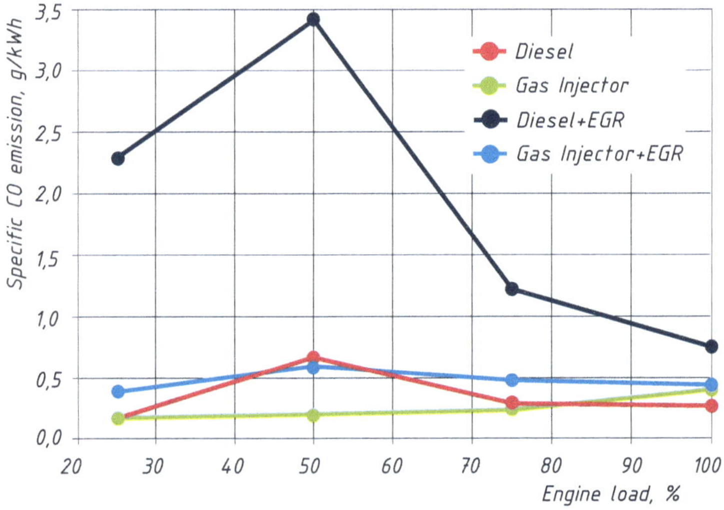

Fig. 7.67 CO emissions during engine operation on diesel and gas-diesel cycles with and without EGR technology (adapted from [41])

the diesel process. Using the EGR unit ensures compliance of NO_x emissions with Tier III standards over the entire range of engine loads.

In this case, the deterioration of combustion associated with a decrease in the proportion of oxygen in the charge of the working cylinder does not significantly affect the emissions of other harmful components, primarily incomplete combustion products. Figure 7.67 shows the change in carbon monoxide emissions for cases of engine operation on diesel and gas-diesel cycles with and without the use of an EGR unit. It clearly seen that the CO content increases when using EGR. In

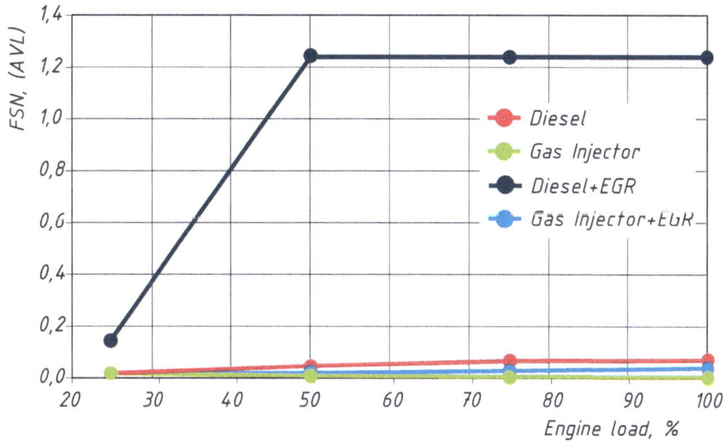

Fig. 7.68 Emissions of particulate matter during engine operation on diesel and gas-diesel cycles with and without EGR technology (adapted from [41])

partial load modes, it can be twice as high as the same indicator for an engine operating on a gas-diesel cycle without EGR. However, this indicator is several times lower than in the case of an engine with EGR operating on a diesel cycle. That is, the effect of EGR on the combustion quality when operating on a gas-diesel cycle is insignificant. These conclusions confirmed by the data on particulate matter emissions shown in Fig. 7.68. Here, an insignificant increase in the particulate matter content in the combustion products is also noted, but it is several times lower than the same indicator for the diesel cycle using EGR.

The greatest deterioration in the engine's environmental performance when using EGR is associated with an increase in hydrocarbon emissions (methane slip), which are already higher in the gas-diesel process than in the diesel cycle, and when EGR is connected, they increase even more (Fig. 7.69).

In addition, as already noted, the use of EGR leads to an increase in the specific fuel consumption of the engine, which is associated with lower combustion temperatures and a more extended heat supply process.

This statement is also true for engines operating on the gas-diesel cycle, which clearly seen in Fig. 7.70.

It should note that the above results obtained for the same level of exhaust gas recirculation both when operating on the gas-diesel and diesel cycles. It is obvious that by varying this indicator, it is possible to optimize the engine's operating process to achieve more specific goals, depending on the task. Thus, the use of technologies to reduce emissions of harmful substances allows us to meet not only current, but also future emission standards for high-pressure gas-diesel engines.

Below, we will provide a comparative assessment of both existing methods of gas fuels in low-speed two-stroke engines.

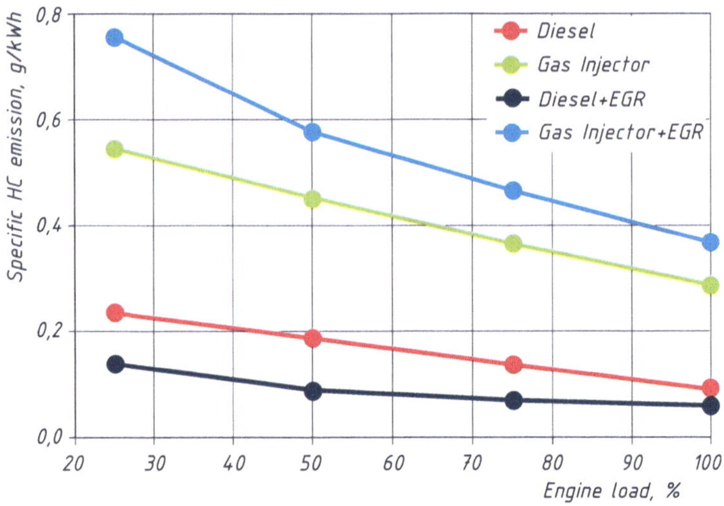

Fig. 7.69 Hydrocarbon emissions (methane slip) during engine operation on diesel and gas-diesel cycles with and without EGR technology (adapted from [41])

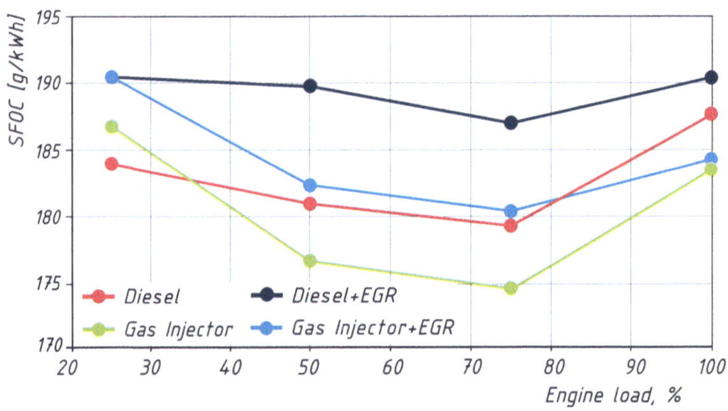

Fig. 7.70 Specific fuel oil consumption given for engine operation on diesel and gas-diesel cycles with and without EGR technology (adapted from [41])

7.8.5 Comparison of Low- and High-Pressure Systems of Dual Fuel Low-Speed Marine Diesel Engines

As shown above, today there are two fundamentally different approaches to organizing the operation of low-speed gas-fueled engines, each of which has its own advantages and disadvantages.

The advantages of low-pressure systems include:

– good mixing of the gas-air mixture during the compression process;

- use of a minimum portion of pilot fuel to ignite the gas-air mixture. Although this advantage has ceased to be relevant with the advent of a new modification of high-pressure engines—ME-GI Mk 2;
- supply of gas fuel to the working cylinder under relatively low pressure, which reduces the likelihood of leaks, simplifies the fuel system, increases the safety of its use, and reduces energy costs for compressing the fuel gas to high pressure;
- cheaper and more reliable screw or centrifugal compressors can be used to supply gas fuel;
- the ability to meet IMO Tier-III requirements for NO_x emissions without using methods to reduce harmful substances in exhaust gases;
- significantly lower content of sulfur oxides SO_x and especially particulate matters (soot) in the exhaust gases compared to engines equipped with high-pressure systems (Fig. 7.71).

Disadvantages of low-pressure systems:

- the possibility of detonation and, as a consequence, higher requirements for the quality of gas fuel;
- inevitable leakage of the gas-air mixture through the piston rings into the sub-piston space during the compression stroke, as well as the likelihood of gas entering the sub-piston space in the event of damage to the gas valve;
- the presence of a fairly large amount of methane in the exhaust gases, the contribution of which to the occurrence of the greenhouse effect is 20–25 times higher

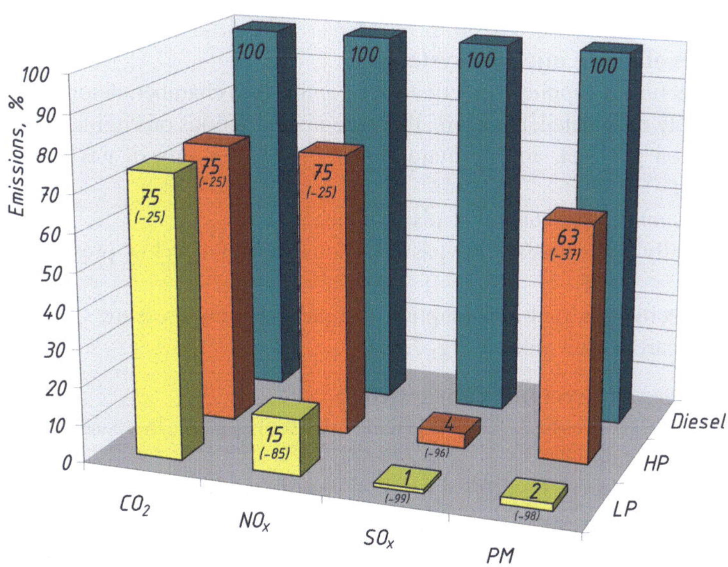

Fig. 7.71 Comparison of emissions of harmful substances from exhaust gases of gas-diesel engines with low-pressure (LP) and high-pressure (HP) systems in relation to a diesel engine (adapted from [43])

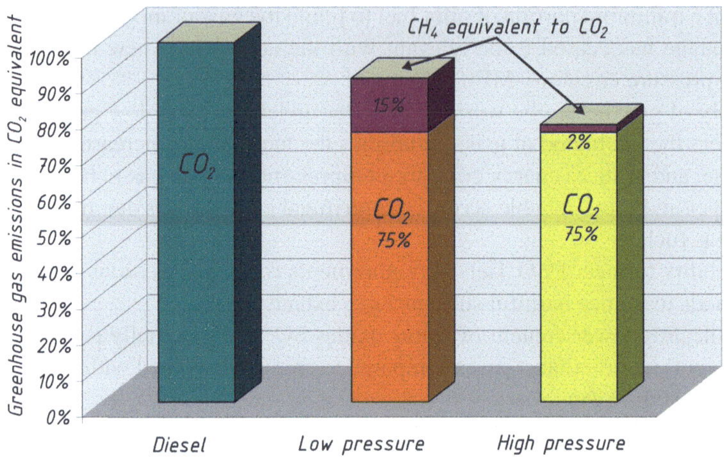

Fig. 7.72 Greenhouse gas emissions (CO_2 + CH_4 in CO_2 equivalent) with exhaust gases of gas-diesel engines with low-pressure (LP) and high-pressure (HP) systems in relation to a diesel engine (adapted from [43])

than CO_2. This increases the equivalent emission of greenhouse gases by engines, thereby reducing the effect of using gas fuel by 10–15% (Fig. 7.72);
- engine power limitation at 80% of the nominal value when running on gas fuel;
- slow response to load changes;
- higher specific consumption of gas fuel compared to high-pressure systems, especially when the load is reduced (Fig. 7.72).

Advantages of direct injection systems:
- when gas fuel is supplied directly to the combustion chamber, detonation can be completely eliminated, therefore, there are no restrictions on engine power when running on gas fuel, and the requirements for the quality of gas fuel are less stringent;
- gas fuel cannot enter the under-piston space;
- significantly lower greenhouse gas emissions compared to low-pressure systems (Fig. 7.72);
- lower specific gas fuel consumption compared to low-pressure systems, especially in partial load modes (Fig. 7.73).

Disadvantages of direct injection systems:
- the use of high-pressure gas complicates the fuel system, increases the requirements for its safety;
- high consumption of liquid fuel for pilot ignition of the gas-air mixture;
- multi-stage compressors are required to compress natural gas, which reduces the economic performance of the power plant due to the increased costs of compressing gas fuel to high pressure;
- the inability to meet IMO Tier-III NO_x emissions requirements without using additional methods to reduce exhaust gas toxicity, such as: exhaust gas recircula-

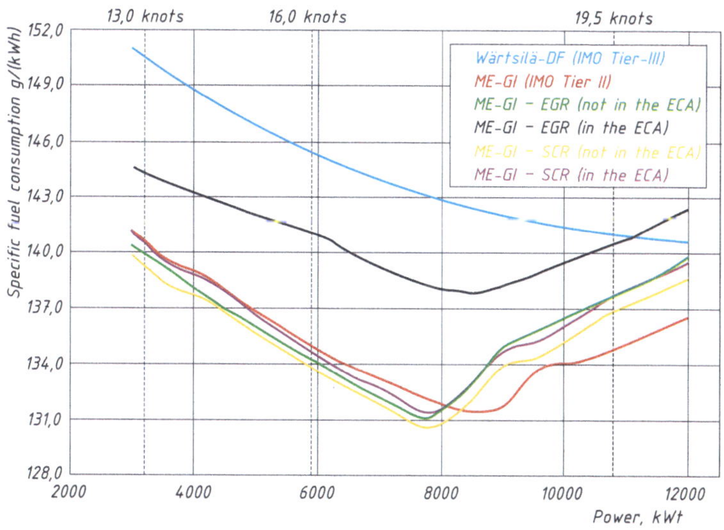

Fig. 7.73 Change in the reduced specific gas fuel consumption for main engines with low-pressure (Wärtsilä DF) and high-pressure (ME-GI) systems, taking into account the pilot fuel converted to gas. *EGR* Exhaust Gas Recirculation; *SCR* Selective Catalytic Reduction; *ECA* Emission Control Area (adapted from [2])

tion (EGR); exhaust gas treatment in a catalytic reactor (SCR), etc., especially for special Emission Control Areas. The use of additional methods leads to an increase in specific fuel consumption, but even in this case it remains lower than that of engines with low-pressure systems (Fig. 7.73).

An important criterion for choosing a particular type of gas-diesel engine is its cost. Thus, according to estimates by Wärtsilä, equipping a basic diesel engine with a low-pressure gas fuel supply system increases its cost by 20–30%, and with a high-pressure system by 35–40%.

For greater clarity, the most significant design differences and features in the organization of the working process of dual-fuel engines with low- and high-pressure systems presented in Table 7.3.

Obviously, both approaches to organizing the working process in gas-diesel engines have their advantages, which for a certain type of installation can be decisive, and the development of designs will eliminate most of the disadvantages inherent today.

The most significant problem for engines using low-pressure technology is the high methane content in the exhaust gases. Considering that in the near future, legislative regulations for emissions of this component expected to introduced, this technology may significantly lose its attractiveness, at least in the form in which it exists now.

In this regard, MAN announced the suspension of its GA series engine production program. However, we can absolutely sure that further accumulation of

Table 7.3 Comparison of features of low-speed gas-diesel engines with low and high pressure systems

Features of the design and organization of the work process	Low pressure	High pressure
Power on gas fuel	80% of N_e^{lf}	100% of N_e^{lf}
Sensitivity to gas quality (methane number)	not lower than 65	no
Sensitivity to air inlet temperature	yes	no
Pilot fuel share (for ME-GI Mk 2), %	≈ 0.5–1	3–8 (0.5–1.5)
Availability of separate pilot injection system	yes	no
Possibility of using HFO as pilot	in the long term	yes
Possibility of using HFO as backup	yes	yes
Possibility of operation on two fuels simultaneously (except pilot), %	no more than 15	yes
Mixture formation	mixing	injection
Pressure of supplied gas fuel, MPa	1.6	30.0
Maximum cycle pressure	increases	no changes
Detonation occurrence with load change	possible	no
Probability of misfire	yes	no
Probability of explosion in the under-piston space	yes	no
Probability of explosion in the exhaust receiver	yes	no
Probability of gas leaks into the engine room	low	high
Effective efficiency on gas fuel	no changes	increases
Effective efficiency on liquid fuel	decreases	no changes
Compliance with IMO emission standards for NO_x	Tier III	Tier II
Emissions of CO_2 with exhaust gases, g/(kW × h)	486	452
Emissions methane with exhaust gases, g/(kW × h)	2.8–3.8	≈ 0.5
GWP (global warming potential in CO_2 equivalent) from liquid fuel, %	not lower than 10	not lower than 23
Possibility of switching from one fuel to another under load	up to 80% of N_e^{lf}	no restrictions
Engine response to load changes	slowed	no changes

experience in designing and operating such engines will help solve this problem. In this regard, it is obvious that in the near future both directions will develop in parallel. Despite significant differences in approaches, the availability of alternative solutions allows the customer to choose the power plant that best meets his requirements.

References

1. Güldner, H.: Das Entwerfen und Berechnen der Verbrennungskraftmaschinen und Kraftgas-Anlagen. Springer Berlin Heidelberg, 1914. 829 p.
2. Bilousov, I.V.: Internal combustion engines of modern marine vessels. Textbook for higher maritime educational institutions. Bilousov, I.V. Bulgakov, M.P.: Astroprint publishing house – Odesa, 2024. – 512 p. ISBN 978-617-7797-54-7

3. Stenersen, D.: Gas Fuelled ships LNG-Fuelled Engines and Fuel Systems for Medium- Speed Engines in Maritime Applications. Marintek GTS Technical Seminar Series, 28.09.2011. 38 p.
4. Project Guide. Bergen engine type C26:33 Gas. Marine. Bergen Engines AS, Rolls-Royce Power Systems Company. Bergen, Norway. 2016. 85 p.
5. Wärtsilä 18V34SG. Instruction Manual. Wärtsilä Finland Oy. Vaasa, Finland. 397 p.
6. Caterpillar Gas Engines Application and Installation Guide G3600-G3300. Caterpillar Inc. USA. 1997. 50 p.
7. MAN dual-fuel GenSets, L23/30DF and L28/32DF. MAN Energy Solutions. Denmark, July 2018. 9 p.
8. L28/32DF Project Guide – Marine. Four-stroke GenSet compliant with IMO Tier II. MAN Diesel & Turbo. 2014. 363 p.
9. MAN L35/44DF Project Guide – Marine Four-stroke dual fuel engine compliant with IMO Tier II. MAN Energy Solutions. 2018. 440 p.
10. Kuiken, K.: Gas- and dual-fuel engines for ship propulsion, power plants and cogeneration Prom 0 to 100.000 kW. Principles. 481 p.
11. Wärtsilä 50DF. Technology Review. Wärtsilä NSD Finland Oy. Vaasa, Finland. 14 p.
12. Technician guide. Series 4000 Common Rail Fuel System. Detroit Diesel Corporation. 2001. 134 p.
13. Wärtsilä W32DF. Instruction Manual. Wärtsilä Finland Oy, Vaasa, Finland. 256 p.
14. Thijssen, B.: Dual-fuel-electric LNG carriers. LNG Shipping Operations. Wärtsilä Finland Oy. Ship Power Solutions. Hamburg, September 27, 2006. 49 p.
15. Heinla, P.: Gas diesel power plant running on shale oil retort gas. Wärtsilä Power Plants. 2014. 23 p.
16. Wärtsilä 32GD. Technology Review. Wärtsilä Corporation. Helsinki, Finland. 16 p.
17. Определение углов открытия и закрытия газовых клапанов при организации внутреннего смесеобразования в судовых малооборотных газодизельных двигателях. / Е.В. Белоусов, И.В. Грицук, В.С. Вербовський: Двигатели внутреннего сгорания. Харьков: изд. НТУ(ХПИ) – 2018. – № 2. – С 21–25.
18. Исследование процессов топливоподачи в газодизельных малооборотных двухтактных двигателях низкого давления. / Варбанец Р.А, Савчук В.П., Грицук И.В., Вербовский В.С. // Двигатели внутреннего сгорания. Харьков: изд. НТУ«ХПИ» – 2019. – № 1. – С 27–33.
19. Белоусов, Е.В.: Организация внутреннего смесеобразования в судовых малооборотных газодизельных двигателях. / Е.В. Белоусов, В.П. Савчук, И.В. Грицук, Т.П. Белоусова: Двигатели внутреннего сгорания // Научно-технический журнал. Харьков: НТУ (ХПИ). – 2017. – №2. – 68 с., с. 13–16.
20. W-X62DF Maintenance Manual “Marine”. Winterthur Gas & Diesel Ltd. Winterthur, Switzerland. 2018. 560 p.
21. WiCE-DF Training. Function of WinGD integrated Controll Electronics. WinGD Training Centre. 12 April 2021. 111 p.
22. MAN B&W ME-GA: Service experience. MAN Energy Solutions. 2024. 42 p.
23. The 2-stroke Low-Pressure Dual-Fuel. Technology: From Concept to Reality. 02 Product Development – Gas & Dual Fuel Engines / M. Ott, I. Nylund, R. Alder, T. Hirose, Y.Umemoto, T. Yamada // CIMAC Congress 2016, Helsinki, Paper No. 233. 12 p.
24. ME-GA – The latest dual-fuel MAN B&W two-stroke engine. MAN Energy Solutions. Copenhagen, Denmark. 2022. 24 p.
25. The ME-GA dual-fuel engine. MAN Energy Solutions 199 14 58–2.0. 2022. 3 p.
26. Ott, M.: Wärtsilä 2-Stroke Dual Fuel Technology. Dual fuel technology update. Technical Seminar 2013. 32 p.
27. X-DF Engines. Excellence built in Low-pressure X-DF Engines FAQ. Winterthur Gas & Diesel Ltd. 2018. 20 p.

28. Aabo, K.: Høytrykks saktegående totakts gassmotor for større skip. Den Norske Gasskonferansen LNG operation 2 stroke and 4 stroke Genset Marine Engines. 27. of March 2014. 66 p.
29. ME-GI Dual Fuel MAN B&W Engines. A Technical, Operational and Cost-effective Solution for Ships Fuelled by Gas. MAN Diesel & Turbo. Copenhagen, Denmark. 2012. 36 p.
30. LNGC-optimised designs of ME-GI engines 4 and fuel gas supply systems. MAN Energy Solutions. Copenhagen, Denmark, 2020. 36 p.
31. Winkel, O.: LNG for deep sea operation. Low Speed – Promotion & Sales. MAN Diesel & Turbo. 2012. 30 p.
32. Dual-fuel conversion. MAN Energy Solutions. Copenhagen, Denmark. 9 p.
33. The 4T50ME-GI Test Engine at MAN Diesel & Turbo's Diesel Research Centre Copenhagen. MAN Diesel & Turbo. Copenhagen, Denmark. 2011. 9 p.
34. LNG Carriers with ME-GI Engine and High Pressure Gas Supply System. MAN Diesel A/S, Copenhagen, Denmark. 37 p.
35. ME-GI Design, Presentation for Croatian Register, 30'th August 2012, MAN Diesel & Turbo, 2012. 64 p.
36. Service experience. Future in the making. MAN B&W two-stroke engines. MAN Energy Solutions, Copenhagen, Denmark. 2023. 36 p.
37. Introduction of ME-GI MK.2. Hyundai Global Academy. Hyundai Global Service. 2022. 82 p.
38. Service experience. Future in the making. Two-stroke technologies. MAN Energy Solutions. Copenhagen, Denmark. 2019. 52 p.
39. 代替燃料と主機関について(次世代形機関；二元燃料機関の紹介)株式会社三井 E&S マシナリー 白井俊範. 2021. 32 p.
40. MAN B&W ME-GI installation in very large or ultra large container vessels. MAN B&W dual-fuel, two-stroke engines. MAN Energy Solutions, Copenhagen, Denmark. 2018. 20 p.
41. Kjemtrup. N.: Performance and Emissions – Current Research. Paper No.3. Licence Days. 2014. 28 p.
42. Juliussen, L.R.: The MAN ME-GI engine: From initial system considerations to implementation and performance optimisation / L.R. Juliussen, S. Mayer, M. Kryger // CIMAC Congress 2013, Shanghai, Paper No. 424. 14 p.
43. Ott, M.: Wärtsilä 2-Stroke Dual Fuel. Technology responding to changing market needs. Wärtsilä. 5TH CIMAC Cascades, Busan. 2014. 25 p.

Chapter 8
Use of Alternative Liquid Fuels with Low Flash Point: In Marine Engines

8.1 Low Flash Point Fuels

Promising fuels with a low flash point include Liquefied Petroleum Gas (LPG), methanol and ethanol, dimethyl ether—all types of fuel that behave similarly to traditional liquid fuel when injected.

Liquefied petroleum gas As in the case of LNG, the initial interest in using liquefied petroleum gas as a fuel for marine engines shown by large ship operators engaged in the transportation of this product. The accompanying factors were the availability and prevalence of this resource, the developed system of its bunkering and the experience of handling liquefied petroleum gases in other modes of transport.

As already noted, the basis of petroleum gases are propane-butane mixtures, which are also the basis of Volatile Organic Compounds (VOCs) that released from crude oil during its transportation and loading operations. For this reason, the topic of using these products was of interest to oil tanker operators.

The use of liquefied petroleum gases can significantly improve the environmental performance of a ship's power plant, their storage on board is much easier than storing LNG. These gases in a liquefied state can be stored at normal temperatures with a pressure in the tank not exceeding 1.6 MPa.

Liquefied petroleum gases, like LNG, considered as alternative fuels for the transition period. They are capable of significantly reducing greenhouse gas emissions and other harmful components until other solutions found that would allow the transition to full decarbonization of shipping.

Methanol It is clear that improving combustion engines alone cannot achieve the radical goal of completely decarbonizing shipping. This can only achieved by switching to carbon-neutral or carbon-free fuels. Among all the potential candidates,

I. Bilousov, R. Varbanets, *Marine Internal Combustion Engines and Environmental Protection Issues*, Springer Series on Naval Architecture, Marine Engineering, Shipbuilding and Shipping 27,
https://doi.org/10.1007/978-3-032-20940-5_8

methanol has several properties that make it particularly suitable for the maritime industry. One of the biggest advantages of methanol, compared to other potential candidates for the marine fuel of the future, is the relatively low capital expenditure required to retrofit an existing fleet. Since methanol is a liquid at ambient temperature and pressure, its use eliminates the need for insulated tanks to maintain low temperatures. Bunkering methanol and storing it on board requires only minor modifications to the vessel's fuel system. In addition, methanol is one of the top five chemical products transported worldwide. Of the world's 100 largest ports, 88 already have methanol-bunkering infrastructure. When methanol enters water, it quickly decomposes, which significantly reduces its potential harm to the aquatic ecosystem.

In terms of reducing harmful emissions into the atmosphere, methanol is also a very attractive fuel. Having a high heat of evaporation, it significantly cools the charge in the combustion chamber, which allows for a significant reduction in NO_x emissions. Since the only carbon atom in the methanol molecule is already bound to an oxygen atom, no soot formed during its combustion. Thus, when using methanol, there is no need to search for compromise solutions between reducing NO_x emissions and increasing exhaust smoke, which is typical for almost all types of diesel engines. This provides more opportunities to optimize the engine operating process in order to increase their efficiency. Methanol engines do not have a problem with methane emissions, which is a serious problem for gas and gas diesel engines. Given that methanol does not contain sulphur, it is already a promising fuel for meeting the restrictions in force in SECA zones and has great potential to meet increasingly stringent emission standards, which may include soot and methane in the near future.

However, all of the above only makes sense if methanol for fleet fuel needs is produced using technologies that are not associated with fossil resources. Today, methanol mainly produced by steam reforming of natural gas or from synthesis gas obtained during coal gasification. In this case, the carbon footprint of methanol can be several times higher than the carbon footprint of the original fuel.

Technologies for producing methanol from renewable sources already exist. For example, natural gas can replaced with biomethane, and synthesis gas can produced by gasifying biomass, such as forest waste and the biogenic fraction of municipal solid waste. If the energy, needed for synthesis, obtained from renewable sources, the carbon footprint of methanol can reduced to a minimum or even zero. Of course, the development of such production processes requires significant investment and experimental design work, but work in this direction is currently underway and the results are optimistic. According to forecasts from the International Renewable Energy Agency (IRENA), by 2050 the production of carbon-neutral synthetic methanol will amount to 250 million tons and 130 million tons of methanols from biomass.

Ethanol or ethyl alcohol is a highly purified biodegradable alcohol that can obtain from renewable raw materials such as sugar cane, corn, wheat or sugar beet. Ethanol-based fuel behaves similarly to methanol, with which the marine industry has accumulated considerable experience. This experiment can applied relatively easily when using ethanol.

Of course, major changes in the structure of fleet supply with alternative fuels are not possible without appropriate legislative regulation designed to make investments in the production of carbon-neutral fuels attractive. On the other hand, the successful development of new technologies for the production of carbon-neutral fuels is possible only if there is a stable demand for them. This requires, first, solving technical issues related to the efficient and bladeless use of such fuels in the fleet.

Below we will consider the technical aspects of using low-flash point fuels in marine engines of various types and purposes.

The main problem faced by marine engine manufacturers is the uncertainty of which alternative fuel will be available for use on ships in the near future. It is obvious that the problem of introducing a particular type of fuel goes beyond the ship industry. It is necessary to solve a number of complex problems in related areas—this includes the creation of sufficient energy capacities based on renewable energy sources, and the development of a raw material base for the production of carbon-neutral fuels and the development of new production capacities for their production. Solutions to these problems for each of the proposed alternative fuels can be successful in different ways and require different periods to bring them to a commercially attractive result. It should be borne in mind that the service life of ship engines measured in decades, and during this period, the concepts of using alternative fuels under the influence of external factors can change several times. For this reason, the only correct solution is to create universal platforms that allow, if not to ensure the operation of engines on several types of alternative fuels, then at least with minimal costs to adapt them to a new type of fuel, the use of which, due to circumstances, at a particular stage will be most appropriate. It is in such conditions of uncertainty that leading manufacturers of marine low-speed engines have begun to create a new generation of power plants that must comply with current environmental legislation and be able to adapt to different scenarios for the development of the fuel market in the future. Because of the work done, approaches have been developed that allow using different types of liquid alternative fuels based on the same technical solutions.

8.2 Use of Volatile Organic Compounds (VOC) as Fuel for Low-Speed Engines

Back at the end of the last century, shuttle tanker operators began to be concerned about the problem of using volatile organic compounds released from crude oil during its loading and transportation. In response to this request, MAN developed a dual-fuel engine technology capable of simultaneously operating on VOC and conventional liquid fuels in any ratio. Since in these engines the alternative fuel was supplied to the combustion chamber in liquid form, the engines were designated LGI (Liquid Gas Injection).

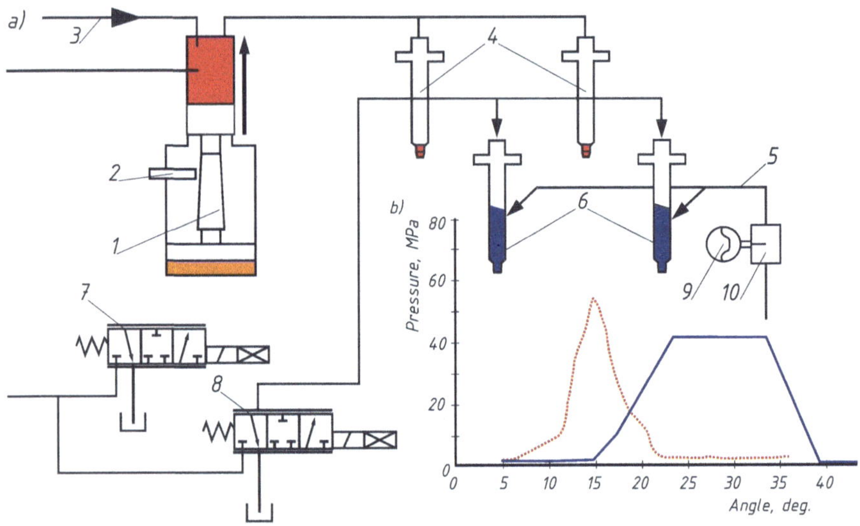

Fig. 8.1 Schematic diagram of the fuel system of the ME-LGI series engine running on VOC (**a**) and the laws for supplying pilot and gas fuel (**b**): (1) high-pressure fuel pump with hydraulic drive; (2) feed rate sensor; (3) liquid fuel supply; (4) liquid fuel injectors; (5) liquefied gas supply; (6) gas injectors; (7) high-pressure fuel pump control spool valve; (8) hydraulic drive control valve; (10) liquefied gas supply control unit; (*Dashed lines*) pilot fuel supply; (*Solid lines*) liquefied gas supply (adapted from [1])

In engines of this type, VOC enters the fuel system in a liquefied state. This circumstance determines a number of design features that will discuss below. The general diagram of the gas system of a gas-diesel engine designed to operate on liquefied gas, shown in Fig. 8.1.

The VOC supply module differs from the design of the LNG supply control module in that liquefied gas is supplied to it under a pressure of 50.0–55.0 MPa. To stabilize the pressure in the liquefied gas control module, a gas accumulator with a separating diaphragm is used. The design of the gas supply control module, shown in Fig. 8.2. Unlike the gas fuel supply control algorithm, when injecting liquefied gas, the main window valve does not close with each injection, but serves to disconnect the high-pressure line of the injectors when the engine is not running. The valve driven by compressed air supplied from the ship's general system.

To clean the system from gas fuel residues, the unit is equipped with a valve for purging the module cavities with an inert gas, which is nitrogen, under, a pressure of 0.3–0.4 MPa.

The injection valve, for injecting liquefied gas, shown in Fig. 8.3.

The injection valve connected to the control unit by high-pressure pipes. Liquefied gas is supplied to the needle valve of the nozzle through the central channel in the needle valve guide and then through the channel in the body of the valve itself. Otherwise, the design and operating principle of the nozzle are similar to the previously discussed design of the nozzle for gas fuel.

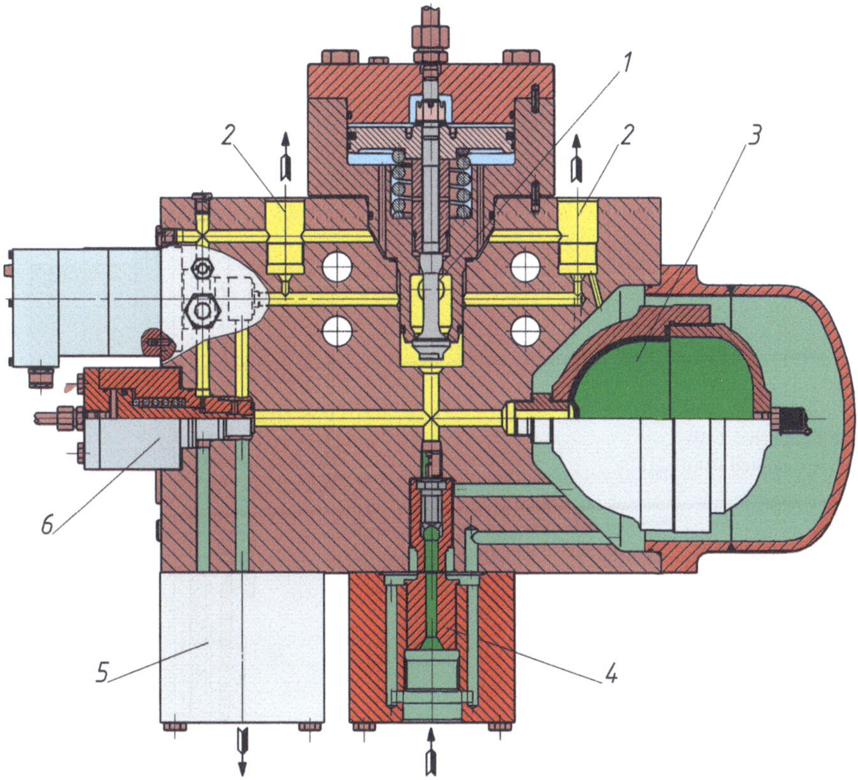

Fig. 8.2 Liquefied gas supply control module: (1) window valve; (2) liquefied gas outlet to gas injectors; (3) gas accumulator with separating diaphragm; (4) liquefied gas supply under pressure of 50–55 MPa; (5) device for connection to the ventilation system; (6) valve for purging the system with inert gas (adapted from [2])

Tests conducted by MAN showed that liquefied petroleum gases could self-ignite in the working cylinder, even without a pilot supply of liquid fuel. At nominal power modes, the nature of the working process without pilot ignition was practically no different from a similar process on traditional fuel. However, when switching to partial loads, the self-ignition delay increased, which led to a significant increase in the rate of pressure increase. This, in turn, increased the loads on the engine design elements, which could lead to a reduction in its service life. For this reason, it decided to retain pilot fuel injection in liquefied gas engines throughout the entire range of operating modes. The standard fuel system used to supply pilot fuel, which also ensures engine operation on liquid fuel. The pilot supply amounted to 5–10% of the full cyclic supply of liquid fuel.

Subsequently, the experience of creating engines designed to operate on VOC used to develop engines capable of operating on liquefied petroleum gas (LPG), which will discussed in the following chapters.

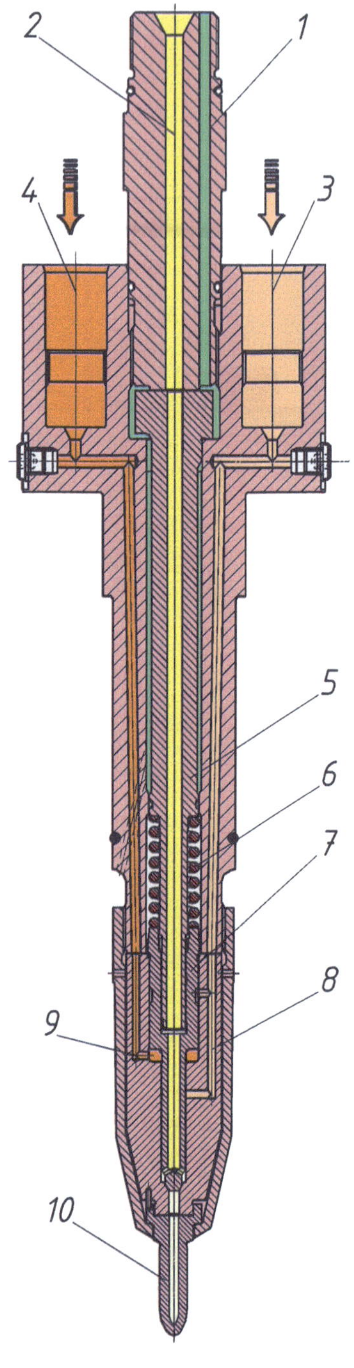

Fig. 8.3 Injector for liquefied gas injection: (1) channel for monitoring gas leaks; (2) channel for supplying liquefied gas to the injector; (3) channel for supplying sealing oil; (4) channel for supplying control oil; (5) needle valve guide; (6) spring; (7) needle valve body; (8) channel for supplying sealing oil to the needle valve; (9) channel for supplying control oil to the cavity of the needle valve hydraulic drive; (10) nozzle (adapted from [2])

8.3 Use of Methanol and Ethanol as Fuels for Two-Stroke Low-Speed Engines

At the beginning of the twenty-first century, the standards for emissions of harmful substances into the environment began to become significantly stricter. In addition, the volumes of transportation of chemical products, primarily methanol, ethanol and LPG, increased significantly. Specialized vessels appeared for the transportation of these goods, which led ship-owners to think about the advisability of using the transported cargo as fuel for the ship's power plant. This would significantly reduce operating costs, in particular for bunkering and fuel processing board. In addition, methanol and ethanol do not contain sulfur, which made it possible to significantly improve the environmental performance of ship power plants, especially when operating them in SECA zones. After the energy efficiency index (EEDI) introduced for ships, methanol and ethanol began to consider as potential carbon-neutral fuels. In response to these demands, leading manufacturers of low-speed two-stroke engines have developed and introduced to the market a new class of dual-fuel engines capable of running on liquid fuels with a low flash point, such as methanol, ethanol, liquefied natural gas, dimethyl ether (DME), gasoline, etc.

8.3.1 MAN Engines

MAN engines developed based on one universal platform, which significantly reduced development costs and the time it took to bring new engines to market. The first engines of this type appeared on ships in 2016. Engines adapted to operate on a specific type of alternative fuel received the appropriate designation:

- for operation on methanol—LGIM;
- for operation on liquefied petroleum gas—LGIP, etc.

The multi-fuel platform was based on the ME series engines with an electronic fuel injection control system and hydraulic drive of the fuel pumps [3]. The standard fuel system ensures engine operation on traditional petroleum fuel, and used to supply a pilot portion when operating on alternative fuel.

For injection of alternative liquid fuel, two or three (depending on the cylinder diameter) Fuel Booster Injection Valve (FBIV) installed on the cylinder cover. The general arrangement of the main elements of the engine fuel system shown in Fig. 8.4.

For the version of engines capable of running on methanol, the FBIV modified, since the use of methanol requires additional cooling and sealing oil, which prevents it from entering the hydraulic drive system. This version of the feed module was designated FBIV-M.

Fig. 8.4 Concept of a two-stroke low-speed engine of the LGI series designed for the use of alternative fuels: (1) hydraulic unit of the exhaust valve drive; (2) high-pressure liquid petroleum fuel lines; (3) control and sealing oil supply lines; (4) power hydraulic drive lines; (5) hydraulic pressure booster for alternative fuel injection (FBIV-M); (6) fuel injectors; (7) alternative fuel injection control unit; (8) liquid petroleum fuel pump; (9) alternative fuel supply lines (adapted from [4])

The FBIV-M feed module is a hydraulic pressure multiplier, made as a single unit with a fuel nozzle (Fig. 8.5). That is, by definition, it is a hydraulically driven pump-injection.

The power drive of the module is hydraulic. Oil supplied under pressure from the control unit installed on the cylinder cover.

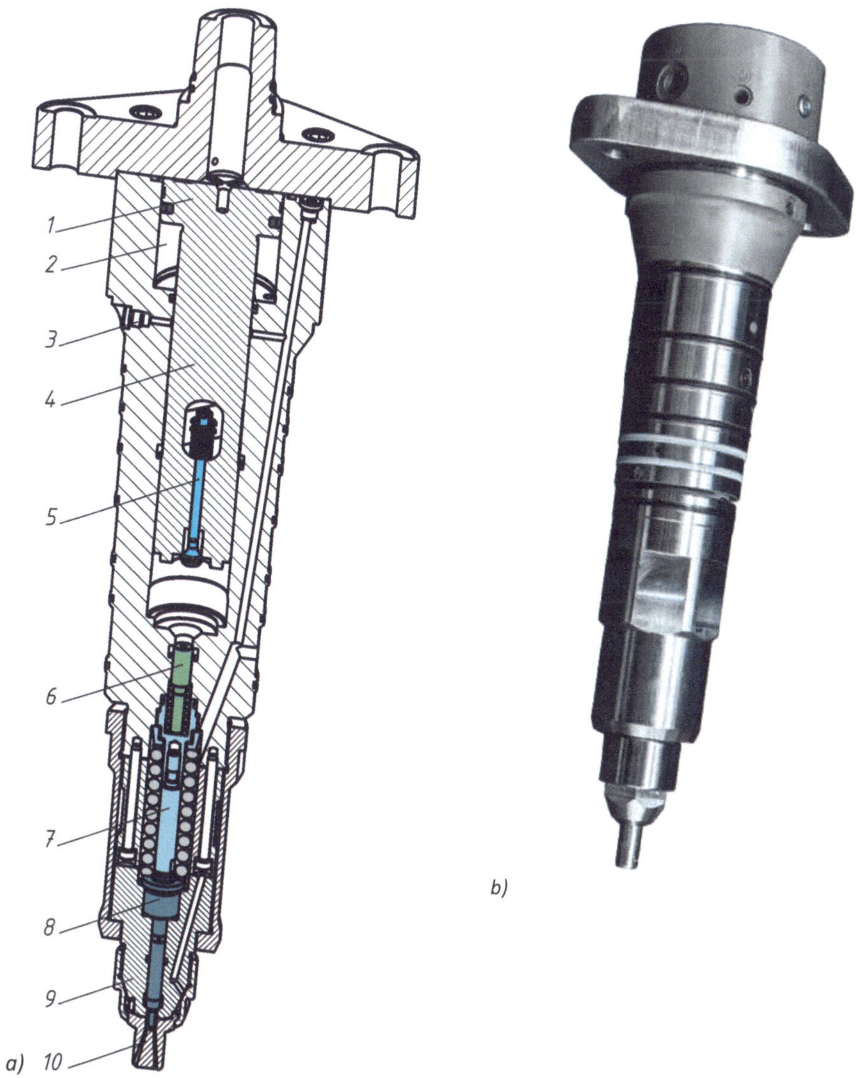

Fig. 8.5 Methanol injection module (FBIV-M): (**a**) device; (**b**) general view; (1) hydraulic actuator piston; (2) hydraulic cylinder; (3) sealing oil supply channel; (4) plunger; (5) filling valve; (6) non-return valve; (7) stop; (8) piston for activating the needle valve of the atomizer; (9) nozzle body; (10) nozzle tip (adapted from [5, 6])

The ratio of the areas of the hydraulic piston and the fuel plunger selected in such a way as to ensure the methanol pressure in front of the nozzle at a level of 60 MPa. Considering that, the calorific value of methanol is 2.15 times lower than that of traditional fuels, its cyclic supply approximately twice as large. Therefore, to ensure injection duration comparable to the supply of petroleum fuels, the cross-sectional area of the nozzle channels had to increase approximately twice.

Methanol has a relatively low boiling point (about 65 °C at 0.1 MPa). When heating the fuel lines and FBIV-M, vapor locks can form in them, which disrupt the supply of liquid fuel. This can lead to unstable engine operation, misfires, and cavitation. The latter can cause erosion of precision parts of the nozzle and plunger pair. Overheating of methanol in the nozzle can contribute to the formation of deposits or polymers, especially if there are impurities in the fuel. Thus, all elements of the fuel system associated with methanol require maintaining a special temperature regime. Cooling helps maintain the cleanliness of the injector and prevents clogging of the nozzle holes, as well as wear of precision pairs of FBIV-M.

To lubricate and cool precision pairs, as well as to prevent methanol from entering the hydraulic drive system, sealing oil supplied to the FBIV-M under a pressure of 5.0 MPa. In addition, the system removes heat from the FBIV-M so that the temperature in it does not rise above 60 °C. To ensure the specified temperature of the FBIV-M, the oil cooled in a heat exchanger, which connected to a low-temperature cooling circuit. The cooling and sealing oil system fully integrated into the engine design, including equipment for continuous monitoring of the presence of methanol in the lubrication system. If methanol detected in the system, the engine will switch to oil fuel mode, and methanol will removed from the engine. At the same time, the cooling oil pump on the supply side will switched to clean system oil, and the oil circuit will be flushed with clean oil. Then the clean oil together with the contaminated oil will collected in the cooling oil tank, and the system will be able to continue working only when no methanol detected in the tank.

Methanol supplied to the engine through double-walled pipes under a pressure of 0.8 MPa. The space between the walls constantly ventilated with air supplied by a special ventilation system. If methanol vapors get into the blown air, the warning system will activated, the sensors of which installed at the outlet of the double-walled pipeline system. In case of too high a content of methanol vapors in the ventilated air, the safety system automatically switches the engine to oil fuel. This switching occurs smoothly and without loss of power.

Considering that methanol an aggressive environment, a selection of materials made for the engine design elements that are in direct contact with it. Measures implemented to prevent corrosion of the combustion chamber fire surfaces. A selection of lubricants made that best meet the new operating conditions.

The subsequent operational experience has shown that the measures taken are sufficient, and the problem of high corrosive activity of methanol is completely solvable.

To activate the methanol combustion process, a pilot portion of liquid fuel supplied to the working cylinder using standard fuel injectors, which is about 5% of the total cyclic supply. In this case, the minimum value of the pilot supply determined by the minimum stable fuel supply of the standard oil fuel injection system designed for engine operation at full load. All systems are controlled using electro-hydraulic valves connected to the general electronic engine management system. The combustion process is monitored using medium pressure sensors (PMI) located in each of the cylinder cover.

8.3.2 Engines from Winterthur Gas&Diesel

Winterthur Gas&Diesel (WinGD), the successor of such well-known manufacturers as Sulzer AG and Wärtsilä, has developed its own version of a methanol injection system for low-speed engines.

Methanol engine versions developed based on the X series diesel engines, which use a common rail fuel injection technology known as RT-Flex. This series of engines was designated X-DF-M.

Like the LGIM series engines from MAN, the standard fuel supply system ensures that the X-DF-M engine operates on liquid fuel and injects a pilot portion when operating on methanol.

For methanol injection, three additional injectors are installed on each cylinder cover, to which methanol is supplied under a pressure of 60 MPa from a hydraulic pressure multiplier (Methanol Booster Units—MBU) (Fig. 8.6).

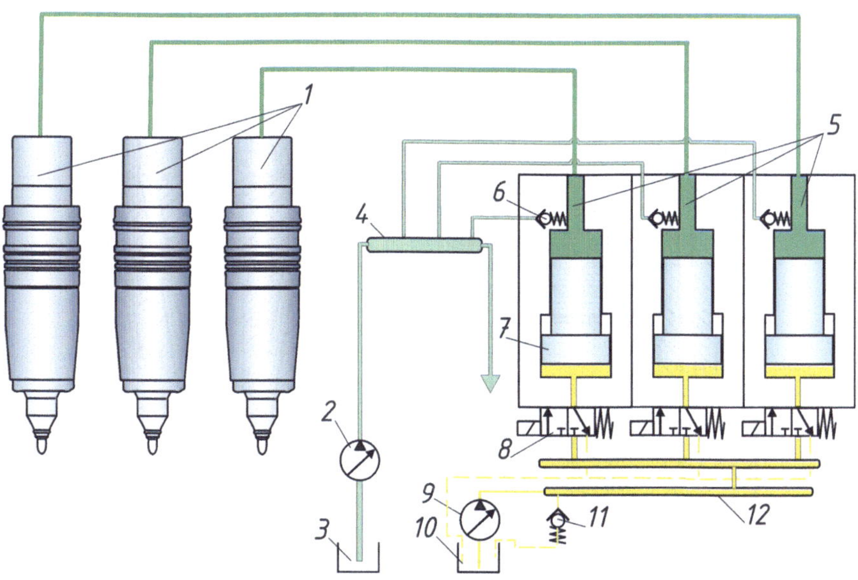

Fig. 8.6 Schematic diagram of the methanol injection system of the X-DF-M series low-speed engines: (1) methanol injection valve; (2) pump for feeding methanol to the hydraulic multipliers; (3) methanol supply tank; (4) intermediate accumulator; (5) Methanol Booster Units (MBU); (6) filling valves; (7) hydraulic pistons; (8) electro-hydraulic control valves; (9) control oil supply pump; (10) control oil storage tank; (11) pressure limiting valve; (12) oil accumulator of the hydraulic drive system pressure (adapted from [7])

Each injector serviced by a separate MBU, assembled into one block (Fig. 8.7). Methanol is supplied to the MBU by a booster pump under a pressure of 1.3 MPa.

The hydraulic drive of the MBU piston carried out from a common hydraulic control system with an oil pressure of 30 MPa.

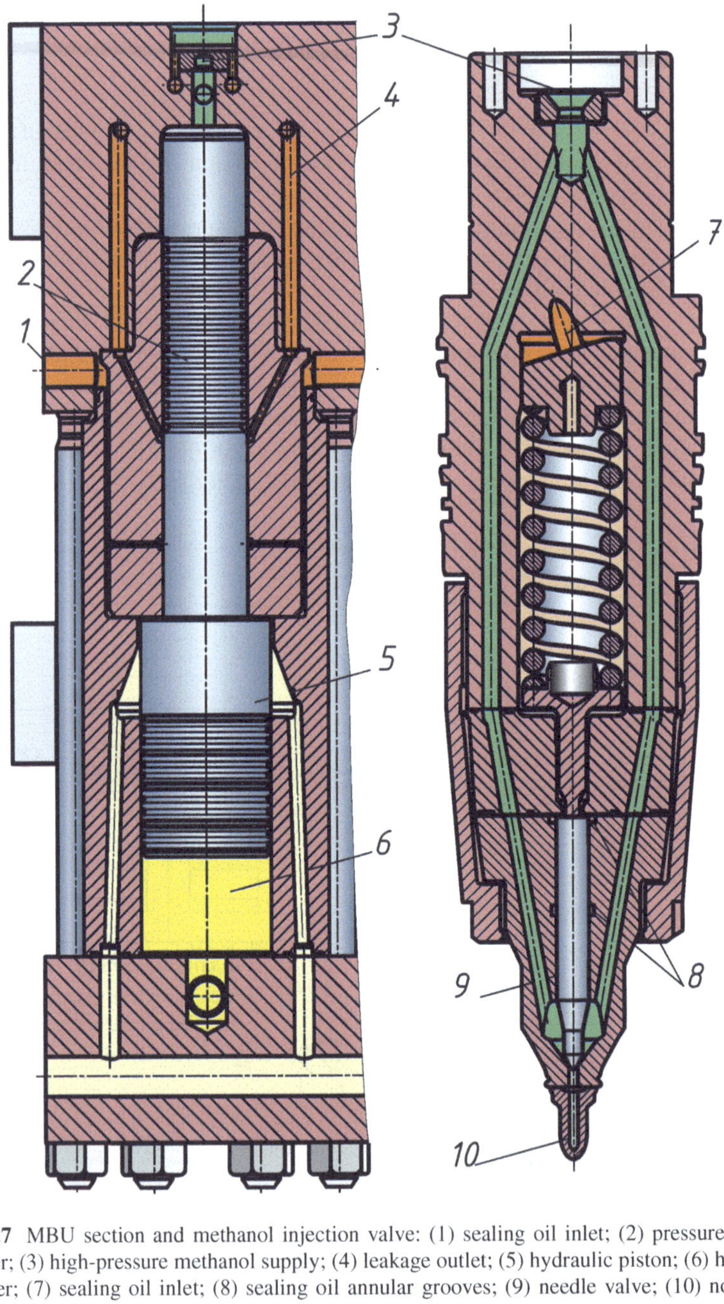

Fig. 8.7 MBU section and methanol injection valve: (1) sealing oil inlet; (2) pressure booster plunger; (3) high-pressure methanol supply; (4) leakage outlet; (5) hydraulic piston; (6) hydraulic cylinder; (7) sealing oil inlet; (8) sealing oil annular grooves; (9) needle valve; (10) nozzle tip (adapted from [7])

The electronic unit controls the opening of the MBU spool valve, which results in the supply of control oil to the hydraulic cylinder. The oil moves the piston, causing the methanol pressure to increase from 1.3 to 60.0 MPa. Methanol under high pressure opens the needle of the spring-controlled injector and enters the combustion chamber through the nozzle channels in the nozzle body. The presence of separate MBUs for each injector allows for very flexible regulation of the methanol supply to the engine working space both through all injectors at once and through each one separately.

8.3.3 Features of the Working Process of Low-Speed Methanol Engines and Their Environmental Performance

Regardless of the manufacturer, engine tests and operating experience have shown that operation on methanol does not differ significantly from operation on petroleum fuel (Fig. 8.8). Analysis of the pressure curves obtained for the laboratory S50ME-T9 engine shows that methanol injection begins earlier, and the combustion pressure reaches a higher maximum value. The process of heat release during methanol combustion is somewhat extended due to the longer injection duration, but at the end of combustion the speed decreases faster, which indicates an earlier completion of combustion. The reasons for this are higher mixing rates due to the greater mass of injected fuel and the initially high oxygen content in methanol, which significantly increases the rate of carbon oxidation. The above features improve the thermodynamic efficiency of the working process when operating on methanol, compared to traditional liquid fuel. As a result, the reduced specific fuel consumption of the engine is 2–3% lower when using methanol compared to traditional fuel (Fig. 8.9).

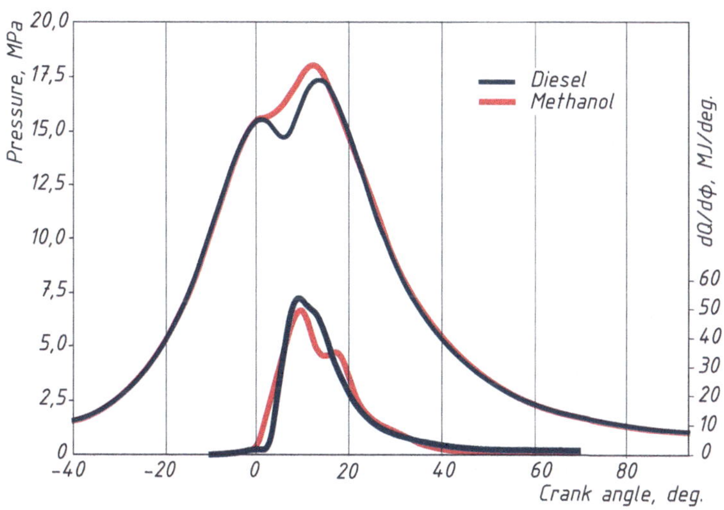

Fig. 8.8 Operating process indicator diagram and heat release profiles for the S50ME-T9 pilot engine at full load when running on petroleum fuel and methanol (adapted from [6])

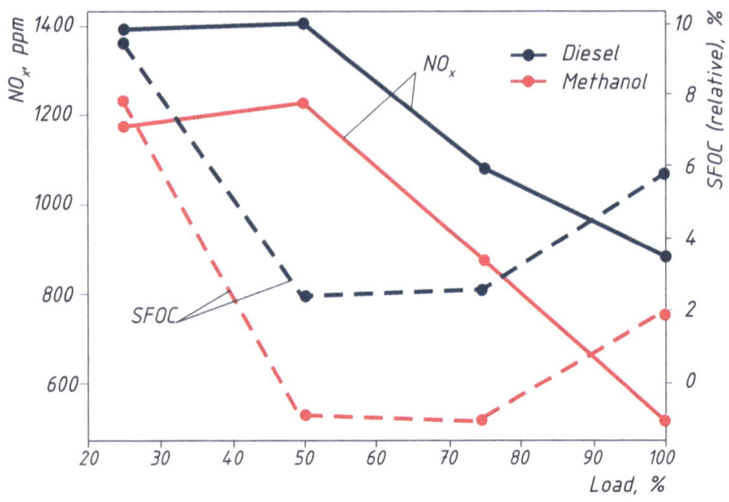

Fig. 8.9 NO_x emissions and relative specific fuel oil consumption on engine load S50ME-T9 (% of MCR) (adapted from [8])

The lower combustion temperature of methanol contributed to the reduction of NO_x emissions. Figure 8.9 shows the NO_x content in the exhaust gases and the relative specific fuel oil consumption (SFOC) at the main operating conditions of the engine when it is running on methanol and petroleum fuel. NO_x emissions reduced approximately 30% when switching from petroleum fuel to methanol, with the same operating parameters.

If we limit ourselves to meeting the Tier II NO_x emission standards, there is a significant margin for optimizing the engine's fuel economy for the operating process-using methanol as a fuel.

To comply with Tier III NO_x emission standards, the LGIM series engine can be equipped with a low- or high-pressure EGR system. The mechanism of the effect of exhaust gas recirculation on NO_x emissions discussed in previous chapters. It should noted that for engines running on methanol, a decrease in oxygen concentration has a much smaller effect on the efficiency and completeness of combustion than is typical for diesel engines. The presence of an oxygen molecule in methanol, already bound to a carbon molecule, prevents the formation of soot. It also promotes more complete combustion, despite the lower flame temperature. Nevertheless, the overall slowdown in reaction rates leads to a decrease in the maximum cycle pressure and, as a consequence, an increase in the specific fuel consumption of the engine, which can see in Fig. 8.10. The same figure shows the results of measuring the levels of NO_x formation for the G95ME-LGIM engine at the main operating load modes—75–100% of MCR. As tests have shown, with EGR disabled, the NO_x emission level corresponds to Tier II, and with it enabled, it is significantly lower than the permissible value of 3.4 g/(kW × h) for Tier III. Thus, the use of EGR in engines running on methanol is a technology that allows achieving a reduction in NO_x to the Tier III level, without a significant deterioration in thermodynamic efficiency.

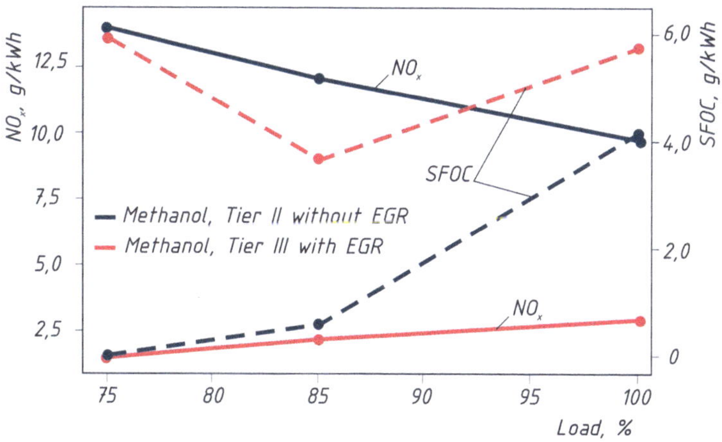

Fig. 8.10 NO_x emissions and specific fuel oil consumption when operating on methanol in Tier II and Tier III modes using EGR for the G95LGIM engine in the range of main operating modes (adapted from [6])

Similarly, to achieve Tier III, a methanol engine can be equipped with a catalytic reactor (SCR) for exhaust gas treatment. In this case, the reactor operating conditions will be much more favorable, since the exhaust gases are practically free of sulfur combustion products, except for a small amount that got there because of combustion of the pilot fuel.

Another promising method for achieving the level of NO_x emissions in accordance with Tier III is the use of not pure methanol as a fuel, but its mixtures with water.

Earlier, we considered the effect of water on the working process of diesel engines, including in the composition of water-fuel emulsions.

Unlike petroleum fuels, methanol dissolves with water in any proportions, forming a homogeneous solution that does not stratify and can be stored for a long time without any problems, if simple rules for its storage observed. Research conducted by MAN has shown that adding water to methanol the amount of 25–40% allows reducing NO_x emissions to the Tier III level without using additional emission control methods such as EGR and SCR (Fig. 8.11).

Approximately 5% of the pilot fuel effectively ignites the water-methanol mixture supplied to the engine combustion chamber.

Engines adapted to this technology are designated LGIM-W.

The main disadvantage of this technology is the limited possibility of using water-methanol mixtures at load modes close to 100% of MCR. This is because their use requires an increase in the cyclic feed by an amount approximately equal to the volume of water added to methanol. However, at full engine power modes, the feed of the pump sections is almost equal to their design capacity, with the exception of a small supply reserve. Thus, to increase the feed at modes close to MCR, it is necessary to increase the size of the pump sections. At the same time, when the load decreases, the supply reserve of the pump sections increases, which allows a proportional increase in the water content in the mixture.

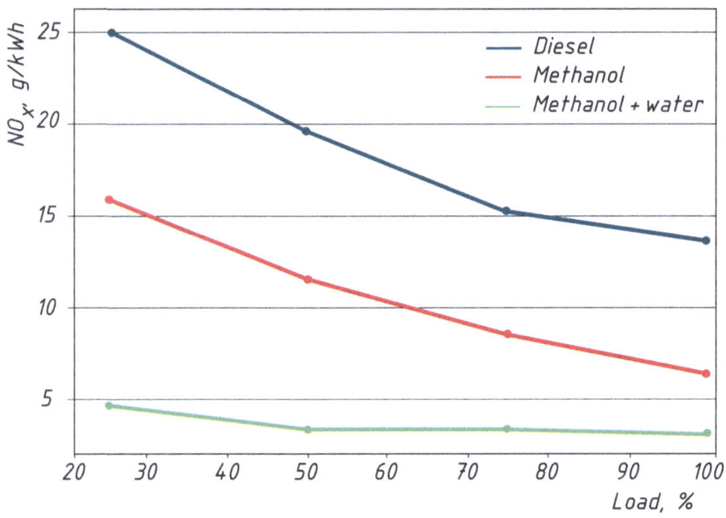

Fig. 8.11 NO_x content depending on engine load running on petroleum fuel, methanol and a mixture of water and methanol (adapted from [9])

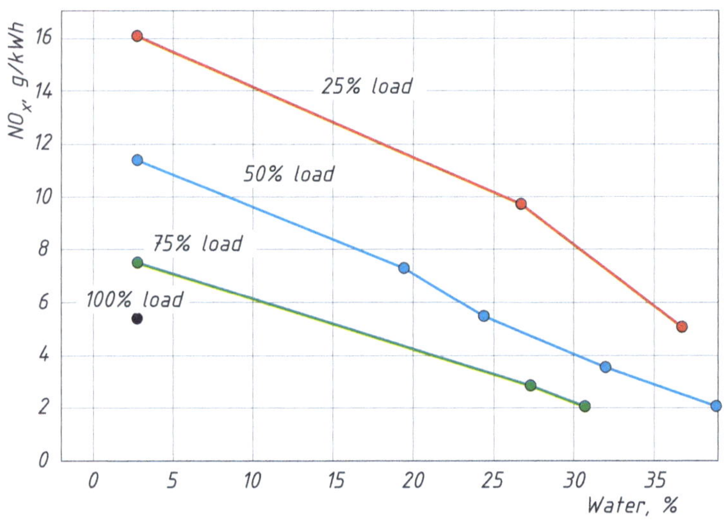

Fig. 8.12 NO_x content depending on the concentration of water in a mixture with methanol at different engine loads (adapted from [10])

The conducted studies have shown that the NO_x content when operating on water-methanol fuel decreases almost in direct proportion to the amount of water in the mixture (Fig. 8.12). In the load range of 50–75% of MCR, the NO_x content can be reduced to 2.0 g/(kW × h), and even at loads close to 25% of MCR it does not exceed the limit established by Tier III.

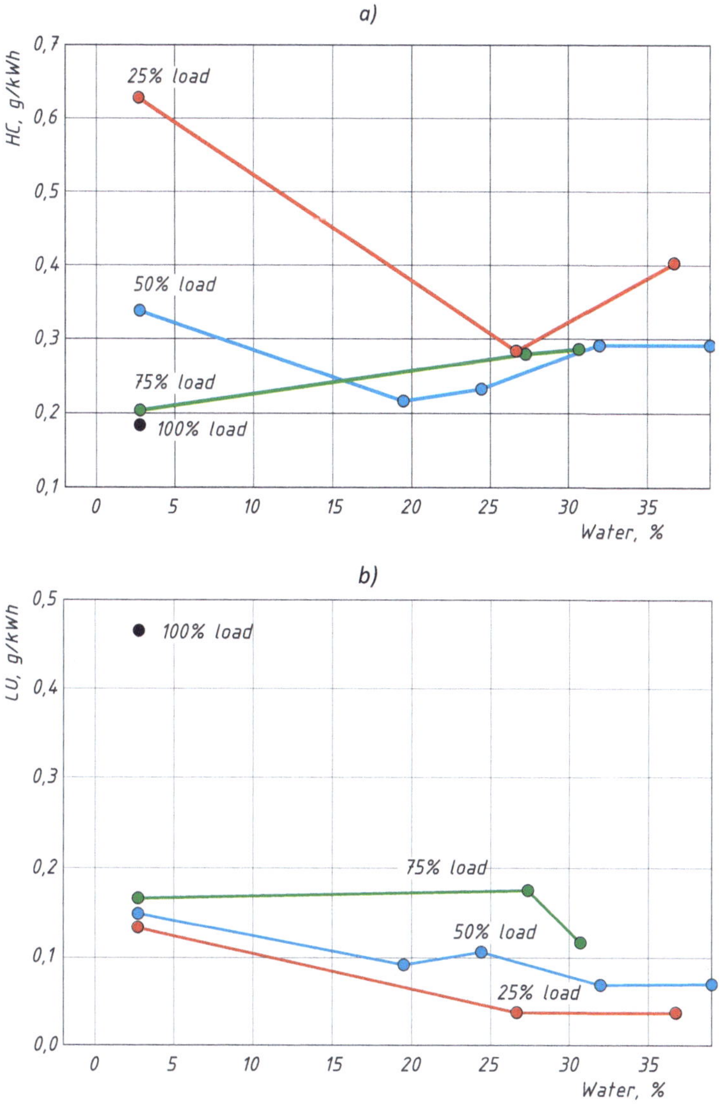

Fig. 8.13 Content of hydrocarbons (CH) (**a**) and carbon monoxide (CO) (**b**) depending on the concentration of water in a mixture with methanol at different engine loads (adapted from [10])

The effect of water in methanol on other indicators of harmful emissions is practically the same as in engines running on water-fuel emulsions (Fig. 8.13).

In general, these indicators remain significantly lower those of diesel engines running on traditional petroleum fuels. This makes the technology of using water-methanol fuels a promising direction in reducing harmful emissions. In addition, given the tendency to reduce the comic speeds of transport vessels, even the impossibility of greatly increasing the water content in methanol does not seem to be a serious problem.

8.4 Use of Liquefied Petroleum Gas as a Fuel for Low-Speed Engines

The first half of the twenty-first century marked by an increase in gas fuel consumption in all areas of human economic activity, and in response to this demand, the volume of its transportation by sea began to grow sharply. This trend affected not only natural gas, but also petroleum gases, which are mainly propane-butane mixtures. Around 2010, it became clear that the volume of liquefied petroleum gas on international markets was growing rapidly, and its prices had become more affordable. The number of specialized vessels designed to transport both LNG and LPG has increased sharply and continues to increase worldwide. As the fleet for LPG transportation developed, ship operators came to understand that using the cargo on board as a relatively cheap fuel for main and auxiliary engines would significantly improve their environmental performance and reduce the cost of bunkering with liquid fuel. This largely facilitated by growing concerns about environmental pollution from ships, which also required quick and effective solutions. In response to this request, manufacturers of low-speed engines continued research and development work on developing methods for using LPG based on their flagship diesel models. Thus, the company Winterthur Gas&Diesel has modified its low-pressure gas engines to use propane-butane mixtures in the gas phase. At the same time, the design of the engine has remained virtually unchanged, only the software and some systems that supply it with gaseous fuel have modified.

8.4.1 MAN Two-Stroke Low-Speed Engines for LPG Operation

Despite the fact that MAN already had engines capable of running on VOC, which are actually petroleum gases in composition, the experience of creating versions for using methanol showed that the use of pump-injectors (FBIV) can significantly improve this technology.

In this case, there is no need to supply liquefied petroleum gas to the gas module under a pressure of 50–55 MPa, and it is enough to limit the pressure to 5.0 MPa. This significantly reduces the load on all structural elements of the gas module and virtually eliminates the need for high-pressure fuel supply pipelines.

To inject liquefied gas into the combustion chamber, hydraulic pump-injectors adapted for this type of fuel are used, designated—FBIV-P (Fuel Booster Injection Valve for Propane). The engines themselves, created because of ME series diesel engines, received the designation—ME LGI-P (Fig. 8.14).

To enable the use of LPG as a motor fuel, standard engines are equipped with additional elements and systems:

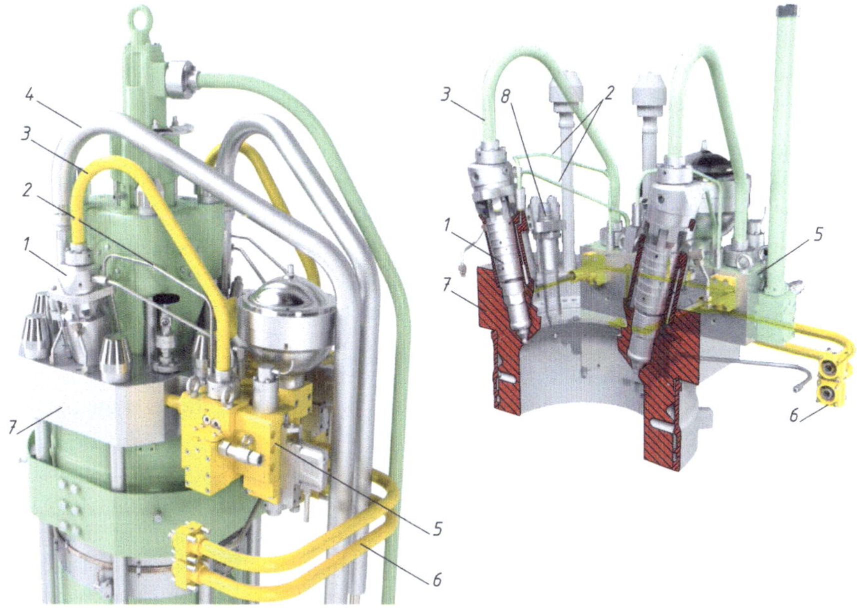

Fig. 8.14 General arrangement of the main components of the fuel systems of ME-LGIP engines: (1) pump-nozzle for supplying liquefied gas (FBIV-P); (2) sealing and control oil supply; (3) FBIV-P power hydraulic drive oil supply; (4) liquid petroleum fuel supply to the injectors; (5) liquefied petroleum gas supply control module; (6) double-walled liquefied gas supply and discharge pipelines; (7) working cylinder cover; (8) fuel injector valve (adapted from [11])

- hydraulically driven pump-injectors (FBIV-P) to increase pressure and inject liquefied petroleum gas into the combustion chamber at a strictly defined angular interval of the working process;
- gas modules used to supply liquefied gas to the pump-injectors. The gas module is equipped with a window valve (Electronic Window valve—ELWI), which sets the time interval during which fuel injection is performed (Fig. 7.73), as well as an electronic injection valve (Electronic Gas Injection—ELGI), which controls the flow of hydraulic oil of the FBIV-P drive;
- a sealing oil system that prevents gas fuel from penetrating into the hydraulic control and drive system.
- a system of double-walled pipelines for supplying and discharging LPG to gas modules with a ventilated space between the outer and inner shells. The pressure in the intermediate space of the fuel pipeline maintained below the pressure in the engine room;
- a ventilation system of the inter-wall space with built-in sensors for detecting the presence of hydrocarbons;
- the engine management system (Electronic Control System—ECS) is a general control system consisting of the ME-ECS core and an additional LPG supply control subsystem. The system controls all functions typical for basic ME

engines, as well as LPG injection and additional functions and corresponding auxiliary systems that ensure efficient engine operation on LPG.
- a safety system when using LPG (Engine Safety System—ESS), monitoring the pressure of liquefied petroleum gas and the combustion process.

In case of emergencies, ESS ensures the transition of the engine to the liquid fuel mode. At the same time, it purges the liquefied petroleum gas pipeline system and all other elements of the gas system using compressed nitrogen.

Nitrogen enters the system when the fuel valve train (FVT) switches and displaces the liquefied gas located there back into the supply tank.

The safety system is independent of the engine management system (ECS). This is critical for reliable and safe engine operation. While the ECS is responsible for optimal engine operation (injection control, boost, valve timing, etc.), the ESS is solely responsible for protecting the engine and preventing accidents if the parameters go beyond critical limits.

In general, the set of systems is typical for all other MAN engines designed to operate on alternative fuels. For safety reasons, many functions duplicated by the control and safety systems, which significantly reduce the likelihood of emergencies. The ECS system controls both pilot and gas fuel injection. For ease of identification, pilot fuel sometimes called the first fuel, and liquefied petroleum gas is the second fuel. We will need this later to understand the abbreviations used.

LPG injection is controlled by sequentially opening the electronic window valve (ELWI) and the electronic gas injection valve (ELGI), which are installed on the liquefied gas supply control module.

The Electronic LPG Injection Valve (ELGI) controls the injection timing and injection rate of LPG. Its operation directly controlled by the ECS, of which the Fuel Gas Supply System (FGSS) is a part. The ELWI sets the period during which LPG can inject and limits the maximum volume of cyclic injection. It connected to the Second Fuel Cylinder Safety Unit (SCSU), which is part of the ESS. The SCSU in turn connected to the Data Acquisition and Safety Unit (DASU), which monitors and analyses combustion processes in real time. The SCSU monitors information from the pressure sensors of specific cylinders and controls the injection and combustion of LPG. In the event of misfires or leaks in the LPG injection system, the ELWI acts as a shut-off device, this immediately shuts off the LPG supply. In this case, the engine automatically switched to liquid fuel. The main LPG supply control unit (Second Fuel Plant Control Unit—SPCU) and the auxiliary LPG supply control unit (Second Fuel Auxiliary Control Unit—SACU) are responsible for this process. These two units monitor the sensors of all safety-related systems. If deviations in the operation of these systems detected, they transmit information to the ECS to stop the supply of gas fuel and switch the engine to liquid fuel. This process occurs without reducing power, for a time commensurate with the intervals between working processes in individual cylinders.

As already noted, the LPG injection system is generally similar to the system used for methanol injection, but there are some differences. The most significant changes affected the design elements of the FBIV-P pump-injectors, associated with

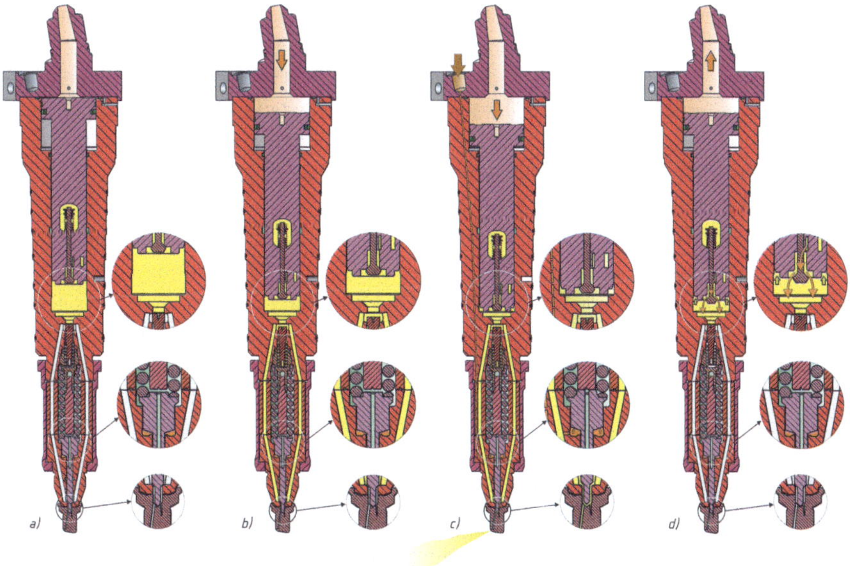

Fig. 8.15 Operating procedure of the FBIV-P module: (**a**) initial state; (**b**) fuel supply to the nozzle; (**c**) injection; (**d**) return to the initial state (adapted from [12])

the physical characteristics of liquefied petroleum gas. Firstly, it has twice the calorific value, which allows to significantly reducing the cyclic feed and dimensions of the device, as well as the flow section of the nozzle channels of the sprayer. Secondly, petroleum gas has greater compressibility than methanol, which requires higher injection pressures to avoid phase transitions.

Figure 8.15 shows the operating procedure of the FBIV-P pump-injectors.

The designation of the main elements corresponds to the positions shown in Fig. 8.5.

When switching the engine to operate on liquefied gas, the safety system (ESS) activates the opening of the window valve (ELWI), because of which liquefied gas enters the pump-injectors. Fuel under pressure of 5.0 MPa opens the automatic filling valve of the suspension type, filling the space under the plunger.

Under the action of pressure, the plunger together with the power hydraulic piston raises upward, squeezing oil into the drain line through the open spool control valve. The engine control system (ECS) through the corresponding (Cylinder Control Unit—CCU) controls the movement of the plunger.

After the power piston reaches the extreme upper position, the FBIV-P goes into a state of readiness for a new injection (Fig. 8.15a). On command from the LPG supply control system (Fuel Gas Supply System—FGSS), the control valve supplies hydraulic oil into the space above the power hydraulic piston (Fig. 8.15b). Oil pressure on the hydraulic piston actuates the fuel plunger, because of which the pressure in the sub-plunger cavity increases. A non-return pressure relief valve (Mon-Return Valve—MRV) is installed between the needle valve of the nozzle and the space under the plunger, adjusted to open at a pressure of 8.0 MPa. It is necessary to prevent liquefied gas from entering the cavity of the working cylinder under the pressure in the discharge line of the fuel supply to the FBIV-P (5.0 MPa), which can happen, for example, when the needle valve of the nozzle hangs. After the pressure under the plunger exceeds the opening pressure of the MRV, the valve moves down, opening a passage for liquefied gas into the chamber of the needle valve of the nozzle (Fig. 8.15b).

At a pressure in the power hydraulic drive system of 30.0 MPa, the ratio of the areas of the hydraulic piston and fuel plunger of 2.4 allows obtaining a pressure in front of the nozzle of approximately 70.0 MPa. However, this pressure is not enough for the needle valve to overcome the force of the spring that holds it in the closed state. The opening of the nozzle needle valve is carried out by supplying control oil under the hydraulic injection activation piston (Fig. 8.15c), for which the ELGI valve installed on the gas module and controlled by the ECS engine management system is responsible. Oil supplied to the cavity under the activator piston opens the needle valve by the amount of the gap between the end of the valve and the stop, opening the passage of fuel to the nozzle channels of the atomizer. Liquefied gas injected into the engine combustion chamber.

At the end of the injection, the control oil from under the needle valve activator piston drained, and the cavity of the power hydraulic cylinder connected to the return line. Fuel under pressure enters the sub-plunger space through the automatic filling valve, lifting the plunger together with the hydraulic piston, pushing the oil out of the power drive cylinder (Fig. 8.15d). The system returns to its original state.

Separating oil supplied to the gaps between the plunger and the needle valve guide under a pressure of 8.0 MPa, which prevents gas fuel from entering the hydraulic drive system and simultaneously lubricates the rubbing pairs. The sealing oil system is consumable, i.e. oil that has leaked into the gaps does not return to the system, but burns together with the fuel.

To maintain the temperature regime in the liquefied gas injection system, its circulation provided with the return of part of the fuel back to the supply tank. An automatic circulation valve allows fuel to circulate through the high-pressure circuit when the engine stopped.

The Hydraulic Cylinder Unit (HCU) consists of a distributor block mounted on a base plate. It has one or more gas accumulators to provide the required peak hydraulic oil flow during fuel injection.

8.4.2 Winterthur Gas & Diesel Two-Stroke Low-Speed Engines for LPG Operation

Despite the presence of ready-made solutions in the line of manufactured engines that allow the use of propane-butane fuels in the gas phase, WinGD has developed a universal platform of the fuel system that can used for adaptation to a certain type of alternative fuel. In this case, the system can use to build an engine with adaptation

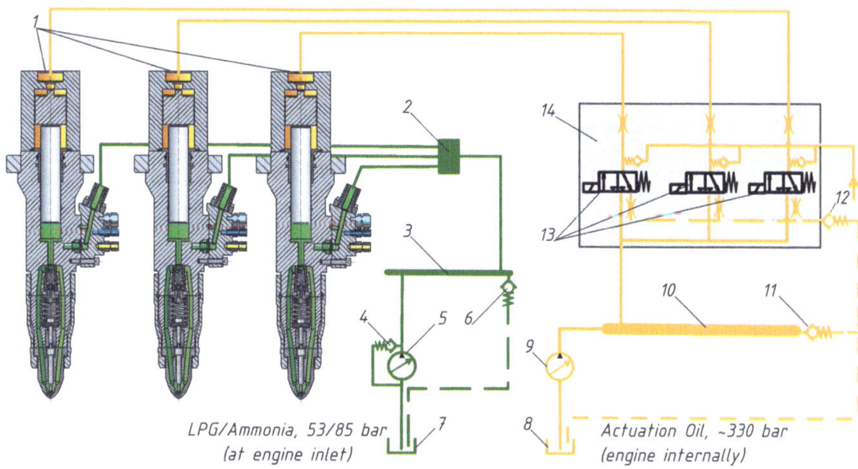

Fig. 8.16 Schematic diagram of the fuel system of the X-DF-P series engines from WinGD for LPG injection: (1) hydraulically driven pump-injectors; (2) distributor; (3) fuel accumulator; (4) pressure limiting valve; (5) fuel priming pump (LPG); (6) safety valve; (7) fuel tank; (8) hydraulic oil tank; (9) hydraulic system pump; (10) oil pressure accumulator; (11) hydraulic system safety valve; (12) hydraulic oil drain line pressure control valve; (13) electromagnetically driven spool control valves; (14) hydraulic distributor block (adapted from [13])

to the fuel that is currently available on the market, and in the future, this same engine, through moderate modernization, can convert to the type of fuel that will eventually become the most competitive, taking into account economic and environmental requirements.

For marine engines, LPG currently considered as a promising fuel, as is ammonia, a promising fuel for the near future. In addition, the engines must be equipped with a standard injection system for traditional liquid fuel to ensure their dual-fuel capability.

When creating such a universal platform, the concept of pump-injectors with a hydraulic drive used, similar to the concept used by MAN.

The general diagram of this system, shown in Fig. 8.16. Engines adapted to operate on LPG as a second fuel are designated X-DF-P.

The system adapted for LPG differs primarily in the performance of the pump section, which is approximately 2.2 times less than when using ammonia. Accordingly, the flow sections of the nozzle channels are also smaller, and selected in such a way as to ensure the duration of supply commensurate with the supply of traditional liquid fuel.

The injection of the pilot portion provided a standard liquid fuel injection system of the Common Rail type, which also ensures operation on traditional fuel.

The materials from which the components used for the ammonia injection system made are completely suitable for operation on liquefied petroleum gas.

The performance of the hydraulic pumps of the drive system, designed for the use of ammonia, significantly exceeds the need for hydraulic oil to drive the pump-injectors designed for LPG, so they can used without replacement.

8.4.3 Features of the Working Process of LPG Engines and Their Environmental Performance

The operating process of an engine running on LPG does not differ significantly from the operating process of the same engine running on traditional liquid fuel (Fig. 8.17).

With the same engine settings (actual compression ratio, maximum cylinder pressure, turbocharger settings, etc.), the heat release profiles in the cylinder have almost the same intensity and duration. The discrepancies in the TDC area are the result of the pilot injection occurring slightly earlier than the liquefied petroleum gas injection. The ascending section of the release rate curve for liquefied petroleum gas has a less steep slope than for liquid fuel, which indicates lower combustion intensity in the initial phase.

As can be seen from Fig. 8.17, the combustion rate can increase by increasing the injection pressure. Although even without optimization, the efficiency of the working process, and when running on LPG turned out to be slightly higher than when running on traditional liquid fuel. Figure 8.18 shows the Specific Fuel Oil Consumption and LPG together with the pilot portion, reduced according to the ISO standard to the reference fuel with a calorific value of 42.7 MJ/kg. It is evident that the most significant reduction in fuel consumption observed in the main operating modes of the engine, in the range of 50–90% of MCR.

NO_x emissions when using liquefied petroleum gas do not differ significantly from similar indicators when the engine is running on diesel fuel (Fig. 8.19).

The ME LGI-P and X-DF-P engines, like other dual-fuel engines running on alternative fuels on the diesel cycle, comply with IMO Tier II requirements for NO_x

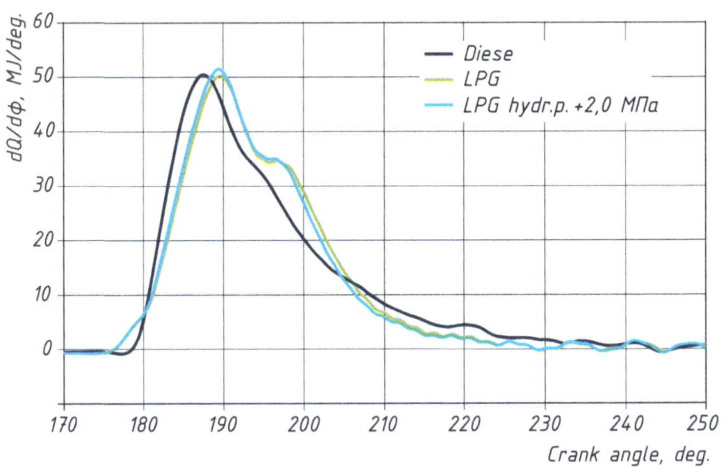

Fig. 8.17 Change in the heat release rate for the working cycle of the ME LGI-P engine on traditional liquid fuel and on LPG at a load of 75% of the MCR at standard and increased injection pressure (adapted from [8])

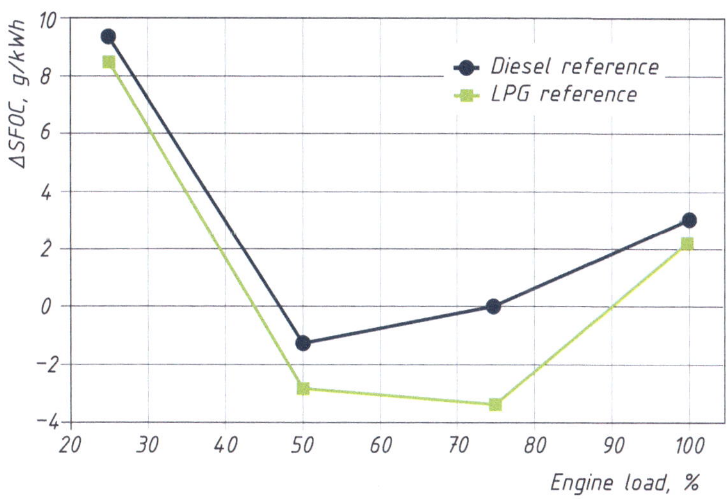

Fig. 8.18 Change in Specific Fuel Oil Consumption of the ME LGI-P engine on traditional liquid fuel and on LPG for the entire range of operating modes (adapted from [11])

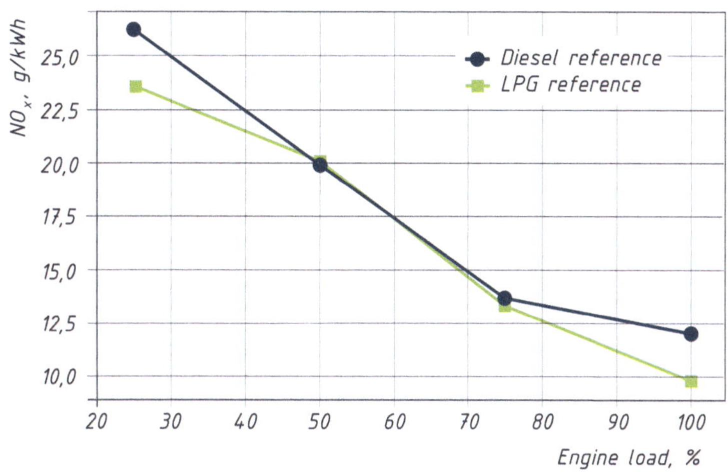

Fig. 8.19 Change in specific NO_x emissions when the ME LGI-P engine is running on liquefied petroleum gas and diesel fuel in the range of operating loads (adapted from [11])

emissions when running on LPG and traditional fuels without additional cleaning and Tier III with exhaust gas recirculation EGR or selective catalytic reduction SCR.

Hydrocarbon emissions when running on liquefied petroleum gas are somewhat higher than similar indicators typical for the working process using liquid fuel (Fig. 8.20). This is primarily due to the presence of two sets of sprayers in the combustion chamber, each of which has sub-needle volume. After injection, hydrocarbons evaporate from the sub-needle volume into the engine working space, and

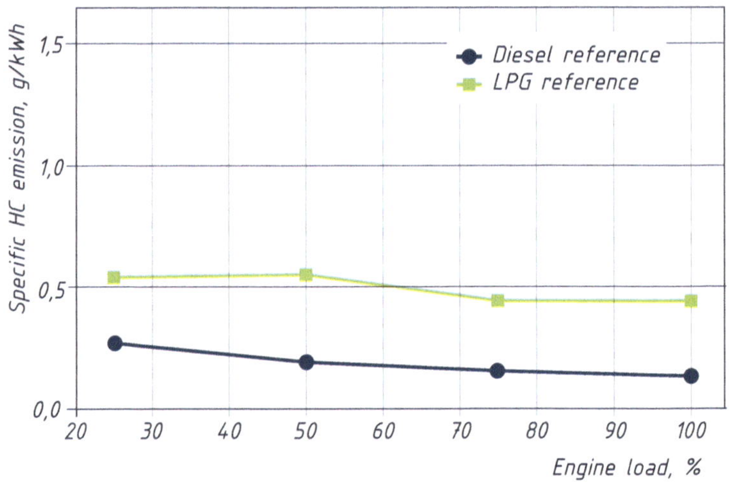

Fig. 8.20 Change in specific emissions of unburned hydrocarbons (HC) during operation of the ME LGI-P engine on liquefied petroleum gas and diesel fuel in the range of operating loads (adapted from [11])

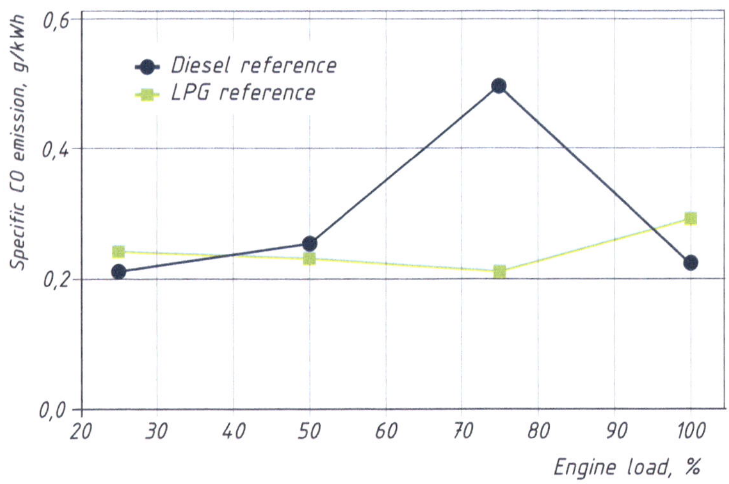

Fig. 8.21 Change in specific CO emissions when the ME LGI-P engine is running on liquefied petroleum gas and diesel fuel in the range of operating loads (adapted from [11])

then, mixing with the already cooled combustion products, pass into the exhaust gases.

CO emissions, when running on liquefied petroleum gas in the range of basic operating modes, are significantly lower than when running on reference diesel fuel (Fig. 8.21). LPG spray plume have a smaller penetration depth, which reduces flame quenching in the wall regions where CO molecules predominantly formed.

Due to the higher calorific value and the higher C/H ratio in liquefied petroleum gas compared to diesel fuel, the level of CO_2 formed during combustion of LPG is approximately 14% lower. However, the level of soot or black carbon is generally somewhat higher when running on liquefied petroleum gas.

ME LGI-P engines can optionally be adapted to run on traditional liquid fuel and LPG in different ratios. This function can be useful for the power plants of oil tankers, which use Volatile Organic Compounds (VOC) as fuel, consisting mainly of propane-butane fractions. With such an organization of the working process, the LPG supply system supplies the engine with all available LPG, and the missing part of the energy compensated by the supply of liquid petroleum fuel.

In general, the key performance characteristics, such as fuel economy and NO_x emissions, of the ME LGI-P and X-DF-P engines are very similar to those of diesel engines of similar size.

8.5 Use of Alternative Liquid Fuels with Low Flash Point in Marine Medium- and High-Speed Engines

In four-stroke engines of marine vessels, the task of using methanol (LPG) arose because of switching the main low-speed engines to this type of fuel.

It is obvious that the use of the same fuel by auxiliary engines as the main engines is economically feasible and extremely attractive. In this regard, leading manufacturers have carried out research and developments work in this direction.

In the course of this work, two fundamentally different approaches based on external and internal mixture formation emerged.

With external mixture formation (Port Fuel Injection—PFI), methanol or LPG in liquid form is sprayed into the intake port of the cylinder cover at the moment of filling the working cylinder with air (Fig. 8.22). During the process of filling the cylinder and during the compression process, the fuel evaporates, forming a homogeneous mixture with air. At the end of the compression stroke, pilot fuel injected into the cylinder, which ignites the fuel-air mixture.

With internal mixture formation (High-Pressure Direct Injection—HP-DI), fuel with a low flash point is supplied to the combustion chamber at the end of the compression stroke using a specially installed injector or through a multi-nozzle injector, which simultaneously supplies pilot and main fuel (Fig. 7.25).

8.5.1 External Mixture Formation

External mixture formation is much easier to implement, since it is only necessary to change the design of the engine intake tract and install the necessary additional equipment. The pressure under which methanol is supplied to the sprayers in

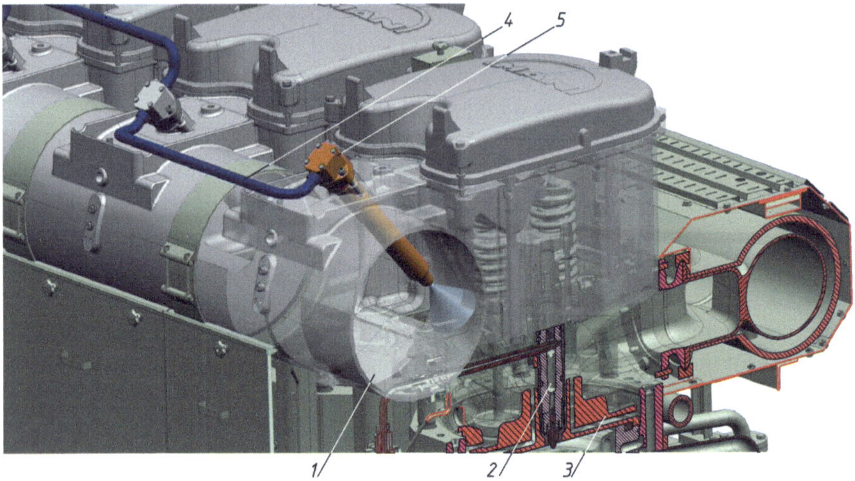

Fig. 8.22 Methanol injection into the intake manifold of the L32/40 engine (PFI): (1) air receiver; (2) fuel injector; (3) cylinder cover; (4) methanol supply line; (5) methanol nozzle in front of the cylinder cover intake ports (adapted from [14])

systems with external mixture formation is much lower than with its direct injection into the combustion chamber, which significantly simplifies the fuel system and increases the safety of its operation. This approach is applicable both to newly created engines and allows for a relatively simple modernization of engines already in operation. For example, MAN offers, based on PFI technology, the modernization of its 48/60, 51/60 or 32/40 series engines (Fig. 8.22) into dual-fuel engines capable of running on both traditional fuel and methanol. In recent years, a new dual-fuel engine with dimensions of 21/31 has been developed and brought to the market, designated DF-M. Following it, a new, more powerful version appeared—L27/38DF-M. Both engines designed for use in auxiliary electric generator sets of ships, using methanol as the main fuel for the main engines. Both engines based on the well-proven diesel engines of the latest generations, which have a modular design, which allows them to easily redesigned for a new fuel. In this case, the design changes affect only individual modules, and have virtually no effect on the reliability of the entire design. In particular, Fig. 8.23 shows the design of a well for a methanol injection valve in a charge air receiver. The injection valves themselves can have different designs depending on the subcontractor that supplies this injection valve. Figure 8.24 shows the design of an injection valve for atomizing methanol into the intake manifold of engines with external mixture formation developed by OMT.

The injector designed for engines with a capacity of up to 1200 kW/cyl. With a multi-point injection scheme, provided that one injector installed per cylinder. It is capable of providing a supply of up to 29,000 mm^3/cycle in engines with a rotation speed of up to 500 min^{-1}, at a methanol supply pressure of 3.0 MPa.

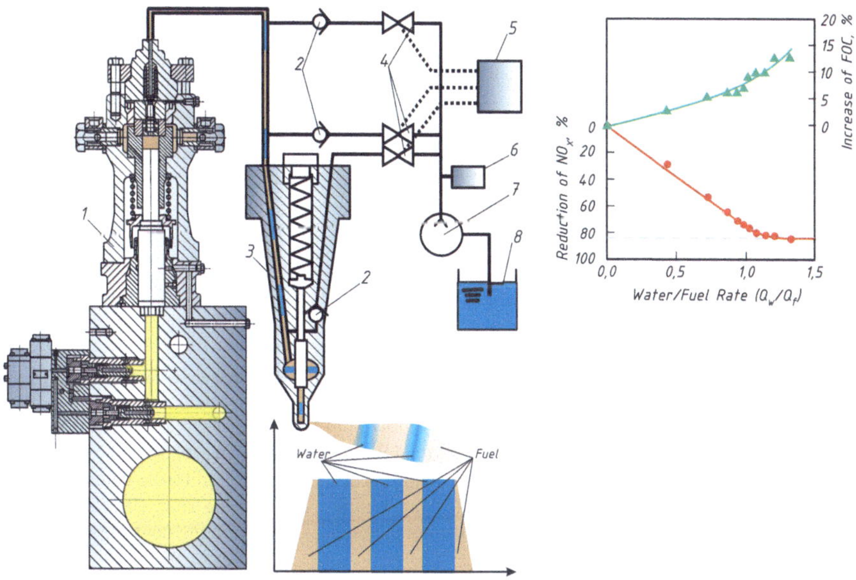

Fig. 8.23 Installation of a methanol injection valve into the intake port of the high-speed L21/31 DF-M engine: (1) nozzle; (2) ventilated cavity; (3) nipple; (4) injection valve connection unit to the methanol supply line (adapted from [15, 16])

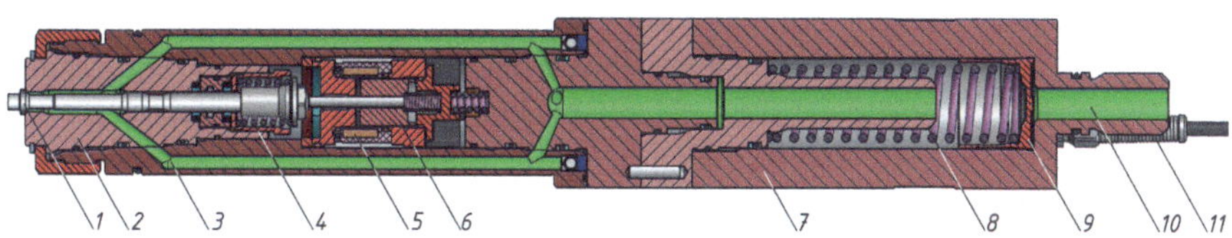

Fig. 8.24 Design of the injector for spraying methanol into the intake manifold of engines with external mixture formation developed by OMT: (1) nozzle plate valve; (2) plate valve body; (3) methanol supply channel to the nozzle; (4) plate valve return spring; (5) solenoid coil; (6) solenoid anchor; (7) maximum fuel flow limiter body; (8) limiter valve seat; (9) maximum flow limiter piston; (10) connecting nipple; (11) electric control cable (adapted from [17])

The injection valve has a direct electromagnetic drive with a built-in solenoid, which eliminates the need for fuel return lines or control oil supply. All potential leaks diverted to a special line with low excess nitrogen pressure or vacuum to eliminate the possibility of an explosive mixture. A built-in maximum supply limiter ensures that the fuel supply to the nozzle shut off in the event of a malfunction. With external mixture formation, the injection duration is approximately three times longer than with direct injection, but due to the significantly lower injection pressure, a significantly larger cross-section of the fuel channel is required. When using a standard nozzle design with nozzle holes, their diameter must be too large to ensure a given throughput. For this reason, in the design under consideration, the shut-off part of the

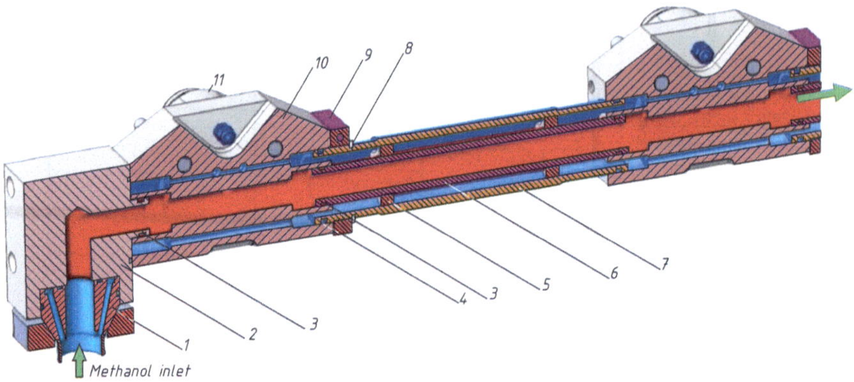

Fig. 8.25 Double-walled methanol supply line to individual nozzles of the L21/31DF-M engine from MAN: (1) connecting flange; (2) methanol inlet unit; (3) sealing ring made of V67F fluororubber with a high fluoroelastomer content (FKM); (4) sealing ring made of nitrile butadiene rubber (NBR); (5) spacer ring; (6) inner pipe; (7) outer pipe; (8) retaining ring; (9) clamp; (10) connection box to the injection valve of the working cylinder; (11) connecting connector (adapted from [15, 16])

nozzle made according to the scheme of a poppet valve opening outward. Thanks to the annular outlet section of the poppet valve, it is possible to form a spray with very fine drops with a valve stroke of only 0.6 mm. Moreover, with this arrangement, virtually all of the available pressure used to impart fluid velocity, whereas in a conventional design some of the pressure is lost in the gap between the needle seat and the needle valve, especially since the maximum stroke of the needle limited by the magnetic force created by the solenoid. Another advantage of the externally opening design is the ability to balance the fuel pressure forces acting on the valve, so that the actuator has to counteract only the spring and inertia forces. This allows the solenoid and, therefore, the injector to be smaller. Moreover, in this configuration, the pressure in the intake manifold acts in the direction of closing the needle, increasing the force on the seat and providing additional protection against fuel leakage from the injector.

A double-walled pipeline, from which a separate connecting box branched off to each of them, connects the individual injectors to each other (Fig. 8.25). An external supply block provides fuel supply to the distribution pipeline. All pipelines within the engine and in the engine room are double-walled. The general requirements for them are the same as for other alternative fuel systems we have discussed earlier. A timing disc, connected to the camshaft, allows the electronic engine management system to track the position of the crankshaft to synchronize the fuel injection process into the engine intake ports.

Since the engine runs on methanol (LPG) according to the Otto cycle, there is a possibility of detonation combustion. Therefore, each cylinder cover is equipped with detonation sensors that transmit signals to the control unit.

Pilot fuel injected by the standard engine fuel system, which also ensures the engine runs on traditional liquid fuel according to the diesel cycle. The share of pilot fuel is 5–15% of the full cycle supply, depending on the engine operating mode.

8.5.2 Internal Mixture

Internal Mixture formation is much more difficult to implement in engines. It requires more complex equipment that must operate at pressures significantly higher than those in external mixture formation systems

The most compelling argument in favor of using internal mixture formation is the ability to significantly reduce the share of pilot fuel to 0.5–5.0% of the full cycle sup-ply.

At this stage of technology development, the use of internal mixture formation is advisable only in powerful medium-speed engines used primarily as the main engines of ships. The first experience of using methanol as a fuel was the re-equipment of the main engines of the Stena Germanica ferry by Wärtsilä specialists from 2015 to 2017 at the Polish shipyard Remontova in Gdansk. As part of scheduled repairs, four Sulzer 8ZAL40S engines were equipped with the necessary components and systems for using methanol as the main fuel, while maintaining the ability to operate on traditional fuels. The initial goal of Wärtsilä and Stena Line, the owner of the vessel, was to dramatically reduce sulphur oxide emissions and significantly reduce nitrogen oxides, as well as gain experience in operating main engines running on methanol.

Using the accumulated experience, the Finnish company Wärtsilä began designing a dual-fuel engine based on the 32 series diesel engine that can run on methanol. Because of the work carried out, the Wärtsilä 32 engine appeared on the market in 2022, which becomes the basis for a new platform of multi-fuel engines that can run on methanol, LPG, liquid biofuel and heavy fuel oil.

The engine uses the principle of internal mixture formation (Fig. 7.23) with the supply of main and pilot fuel through one combined injector. Methanol injected as the piston approaches TDC and ignited by a small amount of pilot diesel fuel. The methanol injection pressure is 60–65.0 MPa.

The diagram of the injection valve for injection of the main and pilot fuel, as well as its layout, is shown in Fig. 8.26a. The standard fuel system of the engine used to supply the pilot portion, which also ensures its operation on liquid fuel.

Considering that the volumetric supply of methanol should be almost three times greater than petroleum fuel, three nozzle tips with corresponding valves are located in the nozzle body for its injection (Fig. 8.26b). These nozzle tips are peripherally located around the central tip intended for injection of liquid fuel. Methanol is supplied to the sprayers by the Common Rail system with small-volume accumulators made directly in the injector body (Fig. 8.26a).

Methanol combustion in the engine carried out according to a cycle close to diesel. At the same time, NO_x emissions reduced by approximately 50% compared to liquid fuel. The engine, when running on methanol, meets the requirements of IMO Tier II. For operation in Emission Control Areas (IMO Tier III), it can be equipped with the SRC system.

Additional pipelines on the engine provide not only the supply of methanol to the working cylinders, but also the supply of oil for sealing the injector, as well as control of its opening and closing. To ensure safety, all pipelines are double-walled with a ventilated space. In the event of an emergency, all cavities filled with methanol

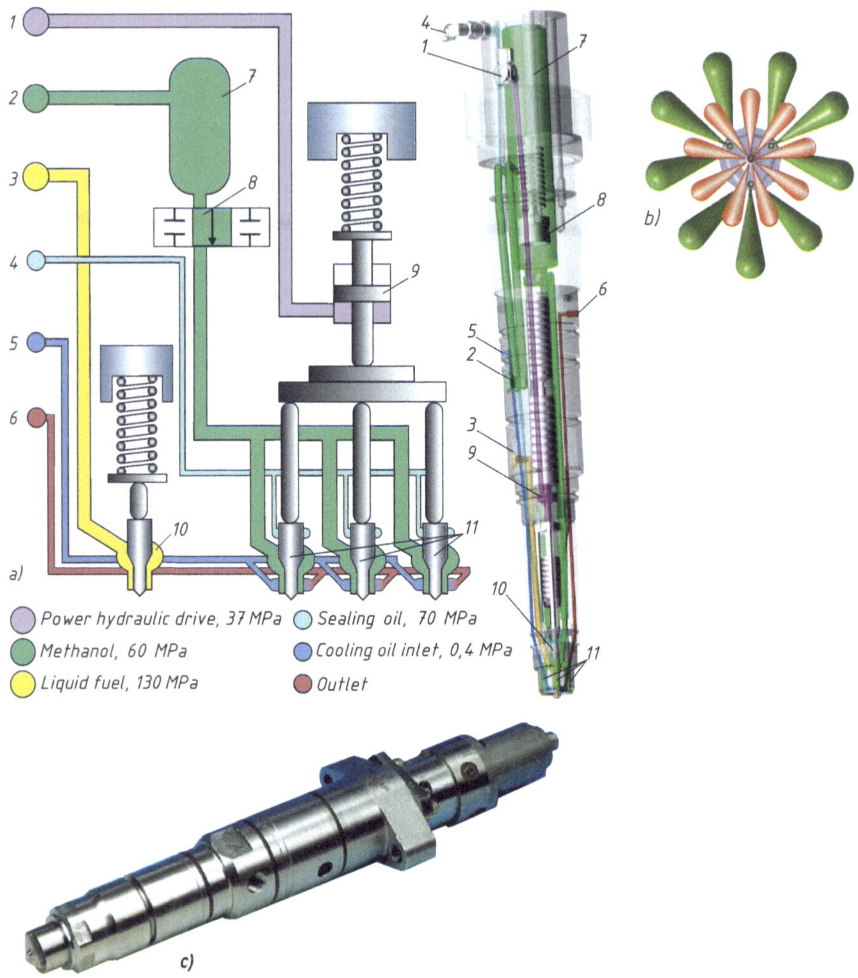

Fig. 8.26 Schematic diagram of the injection module of the Wärtsilä 32 multi-fuel engine and its general arrangement (**a**) diagram of the distribution of methanol and pilot spray plumes (**b**) general view of the nozzle for injection of methanol and pilot fuel (**c**): (1) supply of hydraulic control oil; (2) supply of methanol; (3) supply of liquid fuel; (4) supply of sealing oil; (5) supply of cooling oil; (6) outlet of cooling oil; (7) pressure accumulator; (8) control valve; (9) control piston; (10) liquid fuel nozzle; (11) methanol nozzle (adapted from [18–20])

blown through with compressed nitrogen, which displaces methanol into the supply tank. In this case, the engine control system automatically switches to liquid fuel without reducing the load.

Since the exhaust gases formed during the combustion of methanol contain fewer lubricating particles, it became necessary to modify the exhaust valves to prevent their excessive wear. Methanol is supplied to the engine under pressure from a high-pressure pump unit from Hammelmann (Methanol Fuel Pump Unit—MFPU), which has an independent electric drive (Fig. 8.27).

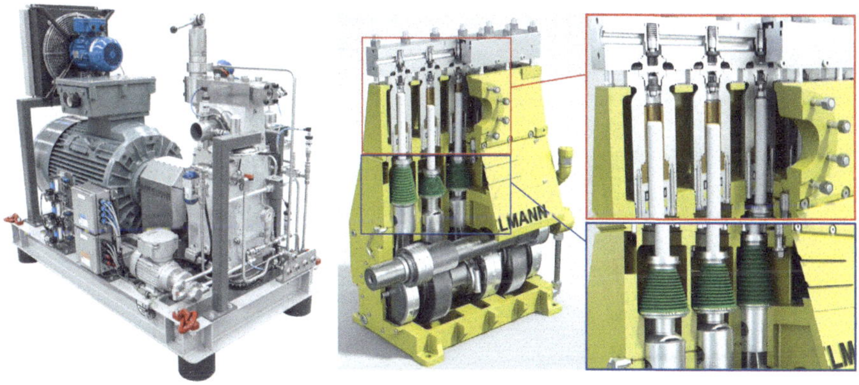

Fig. 8.27 General view and design of the Hammelmann high-pressure pump used to supply methanol to the injection system of the Wärtsilä 32 engine (adapted from [19, 20])

The pump's electric motor controlled by a frequency controller connected to the engine control system. Thus, the pump's performance synchronized with the engine's operating mode.

The oil pump unit supplies hydraulic oil, which used both to control the injection and to ensure an effective seal against methanol leaks during the injection process. In the future, Wärtsilä plans to extend the methanol technology to its entire range of medium-speed engines with cylinder diameters from 20 to 46 cm.

8.5.3 Features of Operation of Four-Stroke Engines Running on Methanol and Their Environmental Performance

Regardless of the method of organizing the mixture formation, the manufacturers of most dual-fuel engines managed to ensure the same performance as when running on traditional liquid fuel. Thus, the conversion of engines to methanol occurs without reducing the output power and without changing the dynamics of response to the load. The use of automatic control and monitoring allows the conversion of engines from one fuel to another without stopping them, without reducing speed and losing power. Due to these features, a complete backup of both fuels is ensured, on which the engine can operate depending on the requirements set by the operator, and in some cases, according to a mixed fuel cycle with a different ratio of fuels supplied to the combustion chamber. Methanol has very low self-ignition ability (cetane number < 5), therefore, in engines with both external and internal mixture formation, the combustion process is initiated by injection of a pilot portion of traditional liquid fuel or biofuel, which can be from 5 to 50% of the full cycle supply. From Fig. 8.28 it is evident that the nature of the methanol combustion process depends significantly on the proportion of pilot fuel, however the duration of the

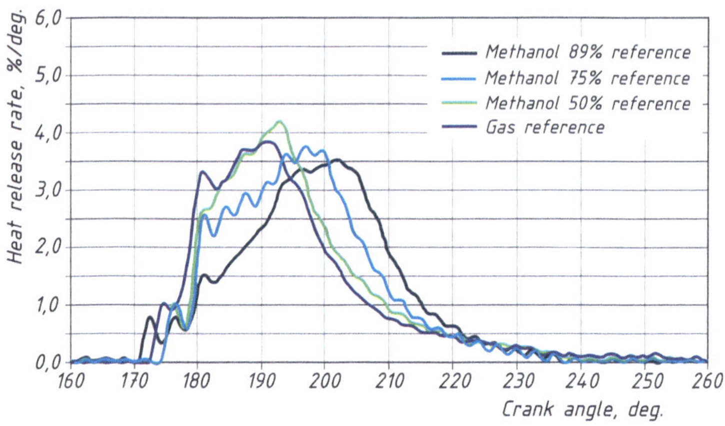

Fig. 8.28 Heat generation dynamics in the Wärtsilä Vasa 32 GD engine when operating on gas fuel and methanol with different proportions of methanol at loads of 75% of MCR (adapted from [20])

main combustion phases for different methanol/pilot fuel ratios differs insignificantly. With an increase in the proportion of pilot fuel, the combustion process becomes more dynamic, approaching in this indicator the basic working process of the engine on compressed natural gas.

The correct choice of the direction of the spray plume of the main and pilot fuel in the combustion chamber, the feed delay time and other parameters allow to significantly optimize the operating process of engines running on methanol, bringing it as close as possible to the cycle of the base engine. At the same time, an increase in the efficiency of engines is noted when they run on methanol, which leads to a reduction in the reduced Specific Fuel Oil Consumption. Figure 8.29 shows the test results of the Wärtsilä Sulzer ZA40S-MD engine when it runs on various types of fuel, including methanol. The overall reduction in fuel consumption when running on methanol is about 2%, while the operating process is quite sensitive to the methanol injection pressure, an increase in which allows improving the fuel efficiency of the engine. Since methanol does not contain sulfur, in the exhaust gases of an engine running on it, the amount of sulfur compounds determined by the sulfur content in the pilot fuel and the proportion of this fuel supplied to the combustion chamber. For engines with external mixture formation, the proportion of pilot fuel is usually 5–50%, while for engines with internal mixture formation it does not exceed 1–5% of the full cycle supply. Considering this, the total emissions of sulfur compounds when operating on methanol can reduced to 99% compared to the use of traditional fuels.

The presence of an oxygen atom in the methanol molecule, bound to a carbon atom, eliminates the formation of soot during its combustion. In fact, the source of soot formation during engine operation on methanol is the pilot fuel. Thus, its quantity, which varies significantly depending on the method of mixture formation,

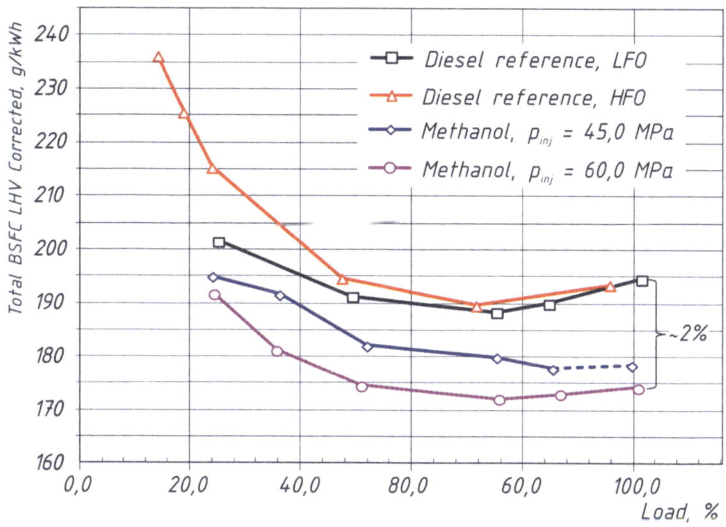

Fig. 8.29 Specific fuel consumption of the Sulzer ZAL40S-MD engine in the range of operating loads when running on light (LFO) and heavy fuel (HFO), as well as on methanol at different injection pressures (adapted from [21])

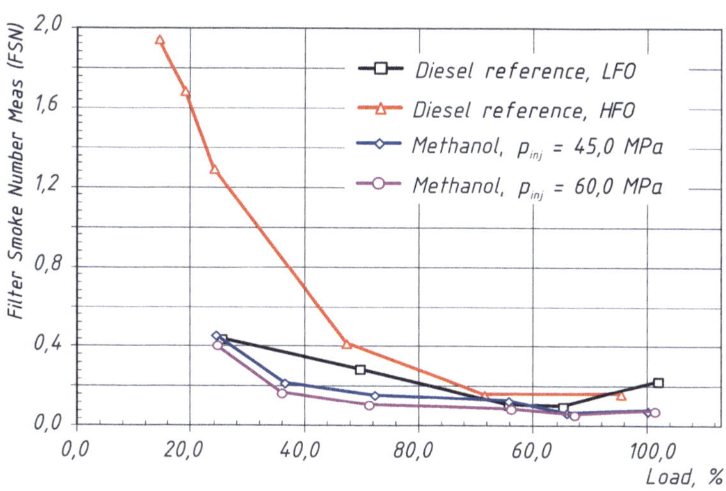

Fig. 8.30 Exhaust smoke of the Sulzer ZAL40S engine in the range of operating loads when running on light (LFO) and heavy fuel (HFO), as well as on methanol at different injection pressures (adapted from [21])

determines the smoke content of the engine exhaust gases (Fig. 8.30). The content of such products of incomplete combustion as CO and CH does not exceed 1.0 g/(kW × h), which generally corresponds to a similar indicator for traditional diesel engines.

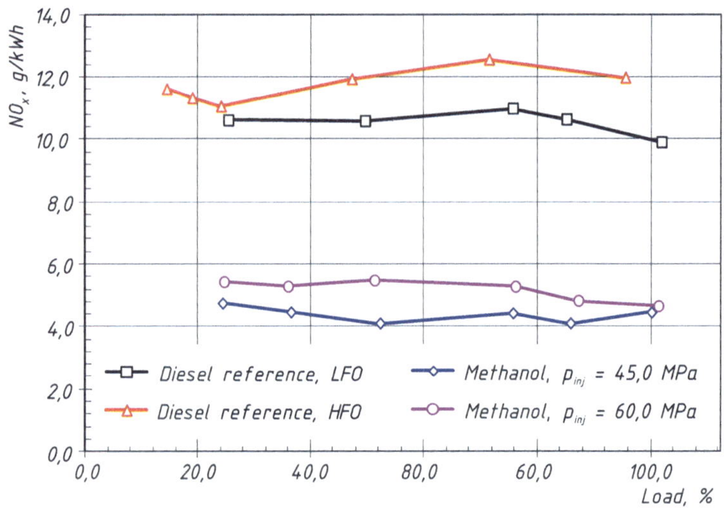

Fig. 8.31 Specific emissions of nitrogen oxides (NO_x) of the Sulzer ZAL40S engine in the range of operating loads when running on light (LFO) and heavy fuel (HFO), as well as on methanol at different injection pressures (adapted from [21])

Significant differences in the organization of working processes in engines with external and internal mixture formation lead to some differences in the mechanisms of nitrogen oxide formation. With external mixture formation, by the moment of ignition the combustion chamber contains a virtually homogeneous fuel-air mixture, which burns without the formation of local high-temperature zones. In addition, the combustion temperature of methanol is lower than that of traditional petroleum fuels (Table 6.1). All this helps to reduce NO_x emissions with exhaust gases, which can reach 30% for this type of engine. In engines with internal mixture formation, methanol injected into the combustion chamber immediately before combustion, evaporating significantly cools the charge, since it has a heat of vaporization that is 4.36 times higher than the same indicator for diesel fuel (Table 6.1). In addition, the cyclic feeds of methanol are approximately twice as high as the cyclic feeds of traditional fuel, which is associated with a lower calorific value. The process of methanol combustion greatly extended in time. This is due to the longer duration of pre-flame processes and the duration of the feed. In fact, the combustion rate determined the rate of feed of new portions of fuel into the combustion chamber. Because of the action of all the listed factors, taking into account the lower combustion temperature of methanol, the reduction of NO_x emissions with exhaust gases for engines with internal mixture formation can reach 60%. Figure 8.31 shows the relationship between NO_x emissions and the nature of the internal mixture formation process. With an increase in methanol injection pressure from 45 to 60 MPa, the NO_x content in the exhaust gases increased over the entire range of operating modes. This can explain by a reduction in the combustion process time due to a reduction in fuel injection time and an improvement in the quality of mixture formation. As a result, the rate of heat release increases, which leads to an increase in the temperature in the combustion chamber and an intensification of the thermal mechanism of NO_x formation. A more intensive increase in temperature with an increase in injection pressure also indicated an increase in the efficiency of the working process, shown in Fig. 8.28. This indicates the possibility of optimizing the operating process according to the criteria of fuel economy or NO_x emissions.

Regardless of the method of mixture formation, engines running on methanol are capable of meeting the requirements of IMO Tier II without the use of additional emission control methods. The total NO_x content usually does not exceed 3–5 g/(kW × h), and the use of control methods such as SCR or EGR make it possible to achieve the IMO Tier III level.

A significant reserve for reducing NO_x emissions, as for two-stroke engines, can be the use of water-methanol mixtures. Adding 20–40%, water to methanol allows you to meet IMO Tier III standards without using additional emission control methods. This is especially true for partial load modes, where an increase in the volume of the mixture can achieve due to the supply reserve of the fuel pumps. The formaldehyde content for most engines using methanol as fuel does not exceed the same indicator for diesel engines—6–14 ppm.

References

1. Winkel, O.: LNG for deep sea operation. Low Speed – Promotion & Sales. MAN Diesel & Turbo. 2012. 30 p
2. Utilisation of Volatile Organic Compoundsin Shuttle Tankers and Crude Oil Carriers. MAN Diesel. 24 p
3. Bilousov, I.V.: Internal combustion engines of modern marine vessels. Textbook for higher maritime educational institutions / I.V. Bilousov, M.P. Bulgakov // Astroprint publishing house – Odesa, 2024. – 512 p. ISBN 978-617-7797-54-7
4. MAN B&W ME-LGI engines. Liquid gas injection – methanol and LPG. MAN Energy Solutions. Copenhagen, Denmark. 2018. 5 p
5. MAN B&W ME-LGIP dual-fuel engines. MAN Energy Solutions. MAN Energy Solutions. Copenhagen, Denmark. 2018. 29 p
6. The methanol-fuelled MAN B&W LGIM engine. MAN Energy Solutions. MAN Energy Solutions. Copenhagen, Denmark. 2023. 32 p
7. Concept Guidance for X-DF-M. Winterthur Gas & Diesel Ltd. 2023. 29 p
8. The Methanol fuelled MAN B&W LGIM Engine. MAN Energy Solutions, Copenhagen, Denmark. 2021. 24 p
9. Market Update Note. Introducing the LGIM-W principle. MAN Energy Solutions. Copenhagen, Denmark. 2020. 1 p
10. Laursen, R.S.: The 2-stroke ME-LGIM. The dual fuel engine designed for operation on Methanol. Methanol Vessel Fuel Blending. Washington, June 13, 2017. 15 p
11. Mayer, S.: Performance and Emission Results from the MAN B&W LGI-P Low-Speed Engine Operating on LPG. Paper No. 169. Mayer, S., Cong, S., Juliussen, L.: CIMAC Congress, Vancouver. 2019. 13 p
12. Jeppesen, M.: Future-proof with ME-LGIP (Liquid Gas Injection Propane). Dual fuel done right, again. Greener Shipping Summit, 13th November 2018. 27 p
13. Announcement Dual-Fuel LPG Engine X52DF-P-1.0, X62DF-P-1.0. Technical Information Note O38. WinGD., 2024. 4 p

14. Methanol in shipping. Marine Four-Stroke. Feindt, A., Frank, F., Mögele, H.: MAN Energy Solutions, Augsburg, Germany. 12 p
15. Set your course to net zero with methanol. New MAN L21/31DF-M GenSet based on proven technology. MAN Energy Solutions. 2023. 26 p
16. Використання метанолу у якості моторного палива для суднових двигунів внутрішнього згоряння. Білоусов, Є.В., Марченко, А.П., Савчук, В.П., Білоусова, Т.П.: Двигун внутрішнього згорання. Всеукраїнський науково-технічний журнал. – Харків: вид. НТУ «ХПІ» – 2024. – № 1. – С 43–51. doi:10.20998/0419-8719.2024.1.06
17. Design and characterisation of a port fuel injector for high power medium speed engines. Coppo, M., Ferro, M., Negri, C., Pinkert, F., Niendorf, M.: 8th Rostock Large Engine Symposium 2024. Pp. 320–333
18. Portin, K.: Wärtsilä gas engine development & methanol adaptation class NK seminar., Singapore 3.11.2015. 27 p
19. https://gimsid.ro/pompe-jet-apa-inalta-presiune-plunjer-piston/electrice-hammelmann. html
20. Stojcevski, T.: Methanol as engine fuel: challenges and opportunities. Wärtsilä, Finland 2016. 21 p
21. Haraldson, L.: Methanol as fuel. Wärtsilä R&D, 17 March 2015, Lund. 26 p

Chapter 9
Use of Ammonia as a Carbon-Free Fuel for Marine Combustion Engines

9.1 Characteristics of Ammonia as a Motor Fuel

The emergence of technologies for using fuels with a high hydrogen-carbon ratio has effectively completed the stage of short-term measures to reduce greenhouse gas emissions under a specific IMO strategy aimed at the complete decarbonization of shipping.

With regard to measures planned for the medium term, a serious reserve been made related to the development of technologies for using fuels that can be obtained from renewable raw materials using energy from renewable sources. Such fuels have received the general name "green". In the case of successful scaling of production of such "green" products as methanol, ethanol and fuels based on fatty acids of organic origin without the use of fossil resources, it will be possible to make shipping carbon-neutral.

However, the greatest interest in the context of the long-term prospects on the path to environmentally friendly shipping is the use of fuels in the composition of whose molecules there is no carbon at all. Such fuels include ammonia and hydrogen.

Hydrogen requires a detailed discussion, which we will do separately, and in this section, we will focus on the use of ammonia as a fuel for marine combustion engines.

The term "carbon-free fuel" refers to fuels that do not produce carbon compounds, primarily carbon dioxide—CO_2, when burned. It is in this context that ammonia considered as one of the potential candidates for the decarbonization of the shipping industry. In addition, ammonia does not contain sulfur, so its use does not require the use of SO_x removal systems from exhaust gases to comply with environmental restrictions. Nitrogen oxides (NO_x), formed during the combustion of

I. Bilousov, R. Varbanets, *Marine Internal Combustion Engines and Environmental Protection Issues*, Springer Series on Naval Architecture, Marine Engineering, Shipbuilding and Shipping 27,
https://doi.org/10.1007/978-3-032-20940-5_9

ammonia, can be removed from the exhaust gases using selective catalytic reduction (SCR) technology.

An important advantage of ammonia over other alternative fuels is that it already produced on a large scale, and most of this product transported around the world by sea. There is already an infrastructure for bunkering ammonia in most major ports, which means there is experience in handling it. It should note that the bulk of ammonia produced today obtained from fossil resources as a raw material for the production of fertilizers. The availability of sufficient ammonia in the markets allows it to use today as a fuel to reduce SO_x emissions. By gradually switching from fossil-based ammonia to "green" ammonia, it is possible consistently reduce CO_2 emissions, thereby minimizing the risks for shipowners.

Ammonia (NH_3) is a chemical compound consisting of nitrogen and hydrogen. In a gaseous state, ammonia is a colorless toxic substance with a pungent odor. At ambient temperature and atmospheric pressure, it is lighter than air and tends to accumulate in the upper parts of spaces. When used on board, it can cause corrosion of the hull surface due to rapid interaction with atmospheric moisture, as it has high hydrophilicity. When absorbing water, the density of ammonia becomes greater than that of air, causing it to settle downwards, forming white clouds. Repeated exposure to the smell of ammonia causes olfactory fatigue or adaptation, which can significantly reduce the olfactory ability. This can lead to the inability to detect the smell even at a high concentration of 300 ppm.

Contact of ammonia with strong oxidizing agents such as chlorine and sodium hypochlorite can lead to the formation of explosive mixtures.

When ammonia interacts with mercury, explosive compounds also formed. For this reason, mercury should not used in devices that may exposed to ammonia. To prevent leaks of ammonia, appropriate safety devices must provided and rules for its safe use must developed.

Gaseous ammonia is a flammable gas with a narrow flammability range and a high lower explosive limit. In addition, the autoignition temperature of ammonia is relatively high, at about 651 °C, and the flame propagation rate is significantly lower than that of other fuels. This significantly reduces the risk of its ignition in a closed space, but these features complicate its combustion in internal combustion engines and can lead to ammonia slip. Unburned ammonia in the engine, entering the exhaust gases, can react with SO_x and NO_x to form secondary air pollutants such as ammonium sulfate and ammonium nitrate, respectively. Therefore, when used in internal combustion engines, pilot ignition used effectively ignite ammonia. At atmospheric pressure, ammonia liquefies at a temperature of −33 °C, so it can be stored in a refrigerated state using low-pressure tanks. This makes it much easier to store on board, but to store a comparable amount of fuel; these tanks would need to be four times larger. This is due to the lower calorific value of ammonia (Table 6.1).

Ammonia is corrosive to some materials such as copper, copper alloys and zinc, and this should take into account when selecting materials for system components in contact with it. It also causes stress corrosion cracking in carbon manganese or nickel steel. Dissolved oxygen in liquid ammonia increases the risk of stress corrosion. The dissolved oxygen level should therefore keep below 2.5 ppm (by weight).

For materials such as rubber, plastic, vinyl or aluminium alloys, only those approved by classification societies should used, taking into account the conditions of use. In summary, ammonia is not an ideal fuel for internal combustion engines; it has a narrow flammability range, low flame speed, high heat of evaporation and high autoignition temperature. Low flame speed leads to a decrease in thermal efficiency and emissions of unburned NH_3. Ammonia combustion also leads to emissions of NO_x and N_2O, which are greenhouse gases with a high global warming potential of 273. New catalytic cleaning technologies may be required to control emissions of these components. Another problem is the toxicity of ammonia and the need to use adequate safety systems on ships running on ammonia. Despite all the listed difficulties, many leading manufacturers of marine engines are actively working on creating versions capable of running on ammonia. Below, we will provide a brief overview of this work.

9.2 Use of Ammonia as a Fuel for Marine Low-Speed Two-Stroke Internal Combustion Engines

Since ammonia can be easily stored and transported in a liquefied state, its use as a fuel for marine engines is possible both in the liquid and gaseous phase.

The saturated vapor pressure of ammonia at a temperature of 20 °C is 0.857 MPa, which allows it to be stored and supplied to the engine in a liquid state at ambient temperatures under a pressure of 1.0–1.2 MPa. Thus, the storage and use of ammonia for engines with internal mixture formation is very similar to the use of liquefied petroleum gas.

At atmospheric pressure, liquid ammonia begins to boil at a temperature of −33.35 °C, so it is easy to convert it to a gaseous state and use it as a gaseous fuel for engines with external mixture formation.

In general, approaches to the design of ammonia fuel systems for its use, both in liquid and gaseous states, are similar to the approaches we have already considered that are typical for other types of alternative fuels. Next, we will dwell in more detail on the features of these systems associated with the use of ammonia as fuel.

9.2.1 The Concept of MAN Ammonia Fuel Engines

The concept of MAN ammonia fuel engines is based on the universal fuel system platform we have previously reviewed, which in turn is based on the well-proven ME diesel engine series [1].

The differences come down to changes in the dimensions of the main elements of the hydraulically driven pump-injectors to obtain the required performance, taking into account the characteristics of ammonia. First, such characteristics include the calorific value, which is 2.26 times lower for ammonia than for petroleum fuels, which requires a corresponding adjustment of the dimensions of the working parts of the pump-injector. In addition, to ensure the required injection time, the diameter of the spray nozzle channels was increased. The materials for the manufacture of the main design elements, as well as the hydraulic sealing oil, carefully selected to ensure compatibility with ammonia. As a result, the pump-injector adapted for ammonia fuel received the designation—FBIV-A, where the letter "A" actually indicates the type of fuel, and the engines themselves received the designation—ME-LGIA. Figure 9.1 shows the main elements of the fuel system of the ME-LGIA engine.

Fig. 9.1 General arrangement of the main components of the fuel systems of ME-LGIA engines: (1) high-pressure liquid fuel pump; (2) double-walled ammonia supply pipes; (3) ammonia supply control unit; (4) diaphragm pressure accumulators; (5) pump-injector for injecting ammonia into the combustion chamber; (6) hydraulic power drive line; (7) flange connection of double-walled ammonia supply lines (adapted from [2])

To guarantee the prevention of phase transitions and cavitation in the fuel system, fuel is supplied to the supply control module and the pump-injector under a pressure of 8.0 MPa, and is injected into the engine combustion chamber under a pressure of 60.0–70.0 MPa. To prevent the accumulation of ammonia vapor in the pump-injectors, a small portion of ammonia continuously recirculated through the supply system. Circulation channels collect fuel around the plunger, as well as along the gap between the rod and the plunger.

The limited volume of the plunger cavity of the pump-injector limits the maximum amount of ammonia injected into the combustion chamber. During a planned or emergency conversion of the engine to oil fuel, the pump-injector and the supply control unit emptied and blown with compressed nitrogen.

For the hydraulic drive of the pump-injectors, oil from the engine control system used, which supplied under a pressure of 30.0 MPa. To prevent ammonia from entering the hydraulic oil, the hydraulic drive system separated from the sub-plunger space by a sealing oil barrier. The sealing oil supplied under a pressure of 10.0–11.0 MPa from a special system that is autonomous and separated from the rest of the engine's hydraulic systems. A small amount of the sealing oil consumed, entering the sub-plunger cavity and mixing with ammonia. From there, it partially injected into the combustion chamber, and partially returned to the supply system. Some of the sealing oil drained from the pump-injector body back into the sealing oil tank. All pipelines and volumes containing ammonia, including the gas block and pump-injectors, made with double walls, having ventilation and a leak detection system. Any ammonia leak from damaged components of the internal circuit is contained within the external pipeline to prevent it from escaping into the engine room. Gas analyzers monitor the ammonia content in the ventilated air, and if a leak detected, they send a signal that causes the engine to switch to petroleum fuel.

To ignite the ammonia, a pilot fuel injection is used, the share of which can reach up to 5% of the total cyclic supply. The standard fuel system of the engine used to inject the pilot fuel, which also ensures its operation on traditional petroleum fuel.

The concept of eliminating anthropogenic carbon dioxide emissions includes considering the use of biofuels from organic fatty acids as a pilot fuel.

9.2.2 The Concept of Ammonia-Fueled Engines from Winterthur Gas&Diesel

The concept of ammonia-fueled engines from Winterthur Gas&Diesel is based on the platform we discussed earlier (Fig. 8.14) designed for use as the main fuel of liquefied petroleum gas, which in turn is based on the well-proven series of diesel engines of the X series [3]. Engines based on this platform were designated X-DF-A.

Conceptually, the approaches to the use of ammonia in X-DF-A engines are practically no different from the approaches used in ME-LGIA engines.

The main element of the fuel system (Fuel Injection System—FIS), which ensures the supply of ammonia to the combustion chamber under high pressure, are hydraulically driven pump-injectors, which are essentially pressure multipliers. Ammonia supplied to them in liquid form under a pressure of 8.3 MPa, which allows avoiding phase transitions and cavitation, and is compressed using a plunger, which affected by the hydraulic piston of the drive. The diameters of the plunger and the hydraulic piston are selected in such a way that the pressure of 30 MPa supplied to the hydraulic chamber of the drive creates a force on the plunger sufficient to inject ammonia under a pressure of 60 MPa. Due to the low energy density of ammonia, its volumetric supply significantly exceeds the supply of traditional fuel. In turn, this significantly affects the consumption of hydraulic oil for the drive of the pump-injectors, which in maximum power modes can reach significant values. Thus, the parameters of ammonia injection directly affect the oil pressure and its consumption. At the same time, when the engine is running on traditional fuels, the consumption of hydraulic oil significantly reduced. For this reason, the oil supply system for the pump-injector drive is a separate system, similar to the oil supply system for the exhaust valve servo drive used in WinGD engines. The pumps for pumping hydraulic oil into the ammonia injection system take oil from the engine's circulating lubrication system. Part of the power taken from the crankshaft through a special transmission used to drive them. The oil under pressure enters the common accumulator, from where it distributed to the individual cylinder control units (Actuation Control Unit—ACU).

To ensure the required level of safety, oil from the accumulator to the individual cylinder control units supplied through double-walled oil lines. The diameter and number of oil lines depend on the engine cylinder diameter and oil consumption.

The electronic control system maintains the optimum pressure level in the accumulator depending on the engine-operating mode, to achieve the most efficient combustion. Each control unit contains electro-hydraulic Oil Control Valves (OCV), one for each pump-injector, allowing individual control of each of them. Each valve regulates the flow of hydraulic oil to the corresponding pump-injector. When power is applied, the valve moves the spool, directing oil to the pump-injector. The duration of the valve opening determines the amount of oil supplied and, therefore, the cyclic portion of ammonia supplied to the engine combustion chamber. When the power is disconnected, the valve closes, directing oil from the hydraulic cylinder of the pump-injector drive to the drain.

Figure 9.2 shows the general structure of the pump-injector of the X-DF-A engines, which includes a hydraulic cylinder, the piston of which drives the fuel plunger, and a needle valve of the atomizer. Ammonia supplied to the cavity under the plunger under a pressure of 8.5 MPa. This pressure ensures the plunger returns to its original position when draining oil from the hydraulic cylinder. At the end of the hydraulic piston stroke, the cylindrical projection on its end surface enters the oil channel, creating an annular gap. This leads to throttling of the oil flow and a decrease in the piston speed at the final stage of its closing, ensuring shock-free placement on the stop.

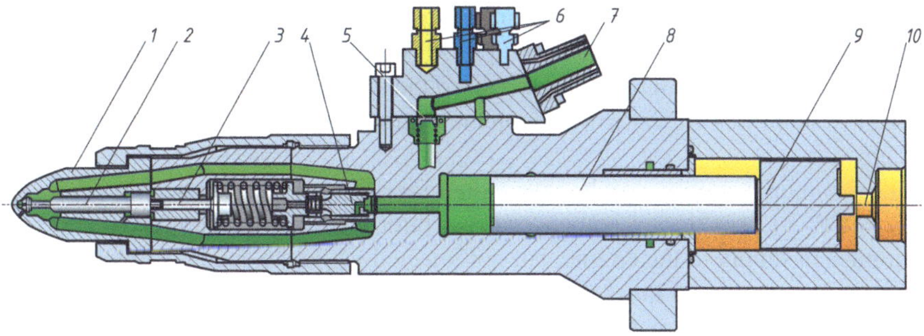

Fig. 9.2 General structure of the X-DF-A engine pump-injector: (1) nozzle body; (2) needle valve; (3) stopper; (4) non-return valve; (5) filler line check valve; (6) sealing oil, coolant and purge nitrogen supply nipples; (7) ammonia supply; (8) plunger; (9) hydraulic drive piston; (10) hydraulic oil supply channel (adapted from [4])

To prevent fuel injection under fuel pressure in the feed line, a non-return pressure-reducing valve installed between the plunger cavity and the nozzle, the opening of which is adjusted to a pressure of more than 8.5 MPa.

When an electric signal supplied from the engine control unit, oil under a pressure of 30 MPa enters the hydraulic drive cylinder. The ratio of the piston and plunger areas is selected in such a way that the injection pressure is 50.0–60.0 MPa. Under the action of pressure, the pressure-reducing valve opens, passing fuel to the nozzle. By acting on the needle valve, the fuel opens it and then through the nozzle holes into the combustion chamber. To reduce of the ammonia slip, the volume of the under-needle cavity of the nozzle minimized as possible.

When the ammonia supply turned off, the purge system ensures its displacement with compressed nitrogen from all volumes of the pump-injector. The X-DF-A engines are equipped with an Ammonia Injector Cooling Water System (AICWS), which used to cool the pump-injectors when the engine is running on petroleum fuel. Sealing rings, low-pressure grooves for collecting leaks, and dirt traps prevent oil contamination, and a leakage monitoring system provides a way to check for issues.

The ammonia pump-injector installed on the working cylinder cover next to the standard petroleum injectors. The cylinder cover is also equipped with special channels integrated into the engine ventilation system, which provides additional safety in the event of a pump-injector failure during operation.

The standard petroleum fuel injection system, characteristic of the X-series engines, used to supply a pilot portion of fuel, ensuring ignition of the atomized ammonia. The same system ensures engine operation when using traditional fuel. The X-DF-A engine can operate continuously on ammonia in the load range from 25–100% or on oil fuel in the load range from 0% to 100%. The engine started on oil fuel. When operating on ammonia, the pilot fuel's share of the total energy balance in the working cylinder is no more than 5%.

In normal operation, liquid ammonia supplied to the engine from the storage system by low and high-pressure pumps. All fuel supplied to the engine is completely consumed, without recirculation into the ammonia supply system. This significantly simplifies the system, both on the engine side and on the ammonia supply system side.

The ammonia supply may be interrupted for a period when power drops below 25% MCR, such as during maneuvering or pilot boarding. In this case, ammonia continues to supply to the engine, but not injected while the engine is operating on oil fuel. This ensures a quick return to ammonia operation. The ammonia supply system stops and is purged if the stop duration exceeds a preset time, the safety system is triggered, or at the operator's request. Liquid ammonia displaced from the system by inert gas. At the same time, the AICWS system activated, and the engine switches to liquid petroleum fuel.

To ensure compliance with safety standards, a vessel equipped with an X-DF-A series engine must be equipped with an Ammonia Vapor Processing System (AVPS). It used to clean gaseous products (e.g. air from the fuel line ventilation system) from ammonia vapors before they are released into the atmosphere.

The piston rod seals have been redesigned to prevent ammonia from entering the engine crankcase. The lower part of the pistons is equipped with a system for monitoring ammonia leaks through the piston rings.

The basis of the safety concept of the X-DF-A series engines is a two-stage purging procedure. It designed to increase the safety of the crew in the event of an ammonia leak or during maintenance work on the engine systems:

- initial purge: inert gas (e.g. N_2) is supplied through the inert gas supply system. All liquid ammonia from the engine and fuel system lines forced out into a special tank;
- residual ammonia vapor removal: At this stage, the AICWS cooling system activated. This system not only cools the pump-injector, but also absorbs, flushes and collects any residual ammonia (e.g. from internal pipe surfaces wetted with liquid ammonia).

Both systems connected via the Fuel Valve Unit (FVU), which includes a series of valves to control the fuel supply and ensure interaction between the engine and auxiliary systems.

The annular space between the pipes mechanically ventilated with a slight vacuum, providing 30 air changes per hour.

Ammonia detectors are located in the annular ventilation system of the pipes, as well as in the lower part of the piston and the engine exhaust system, for early detection of leaks. In the event of a leak, the engine goes into a safe mode, starting the purge process to ensure a safe inspection by the crew.

The fuel lines and components containing ammonia designed to minimize the risk of mechanical damage during vessel operation.

9.2.3 *Ammonia Fuel Engine Concept from Japan Engine Corporation*

Japan Engine Corporation (J-ENG) was founded in April 2017 through the merger of the marine diesel engine division of Mitsubishi Heavy Industries Marine Machineries & Engine Co., Ltd. (MHI-MME) and Kobe Diesel Co., Ltd.

Too efficiently, burn ammonia in the combustion chamber of a low-speed engine, the company used the previously proven "stratified injection" technology. With this technology, various liquids sequentially injected through one injection valve, forming clearly defined layers of the resulting aerosol of different compositions in the spray cone. Previously, this technology used to inject water together with fuel in order to reduce NO_x emissions, which discussed in the relevant chapters of the book.

The UEC series diesel engines were as the base engine for creating the ammonia version.

Engines designed to operate on ammonia as the main fuel were designated—UEC-LSJA. As previously, noted, special conditions must create for the efficient combustion of ammonia, as this fuel is quite resistant to ignition.

Stratified injection, in which a portion of pilot fuel first enters the combustion chamber, followed by a portion of liquid ammonia, alternating with traditional fuel in several layers, allows for maximum control over the combustion process and a significant reduction in NO_x emissions due to the high efficiency of ammonia combustion in mixed fuel.

The diagram of the fuel system for sequential, layered injection of ammonia and pilot fuel, as well as the order of layer formation are shown in Fig. 9.3.

Modeling of layered spraying processes showed that aerosol droplets (shown in red in Fig. 9.4), formed at the beginning of the injection, are significantly slowed down due to resistance relative to stationary air.

During approximately the first 1/8 of the injection time, all the droplets of injected fuel are located at a small distance from the nozzle tip of the nozzle (at $t = 0.9$ ms). The droplets of fuel injected later (in Fig. 9.4, they are marked in yellow) are already moving in the airflow, which has acquired some mobility, having received an impulse from the droplets injected earlier. These droplets move at a higher speed, penetrating the cloud of previously injected droplets, throwing them aside (at $t = 1.5$ ms). Gradually, all the droplets injected into the combustion chamber first are dispersed along the generating surface of the spray plume, as if enveloping it from the outside (at $t = 2.4$ ms).

If at the beginning of injection, a portion of pilot fuel with good self ignition (for example, MDO) supplied to the combustion chamber, and then the flame from its combustion will surround a cloud of ammonia droplets. This will lead to a significant increase in temperature inside the spray plume and intensive evaporation and ignition of ammonia droplets.

The last portions of fuel form an aerosol that remains in the center of the flame, where almost all the oxygen has been used up.

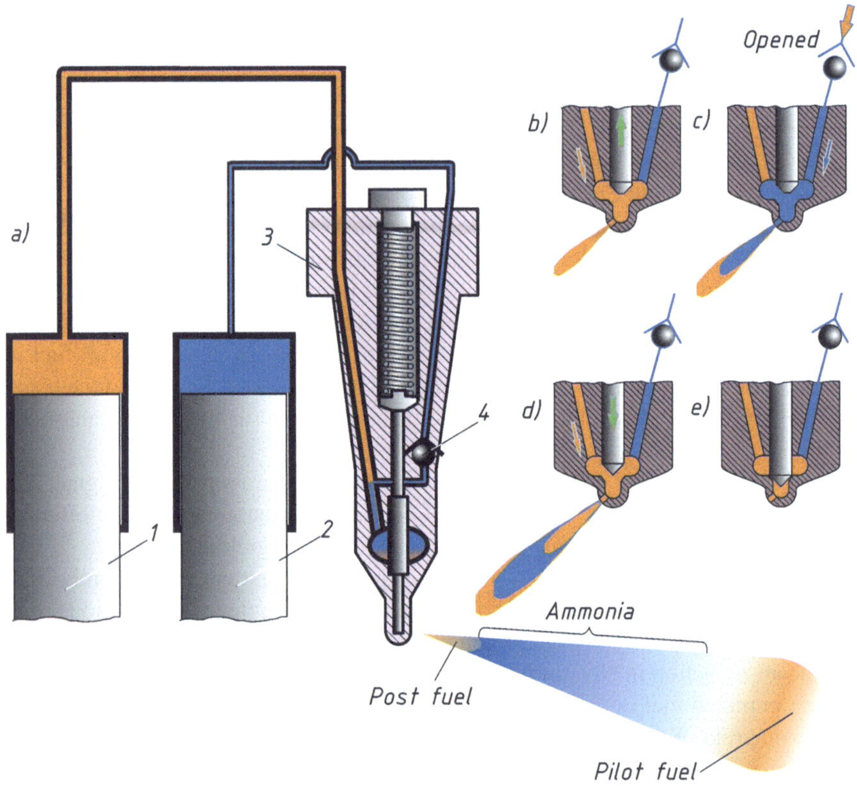

Fig. 9.3 Schematic diagram of the fuel system for forming a stratified injection of pilot fuel and ammonia used by Japan Engine Corporation of the UEC-LSJA series and the diagram of forming a stratified charge: (**a**) general diagram; (**b**)injection of the front portion of pilot fuel; (**c**) injection of ammonia; (**d**) post-injection of pilot fuel; (**e**) end of injection; (1) fuel pump for supplying pilot fuel; (2) fuel pump for supplying ammonia fuel; (3) fuel injection valve; (4) check valve (adapted from [5])

In Fig. 9.4, these droplets are shown in black. They remain in the center of the flame, since there is no more spraying that pushes them to the periphery, to areas more saturated with oxygen. This is the reason for the long combustion and poor afterburning of the last portions of fuel, especially such a low-activity fuel as ammonia. If a highly active fuel, such as MDO, injected into the combustion chamber last, this problem will be much less acute.

The results of this modeling explain the high efficiency of the stratified fuel injection system, especially when using fuels with low reactivity.

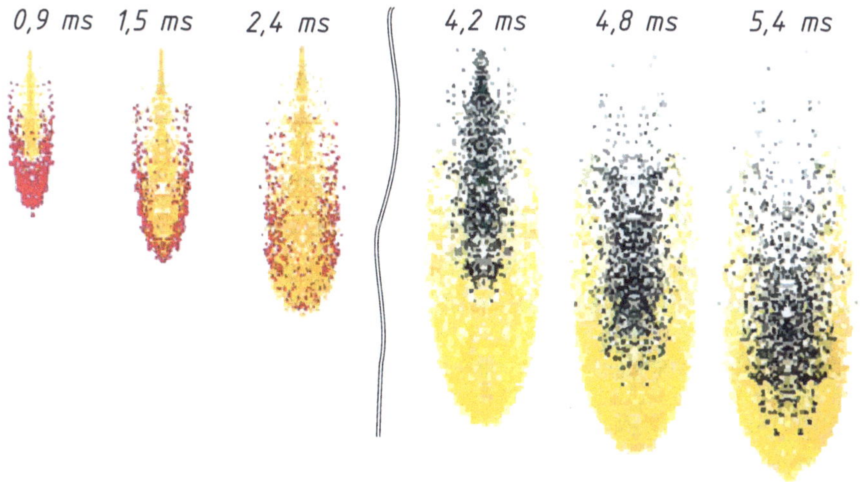

Fig. 9.4 Sequence of aerosol droplet distribution in the spray plume by time of entry into the combustion chamber. Droplets injected during the first 1/8 of the injection time are marked in red. Droplets injected during the last 1/4 of the injection time are marked in black (adapted from [5])

9.2.4 Features of Operation and Environmental Performance of Two-Stroke Low-Speed Engines Running on Ammonia

The main safety issue when working with ammonia is its toxicity and gas-dispersion properties. For this reason, ammonia emissions into the air must be limited to the lowest possible level. Direct discharge of ammonia into the atmosphere prohibited during normal operation, including for pressure control in tanks, except in emergencies when this does not lead to the occurrence of hazardous concentrations.

Regardless of the manufacturer, the concepts of safety systems for engines running on ammonia have been developed in such a way as to ensure the same level of safety as when using traditional fuel. They based on three key principles:

- the principle of gas safety of the engine room, which ensures that a single failure in the ammonia supply system will not lead to its release into the engine room;
- the principle of secondary localization of leaks in the double-walled pipeline system with automatic early detection of any ammonia leaks;
- the principle of compliance with safety requirements when changing the type of fuel, in the event of leaks, as well as in preparation for maintenance. This achieved by purging all volumes filled with ammonia.

In addition to design measures aimed at preventing the ingress of ammonia into the ship's spaces, precautions (e.g. adequate personal protective equipment and procedures) for the crew during engine operation on this type of fuel, as well as special training of the crew, are necessary.

When the engine is running on ammonia or in all other cases when the fuel system is under pressure (e.g. in standby mode), crane work near components handling ammonia should be prohibited. All objects in the engine room should securely fasten to prevent damage by them to components working under ammonia pressure.

At the time of writing this book, the results of performance tests of ammonia engines by manufacturers have not published. The information below based on laboratory studies conducted by various companies and requires further clarification.

9.2.4.1 The Combustion Characteristics of Ammonia

The combustion characteristics of ammonia, and primarily the low flame propagation speed, significantly affect the engine operating process, and at the same time significantly expand the possibilities for its optimization. The key role in the successful development of ammonia technologies played by the engine fuel system, which determines the start time and duration of the main and pilot fuel supply, as well as the injection pressure. By varying these parameters, it is possible significantly influence the combustion character and operating process parameters.

Figure 9.5 shows the heat release characteristics for two different supply durations in a laboratory setup simulating the injection of ammonia and pilot fuel into the combustion chamber of a low-speed engine.

The initial conditions in the setup correspond to the parameters at the end of the compression process with the engine running at 50% MCR load. The pressure at the start of injection is 11.6 MPa, and the temperature is 860 K.

The data obtained for a system with one pilot and one main injector. The heat release rates given for two different ammonia injection durations with the same

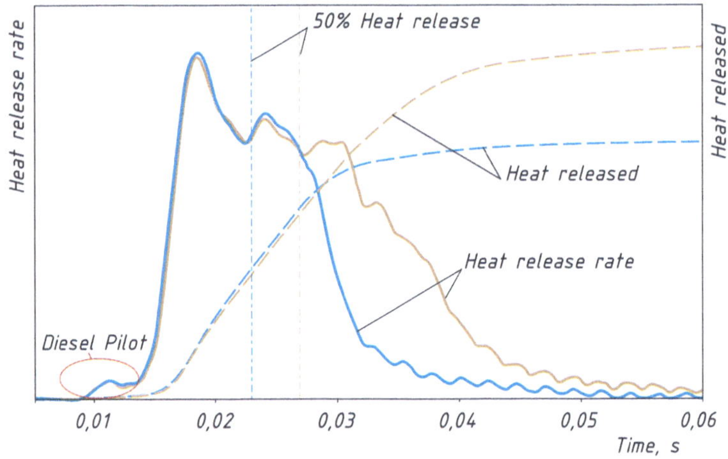

Fig. 9.5 Heat release characteristics in a laboratory setup simulating combustion chamber conditions at two different injection durations with one pilot injector and one ammonia pump-injector. Chamber pressure 11.6 MPa, temperature—860 K (adapted from [4])

proportion of pilot fuel, which is approximately 5% of the total heat supplied. Initially, the combustion of the pilot fuel is clearly visible.

After the interaction of the diesel combustion flame with the ammonia jet, its intensive combustion begins. At the initial stage, regardless of the injection duration, the combustion pattern differs only slightly. With an increase in injection time, the proportion of heat released increases.

To determine the optimal injection parameters for both fuels to achieve high efficiency and low emissions, WinGD carried out numerous variations of the settings of the SCE52 laboratory engine.

One of the key settings for a dual-fuel ammonia engine is the pilot fuel and ammonia injection sequence, which significantly affects its performance and emissions levels. Figure 9.6 shows the qualitative effect of the pilot fuel injection start angle with a fixed ammonia injection angle.

The study conducted at reduced engine load with an increased proportion of pilot fuel. The results showed that there is a clear optimum setting in terms of reducing NH_3 slip and NO_x (including N_2O) emissions. The start time of the pilot fuel injection is a very influential tuning parameter, with its optimum range being wide and stable, ensuring stable operation over a wide range of conditions. These results highlight the significant optimization potential for reducing engine emissions and further confirm the potential of the ammonia dual-fuel engine as a highly efficient, low-emission solution.

Figure 9.7 shows the cylinder pressure and heat release rate curves corresponding to 25, 50, 75 and 100% engine load, respectively, obtained in tests on a 4T50ME-X laboratory engine carried out at the MAN B&W Research Centre in Copenhagen. For comparison, typical pressure and heat release rate curves in the cylinder for a similarly sized commercial engine running on marine diesel fuel

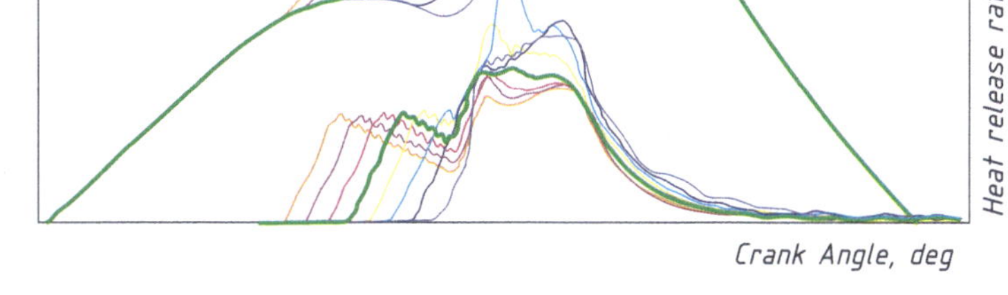

Fig. 9.6 Effect of pilot fuel injection angle on the nature of pressure change and heat release rate in the working cylinder of a low-speed laboratory engine SCE52 running on ammonia (adapted from [4])

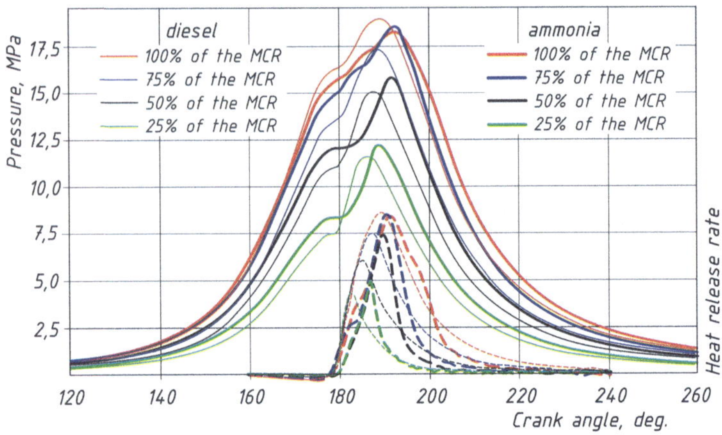

Fig. 9.7 Change in pressures and heat release rates in a laboratory two-stroke low-speed engine MAN B&W 4T50ME-X when operating on ammonia and diesel fuel at different loads (adapted from [6])

presented. During the studies on ammonia combustion in the engine, the share of energy from the pilot fuel was 3–6% of the total energy at full load.

The engine injection system designed to provide ammonia injection duration comparable to that of a similar diesel engine.

An analysis of the curves shows that the pilot fuel portion, which injected shortly before the ammonia injection, self-ignites almost instantly. This contributes to the efficient ignition of ammonia, as indicated by a sharp increase in pressure and heat release rate immediately after TDC. At the same time, the pressure increase does not exceed 4.0 MPa, which is typical for diesel engines of this type. The decrease in the heat release rate also occurs quite quickly, which indicates efficient combustion of the fuel. Thus, the heat release curves show that ammonia burns at rates that are typical of traditional fuels. This suggests that the combustion process occurs in a controlled mixing mode, in which all the fuel burned during the injection. This also indicated by the fact that as the load increases, the combustion duration increases, since the increase in cyclic supply achieved by increasing the injection time. The fuel-air ratio at a load of 25% is about 3, with an increase in load to 100% it decreases to 2.

Tests have shown that the stability of the combustion process from cycle to cycle when running on ammonia is similar to that when running on diesel fuel. For an engine running on NH_3 at 50% load, the relative deviation of the maximum cylinder pressure is about 1.26%. For diesel fuel under similar operating conditions, it is 1.09%.

In general, the main characteristics of the working process on ammonia are quite similar to similar characteristics for engines running on traditional fuels. This also indicated by the relatively high efficiency of the engine running on ammonia, which is comparable to similar indicators for engines running on traditional fuels (Fig. 9.8).

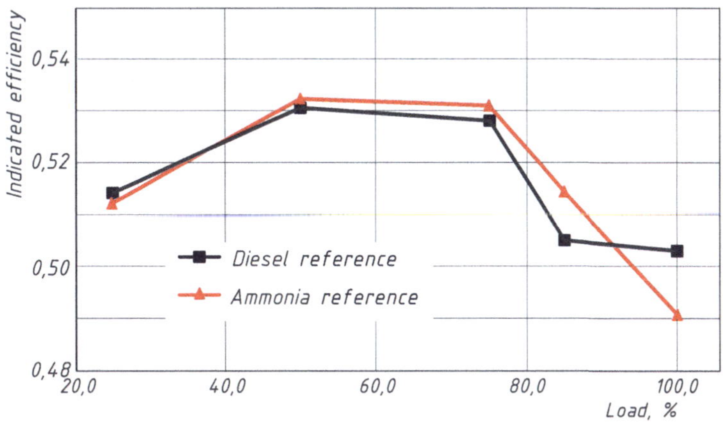

Fig. 9.8 Indicator efficiency of the working process in the range of operating modes of the laboratory two-stroke low-speed engine MAN B&W 4T50ME-X when running on ammonia and diesel fuel (adapted from [6])

9.2.4.2 Environmental Performance of Low-SPEED engines with Direct Ammonia Injection

The absence of sulfur and carbon in ammonia fuel makes it quite environmentally friendly, reducing the volume of emissions of sulfur compounds, soot and carbon monoxide to a minimum, which determined by the amount of pilot fuel used.

At the same time, the use of ammonia fuel leads to the presence of such harmful substances in the exhaust gases as ammonia itself (NH_3), unburned in the engine, dinitrogen oxide (N_2O) better known as nitrous oxide and nitrogen oxides (NO_x).

By analogy with other fuels, the presence of unburned ammonia in the exhaust gases called its slip.

Ammonia slip or unburned ammonia entering the exhaust gases can occur in several ways. These are incomplete combustion, for example, due to a lack of oxygen in certain areas of the combustion chamber, flame extinction near cold walls or in confined spaces, and evaporation of ammonia from the sub-needle volumes of the injectors. The sub-needle volume factor can minimize by choosing the optimal design of the nozzle tip. The low surface-to-volume ratio of the combustion chamber, typical of low-speed engines, and the strong swirl of the charge in the working cylinder help to keep the flame away from the walls. This helps to minimize the slowdown of reactions in the near-wall areas of the combustion chamber. In combination with the sufficiently long time allocated to the combustion process in low-speed engines, even relatively slowly burning ammonia has time to burn out almost completely. For the tested designs of two-stroke low-speed engines with internal mixture formation, the ammonia content in the exhaust gases did not exceed 1–3 ppm. It should note that at the time of the book, there were no current restrictions on ammonia emissions with exhaust gases.

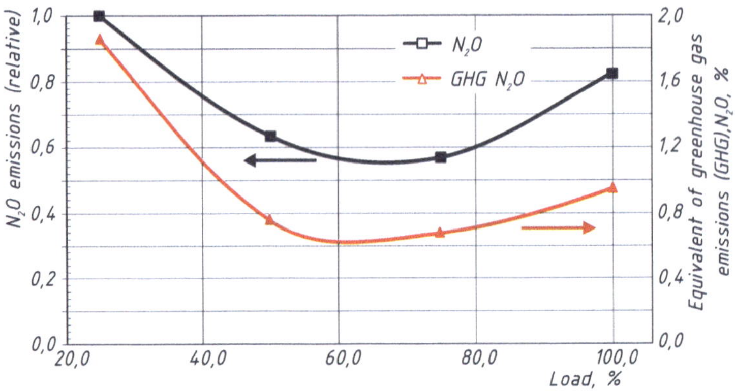

Fig. 9.9 Changes in nitrous oxide emissions and its contribution to the overall greenhouse effect depending on the engine load of the MAN B&W 4T50ME-X when running on ammonia (adapted from [7])

Nitrous oxide, also known as laughing gas, is a greenhouse gas with a Global Warming Potential (GWP) equivalent to 265 times that of CO_2. For this reason, its emissions must keep as low as possible. Otherwise, the effect of using carbon-free fuel will significantly reduce or even eliminated. In the tested engine designs, the nitrous oxide content in the exhaust gases did not exceed 5 ppm. Figure 9.9 shows how nitrous oxide emissions and their contribution to the overall greenhouse effect change depending on the engine load relative to maximum emissions at minimum load.

It is clearly visible that emissions are minimal in the main operating modes, which is especially important to understand when operating ammonia engines in order to minimize their impact on the environment.

Nitrogen oxides. In general, tests have shown that using ammonia as a fuel results in lower NO_x emissions in engine exhaust gases compared to diesel fuel (Fig. 9.10). Since nitrogen oxides are a common problem for all types of fuels used, methods for their reduction well established, as has already been discussed in the relevant chapters. If necessary, these methods can also apply to engines running on ammonia.

Particulate Matter. Although ammonia does not contain any carbon, engines that run on it can emit particulate matter, primarily ammonium salts, into the atmosphere. These particles can form because of the reaction of ammonia with other exhaust components. A small portion of the particulate matter may be carbon-based, which is formed a result of the combustion of pilot fuel.

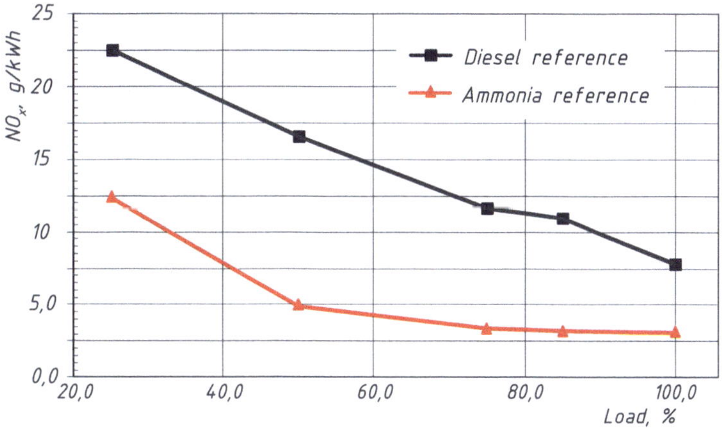

Fig. 9.10 Changes in nitrogen oxide emissions depending on engine load for a MAN B&W 4T50ME-X when running on ammonia and diesel fuel (adapted from [7])

9.3 Use of Ammonia as a Fuel for Marine Medium- and High-Speed Four-Stroke Internal Combustion Engines

As already noted, ammonia is a gas under normal conditions, which, if the necessary precautions are observed, can be used as a fuel for engines with external mixture formation. For four-stroke engines, the most widely used schemes for external mixture formation were previously developed for gas-diesel engines, with the supply of gaseous ammonia into the intake port of the cylinder covers under low pressure. With this approach, changes made to the design of gas engines related to the properties of ammonia. For example, due to the lower calorific value of ammonia compared to natural gas, the cyclic supply of ammonia must increase approximately twice, and the gas valves must ensure this supply.

In addition, the corrosive activity of ammonia leads to significant changes in some engine components that exposed to it. Since ammonia is significantly more toxic than natural gas, at the system level, the main difference is the addition of an ammonia emission control system. To achieve this, additional safety features added to the proven dual-fuel engine technology, taking into account the unique properties of ammonia.

Examples of the implementation of ammonia combustion technology in a four-stroke engine with external mixture formation operating according to the Otto cycle are the W25 engines from Wärtsilä (Fig. 9.11a) or the L28ADF from IHI Power Systems (Fig. 9.11b).

In this case, for example, Wärtsilä offers a complete solution, which in addition to the engine itself includes a fuel supply system, an ammonia emission prevention system, an exhaust gas aftertreatment system and a hybrid battery system.

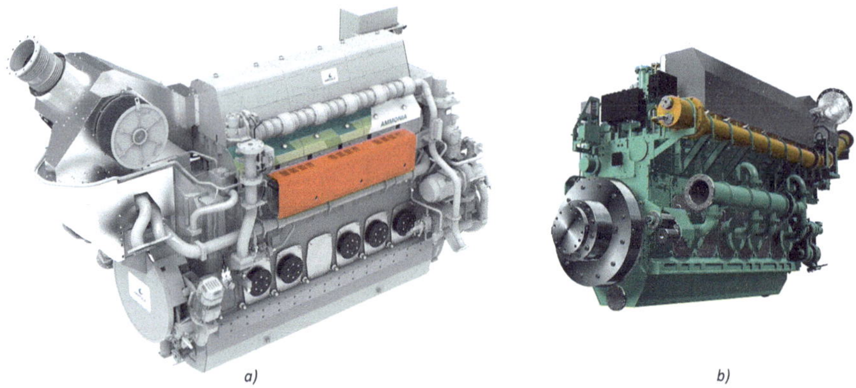

Fig. 9.11 Dual-fuel, four-stroke marine engines designed to operate on gaseous ammonia: 6L25DF from Wärtsilä (**a**) and 6L28ADF from IHI Power Systems (**b**) (adapted from [8, 9])

The supply of ammonia to the intake port of this class of engines carried out similarly to the previously considered gas engines, with external mixture formation. An electronically controlled metering valve regulates the amount of gaseous ammonia supplied to the intake ports of the cylinder cover during the intake stroke. Gaseous ammonia mixed with the air required for combustion and then enters the working cylinders as part of the mixture.

Ammonia supplied to the metering valves from the supply pipeline, the temperature of which constantly monitored and maintained above the dew point to prevent condensation of ammonia. In case of too low temperature of gaseous ammonia or gas pipeline, the automation system initiates the necessary equipment protection.

For injection of the pilot portion, the standard liquid fuel supply system usually used, which also ensures engine operation in diesel mode. The share of pilot fuel in the total energy balance of the engine operating process usually does not exceed 5–10% over the entire range of operating modes.

In Fig. 9.12 shows the nature of the change in the working process and heat generation parameters during operation of the 6L28ADF engine on the diesel cycle, as well as on the Otto cycle on natural gas and ammonia.

The Wärtsilä 25DF engine equipped with a standard Common Rail fuel system can switch to diesel cycle operation either completely or in individual cylinders, which increases reliability without compromising safety.

Tests have shown that the efficiency of engines running on ammonia according to the Otto cycle is no different from the efficiency of dual-fuel engines on liquefied natural gas ($\approx 46\%$) and is slightly higher than the efficiency of an engine operating on the diesel cycle.

During the tests, ammonia proved itself as an effective fuel for engines operating on the Otto cycle. It has high detonation resistance and ensures stable combustion in a wide range of engine settings. Due to the flexibility of automatic control technologies, the effect of the low combustion rate of ammonia can level.

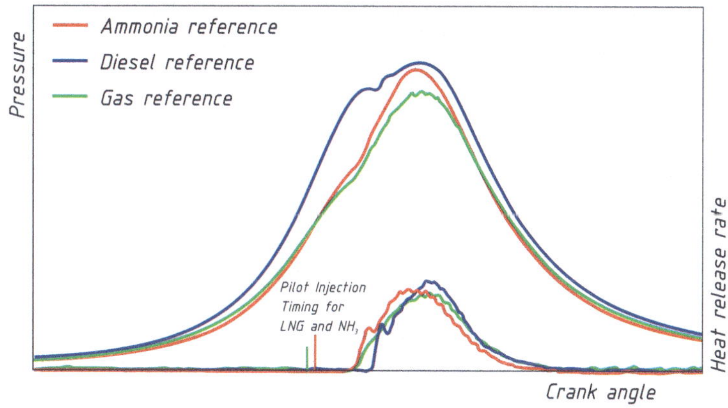

Fig. 9.12 Pressure and heat release rate behavior in the 6L28ADF dual-fuel engine from IHI Power Systems when running on natural gas, ammonia and diesel fuel (adapted from [9])

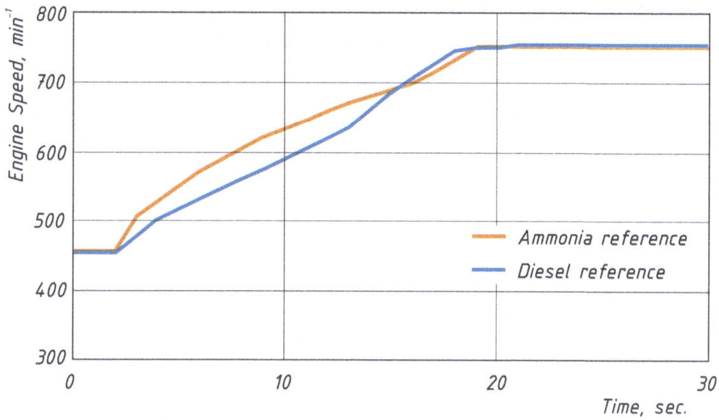

Fig. 9.13 Acceleration dynamics in diesel mode and ammonia mode of operation of the dual-fuel engine 6L28ADF from IHI Power Systems when operating according to the screw characteristic (adapted from [9])

The dynamic characteristics of engines running on ammonia differ little from similar characteristics of the same engine when it is running on traditional fuel.

In Fig. 9.13 presents the results of tests of the 6L28ADF engine installed on a tugboat and operating according to the propeller characteristic.

When using diesel as a pilot fuel, the greenhouse gas emissions of the engines are below 50 g/(kW × h) CO_2 equivalent, calculated from tank to wake. Ammonia slips, for example, for the Wärtsilä 25DF engine does not exceed 50 ppm, which does not pose a threat to the environment or safety.

A standard SCR system used to control NO_x emissions, the only difference being that in addition to NO_x, unburned ammonia in the exhaust gas will act as a reactant, and the external reactant supply must adjusted accordingly. For this purpose, the system is equipped with sensors and analyzers that support advanced combustion balance calculations to determine the correct reactant dosing level. The engine and exhaust gas cleaning system are optimized together, which allows nitrogen oxide emissions to be reduced to the corresponding IMO Tier III level. When running engines with external mixture formation, the concentration of gaseous ammonia in the crankcase increases due to blow-by through the piston rings. Therefore, a slight vacuum maintained in the crankcase by an exhaust fan when operating on this fuel. This is an additional safety measure and to prevent the formation of odors in the engine room. The total flow rate of crankcase gases is usually less than 0.5% of the total flow rate of engine exhaust gases.

Ammonia can penetrate into the lubricating oil, but detailed laboratory tests have not revealed any changes in its properties due to this. The smell of ammonia can persist in engines for up to several hours, so before opening the crankcase hatches, it is necessary to wait a sufficiently long time, continuing to ventilate the crankcase space.

Before starting maintenance or repair, it is important to ensure that systems that may contain ammonia fuel or ammonia vapors are properly purged. Personnel engaged in maintenance and repair should provide with portable ammonia detectors and easily accessible breathing masks. In addition, the engine room must effectively ventilate, and ammonia detection systems must install at various points.

References

1. Bilousov, I.V.: Internal combustion engines of modern marine vessels. Textbook for higher maritime educational institutions. Bilousov, I.V., Bulgakov, M.P.: Astroprint publishing house – Odesa, 2024. 512 p. ISBN 978–617-7797-54-7
2. Jeppesen, M.: Ammonia Engine development update. Greener Shipping Summit Athens, November 2024. 21 p
3. Bilousov I. Modern Marine Internal Combustion Engines. A Technical and Historical Overview. Bilousov, I., Bulgakov, M., Savchuk, V.: Springer Series on Naval Architecture, Marine Engineering, Shipbuilding and Shipping (NAMESS, volume 8). Springer Nature. 2020. 385 p
4. WinGD's ammonia engine technology for the new X-DF-A family New Engine Concepts & Systems. Paper No. 329. Graziano, B., Topaloglou, S., Purayil, A., Avvisati, I., Christen, C., Bouwstra, A., Rotz, B., Heizer, V., Rohner, L.: Zürich, CIMAC Congress 2025. 17 p
5. The latest technological development of the J-ENG UE engine for zero emission and digital transformation New Engine Developments – Diesel. Paper No. 99. Imanaka, K., Kinoshita, Y., Edo, K.: Busan, CIMAC Congress 2023. 12 p
6. High efficiency ammonia combustion in the MAN B&W LGI-A marine two-stroke engine. Fuels – Alternative & New Fuels. Paper No. 82. Hult, J., Sjöholm, J., Westlye, F.R., Ringsted, S.B., Kaltoft, J., Mayer, S.: Zürich, CIMAC Congress 2025. 13 p
7. Ludwig, C.: Ammonia engine development MAN B&W ME-LGIA. Future in the making. MAN Energy Solutions, March 2025. 16 p
8. Ammonia-powered future: introducing Wärtsilä 25. New Engine Concepts & Systems. Paper No. 528. Krook, T., Engsbo, R., Torrkulla, J., Hattar, C.: Zürich, CIMAC Congress 2025. 10 p
9. Development of an ammonia-fueled engine (28ADF) for future marine industries Fuels – Alternative & New Fuels. Paper No. 214. Masuda, Y., Mashima, Y., Kazama, S., Naruse, H., Aiba, K., Hirose, T.: Zürich, CIMAC Congress 2025. 10 p

Chapter 10
Use of Hydrogen as a Fuel for Marine Internal Combustion Engines

10.1 Characteristics of Hydrogen as a Motor Fuel

The attractiveness of hydrogen as a fuel lies in its unique physical and chemical properties, which can provide virtually zero carbon emissions and high efficiency. When it burns, it produces only water (H_2O), making it an ideal fuel for achieving zero CO_2 emissions.

Of all fuels, hydrogen has the highest specific calorific value, which is 2.8 times higher per unit mass than diesel fuel. This ensures high efficiency and the potential for high power. When mixed with air, hydrogen ignites over a very wide range of concentrations, from 4 to 75% by volume, which allows the engine to operate on very lean mixtures. The high flame propagation speed of hydrogen ensures rapid and complete combustion, which increases thermal efficiency and increases engine power.

At the same time, the wide flammability range and high flame propagation speed make hydrogen a complex fuel prone to abnormal combustion phenomena. In addition, despite its high calorific value, hydrogen has a low volumetric energy density, which makes it unsuitable for ocean-going vessels and other large mobile engines.

Storing hydrogen on board is a complex task in itself, which has three main solutions:

- creating high pressure for storage in a gaseous state;
- liquefaction, which is achieved at atmospheric pressure by cooling to a temperature of −253 °C;
- storage in metal hydrides or other solid molecular structure.

Storing hydrogen in a compressed state is the most common and well-proven method in industry. Hydrogen is stored under very high pressure, about

I. Bilousov, R. Varbanets, *Marine Internal Combustion Engines and Environmental Protection Issues*, Springer Series on Naval Architecture, Marine Engineering, Shipbuilding and Shipping 27,
https://doi.org/10.1007/978-3-032-20940-5_10

35.0–70.0 MPa, which provides a density of 23.3–39.3 kg/m^3. This is about two times less than when stored in a liquefied state. In addition, special tanks needed for storage that can withstand high pressures. There are several main types of such tanks:

- all-metal;
- semi-composite - metal tanks covered with an outer shell made of composite materials (such as carbon fiber) in the form of a rim or a solid outer layer;
- all-composite.

Metal tanks can withstand high pressures. They are relatively cheap, but they have a significant weight, which is several times greater than the weight of the hydrogen stored in them.

Semi-composite and all-composite tanks are much lighter, they can operate under higher pressure, but their cost is significantly higher than that of metal tanks.

This method is hardly suitable for on-board storage, since it does not allow making any significant reserves that can ensure the necessary autonomy of the vessel.

The basis of the technology for storing hydrogen in liquid form is its liquefaction by cooling to a boiling point of −253 °C. The advantage of liquefaction is the ability to achieve a high density of hydrogen at atmospheric pressure, which is 775 times higher than that of gaseous hydrogen. However, even with this storage method, to place hydrogen reserve equivalent to traditional fuel on board, 7.2 times more volume will needed, which significantly reduces the space of the ship's cargo spaces. One cubic meter of liquefied hydrogen weighs just over 70 kg, while traditional fuel in the same volume will weigh 860–1050 kg.

Liquefaction and storage of hydrogen in liquid form under high pressure is an energy-intensive process. Thus, approximately 18% of the initial energy of the hydrogen spent during liquefaction.

To store hydrogen in a liquefied state, it is necessary to solve the problem of its evaporation due to external heat influx. Evaporation has two side effects: firstly, it is an increase in pressure inside the tank, and secondly, the loss of energy spent in the process of liquefying hydrogen. Reducing heat transfer through the tank is a key solution to prevent evaporation. The shape of the tank plays a role here, or rather the volume-to-surface ratio, which should maximum. Spherical tanks meet this requirement best. Thermal insulation material is also crucial for creating resistance to heat transfer. For modern well-insulated spherical tanks, the boiling rate is less than 0.1% per day.

This method of storing hydrogen on board can be acceptable for ships making relatively short voyages between permanent ports, in which the necessary infrastructure for bunkering can create. For example, these can be inland waterway vessels or ferries serving regular routes.

There are various methods of storing hydrogen in solid materials, but the most widely used are storage methods based on metal and boron hydrides.

Metal hydrides store hydrogen by chemically binding it to the metal. They have unique properties to adsorb and release hydrogen at different temperatures and pressures, so each of the metal hydrides has different characteristics.

The main solid-state hydride of an elemental metal is magnesium hydride. Since magnesium is widely available, magnesium hydrides are common and attractive, allowing hydrogen to be stored with a density of 1.45 g/cm^3 and a hydrogen storage capacity of 7.6% by weight. However, the process of hydrogen storage and release in magnesium hydride is very slow. To achieve acceptable hydrogenation/dehydrogenation rates, it is necessary to maintain the temperature in the reactor at about 300 °C, which requires additional energy costs and reduces the efficiency of the process as a whole. An alternative to metal hydrides are solid-state hydrogen storage systems based on boron. Boron classified as a metalloid and has both metallic and non-metallic properties. Boron-based storage systems have a hydrogen storage capacity of more than 10% by weight due to the lightweight boron with a high storage capacity. Using hydrides allows storing large amounts of hydrogen in a relatively small volume, but such reactors have a large mass, which makes their use in transport energy less attractive.

10.2 Use of Hydrogen as a Fuel for Marine Low-Speed Two-Stroke Internal Combustion Engines

At the time of writing this book, research and development work on the creation of low-speed dual-fuel engines capable of running on hydrogen was still in progress. Only one manufacturer—J-ENG in a consortium with Kawasaki Heavy Industries, Ltd. and Yanmar Power Technology Co., Ltd. have declared their readiness to begin supplying dual-fuel hydrogen engines of the UEC35LSGH series from 2026.

Considering that, hydrogen enters the engine in a gaseous state, the hydrogen engines based on existing models of low-speed gas-diesel engines, which are in the production program of each manufacturer [1, 2].

Theoretical calculations and the first experimental studies have shown that the concept of preliminary mixing of hydrogen with an air charge on the compression stroke, typical for low-pressure gas engines, is of little use for this type of engine.

The high reactivity of hydrogen will contribute to its premature self-ignition, with subsequent abnormal combustion in the working cylinder.

In Fig. 10.1 shows the results of tests of the laboratory engine 4S50ME-T conducted at the MAN research center in Copenhagen together with Mitsui. One cylinder of this engine converted to operate on hydrogen fuel, with pilot ignition of the hydrogen-air mixture. Hydrogen supplied to the working cylinder using two gas injectors, which allowed both its preliminary injection during the compression stroke and directly at the end of the compression immediately after the injection of the pilot fuel. Understanding that the high temperature and pressure typical of engine operation at full load can provoke abnormal combustion, studies

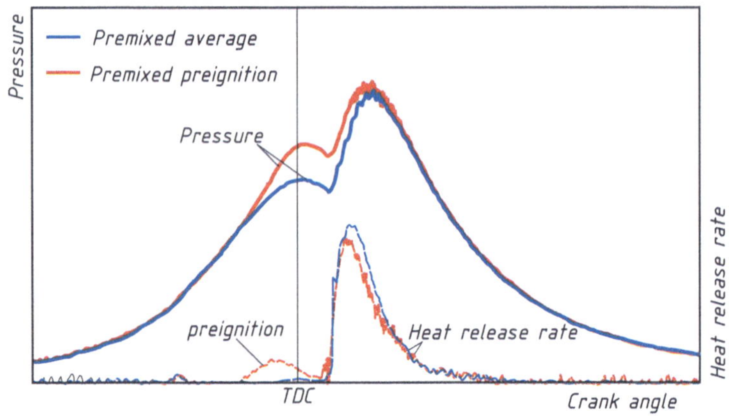

Fig. 10.1 The nature of the change in pressure and heat release rate in the laboratory dual-fuel engine 4S50ME-T when operating on hydrogen at a load of 25% of the MCR in the mode of preliminary mixing of 5% hydrogen with air (adapted from [3])

of mixture formation with preliminary mixing began with a load of 25% of the MCR.

The share of hydrogen in the total energy balance during the tests did not exceed 5%, the rest was diesel fuel supplied by the standard engine injection system. However, even with such low hydrogen consumption, during the thermodynamic cycle, its premature ignition and further abnormal combustion in the working cylinder, accompanied by detonation, periodically occurred.

This study showed that the pre-mixing method is of little use for low-speed engines, which have an order of magnitude more time allocated for the compression and combustion processes than medium- and high-speed engines.

The most rational approach to organizing the working process of an engine running on hydrogen fuel is the injection of gaseous hydrogen into the combustion chamber immediately after the pilot fuel supplied there.

This mixture formation scheme, previously implemented in ME-GI engines from MAN, can be adapted to the use of hydrogen with minor changes.

Figure 10.2 shows a comparison of the working process of the laboratory engine 4S50ME-T when running on diesel fuel and on gaseous hydrogen.

The share of hydrogen in the total energy balance of the working process reached 90%, without any anomalies.

The obtained working process did not differ significantly from a similar process when using natural gas as fuel, and correlated well with the diesel working process.

Tests have shown that after injection, hydrogen burns faster than diesel fuel, which leads to less afterburning. Therefore, hydrogen fueled engines can potentially have a higher thermal efficiency than base diesels. At the same time, their combustion temperature is higher, which contributes to the formation of more thermal NO_x, and this requires additional measures to reduce them.

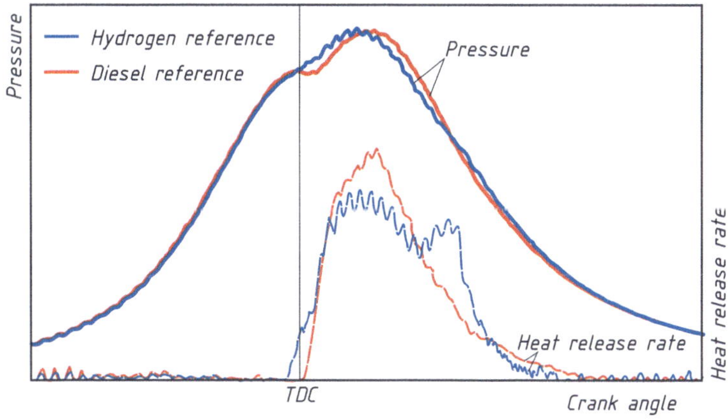

Fig. 10.2 The nature of the change in pressure and heat release rate in the laboratory dual-fuel engine 4S50ME-T when operating on hydrogen and diesel fuel at a load of 100% of the MCR in the mode of direct injection of hydrogen into the combustion chamber (adapted from [3])

10.3 Use of Hydrogen as a Fuel for Marine Medium- and High-Speed Four-Stroke Internal Combustion Engines

As with two-stroke engines, four-stroke hydrogen-powered marine engines are still in the development or trial operation phase. For example, a tugboat equipped with two 12DZD dual-fuel hydrogen engines with a power of 2000 kW each from ABC is undergoing trial operation in the port of Antwerp-Benges.

Almost all major manufacturers of medium- and high-speed marine engines are working in this direction. Companies such as ABC, Wärtsilä, MAN Energy Solutions, Caterpillar (MaK) and Rolls-Royce (MTU) are not only conducting research, but also already have specific products or plans to bring hydrogen versions of their engines to market.

Developers of four-stroke hydrogen engines prefer the concept of external mixture formation, with hydrogen supplied into the cylinder cover's intake port to mix with the incoming air. This scheme been well tested on gas engines and has the necessary element base for its implementation, which can be adapted for hydrogen with minimal costs.

The possibilities of creating both purely hydrogen engines with spark ignition and dual-fuel engines with ignition from a pilot portion of liquid fuel are considered.

In spark ignition engines, even with relatively large cylinder diameters, a scheme with direct ignition of the combustible mixture from an electric spark can effectively used. The high reactivity of hydrogen, a wide range of its flammability limits and a high flame propagation speed allow for the effective combustion of even relatively lean mixtures without resorting to prechamber-torch ignition. This significantly simplifies the engine design compared to other engines using gaseous fuels.

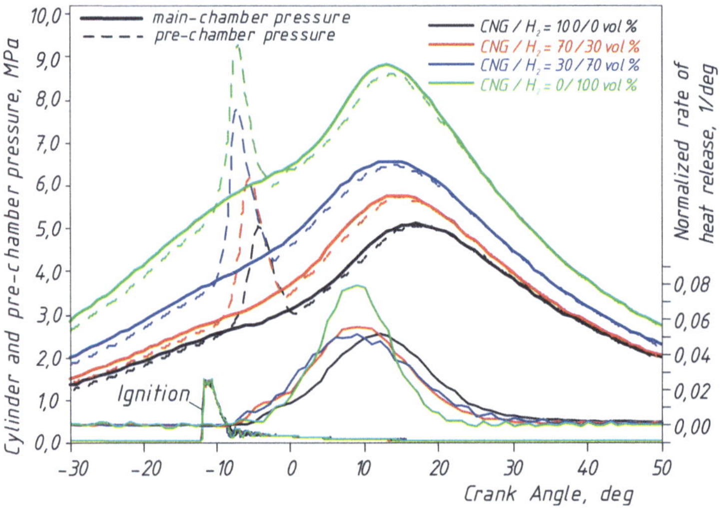

Fig. 10.3 The nature of the change in pressure and heat release rate in a laboratory gas engine with dimensions of 30/43 cm, when operating on a mixture of hydrogen and propane at an average effective pressure of 1.1 MPa (adapted from [4])

At the same time, most studies on the use of hydrogen in four-stroke engines performed on existing samples of gas-diesel or gas engines. The latter were mostly equipped with prechamber ignition. Many studies used hydrogen as an additive to gaseous fuel.

Figure 10.3 shows the results of tests of a laboratory gas engine with a cylinder diameter of 30 cm and a piston stroke of 43 cm, with a rotation speed of 750 min^{-1} and a cylinder power of 573 kW [4]. During the tests, a mixture of propane and hydrogen used as fuel, with the volume fraction of hydrogen, gradually increasing up to 100%. At a given compression ratio of 10.3:1, to improve the stability of the engine and prevent detonation, as the proportion of hydrogen increased, the fuel-air ratio was increased. This achieved by increasing the boost pressure.

In dual-fuel engines, a pilot portion used to ignite the hydrogen-air mixture. The standard engine injection system, which also handles liquid fuel operation, used to inject this pilot portion. The share of hydrogen in the overall energy balance of the working process is usually 90–95%.

External mixture formation for hydrogen engines has several distinct features, despite its simplicity and regardless of how combustion activated. The high combustion rate and low ignition energy of hydrogen make the homogeneous mixture very sensitive to uncontrolled ignition under pressure. To combat detonation, several key strategies used in hydrogen-powered engines with external mixture formation:

- the use of a very large excess of air, i.e. lean combustion at a low concentration of hydrogen in the mixture. This allows to significantly slowing down the flame

propagation rate and peak temperatures in the combustion chamber, which prevents abnormal self-ignition of the fuel-air mixture, leading to detonation;
- charge parameter control. Reducing the temperature of the supplied air directly reduces the temperature of the charge, which reduces the risk of detonation. Too high a boost pressure can lead to an excessive increase in temperature and pressure at the end of compression, which also increases the risk of detonation. Careful control of boost and air-cooling allows maintaining parameters at a safe level;
- charge cooling by injecting water both directly into the combustion chamber and into the intake receiver is an effective way to combat detonation. As shown in the previous chapters, water has a high heat capacity, and its evaporation in the cylinder absorbs a large amount of heat. In addition, water acts as diluents, further reducing peak temperatures in the combustion chamber, which also helps to reduce NO_x emissions;
- optimization of the design and settings, in particular the shape of the combustion chamber, for unimpeded flame propagation without the formation of too hot or stagnant zones in which detonation can occur. The choice of the ignition advance angle allows the combustion process to shift further from TDC, which reduces peak pressures and temperatures in the combustion chamber. Increasing the excess air is the most common method of stabilizing hydrogen combustion, which makes the thermodynamic layout of the engine and its turbocharging system a difficult task. To operate an engine on pure hydrogen, a significant increase in the excess air ratio is required to compare to operation on traditional fuels. A high air mass flow can achieve by increasing the boost pressure, but this will lead to a significant decrease in the exhaust gas temperature and a decrease in turbine power. Thus, in an engine running on pure hydrogen, single-stage turbocharging is associated with difficulties associated with the combination of a large compressor and a small turbine.

Figure 10.4 shows an increase in the excess air ratio, boost pressure, and a decrease in exhaust gas temperature with increasing volume fractions of hydrogen for a laboratory medium-speed engine with dimensions of 30/43 cm.

High excess air coefficients have a positive effect on reducing NO_x emissions and limiting the increase in cylinder pressure.

The indicated efficiency of the engine when running on pure hydrogen increased by 2% compared to running on pure natural gas (Fig. 10.5). The lower part of the diagram shows the pumping losses, which increase due to a decrease in the enthalpy of the exhaust gases, which is associated with a decrease in their temperature.

Figure 10.6 shows the measured emissions of the engine operating at constant load and speed. When running on pure hydrogen, virtually zero carbon dioxide observed. The small amount of carbon monoxide is apparently due to the combustion of lubricating oil in the working cylinder. Zero NO_x emissions observed when running on pure hydrogen.

The proportion of unburned hydrogen in the exhaust gases increased with increasing hydrogen content in the mixture. When running on pure hydrogen, it

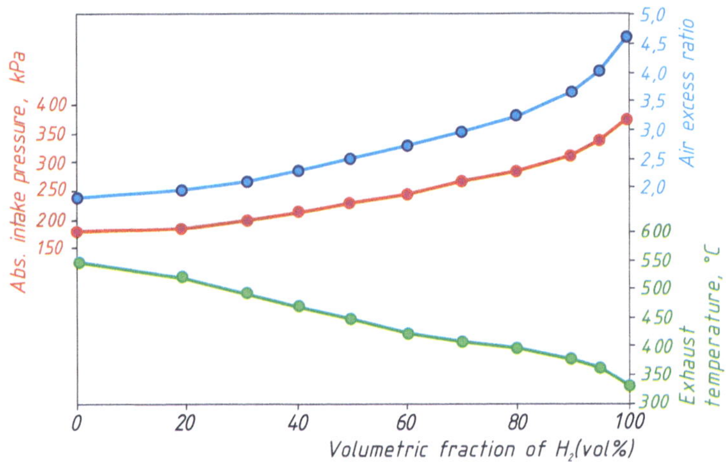

Fig. 10.4 Change in the excess air coefficient, boost pressure and exhaust gas temperature with an increase in the volume fraction of hydrogen in the fuel mixture for a laboratory medium-speed engine with dimensions of 30/43 cm (adapted from [4])

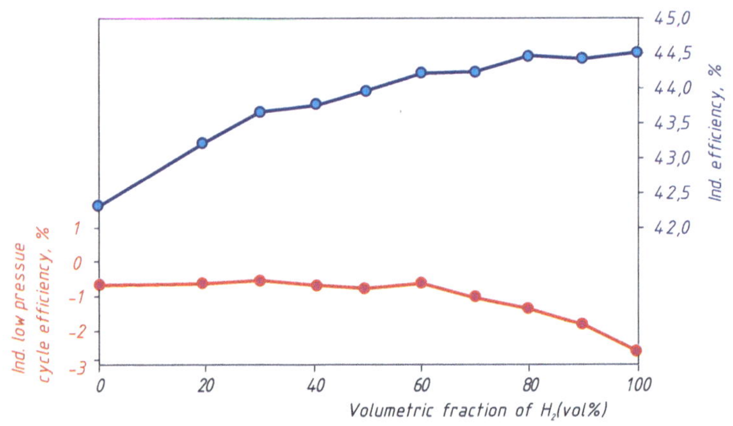

Fig. 10.5 Change in indicator efficiency and pumping losses with increasing volume fractions of hydrogen in the fuel mixture for a laboratory medium-speed engine with dimensions of 30/43 cm (adapted from [4])

reaches values of 2.5 g/(kW × h) (or ~ 1800 ppm). Carbon monoxide emissions remain virtually constant until the hydrogen content in the mixture exceeds 90%, after which they begin to decrease virtually to zero. Total hydrocarbon emissions remain constant until the hydrogen content in the mixture reaches 60%, and then begin to decrease.

Thus, the presented data confirm the prospects of using hydrogen as an environmentally friendly fuel, ensuring high efficiency and environmental friendliness of engines operating on it.

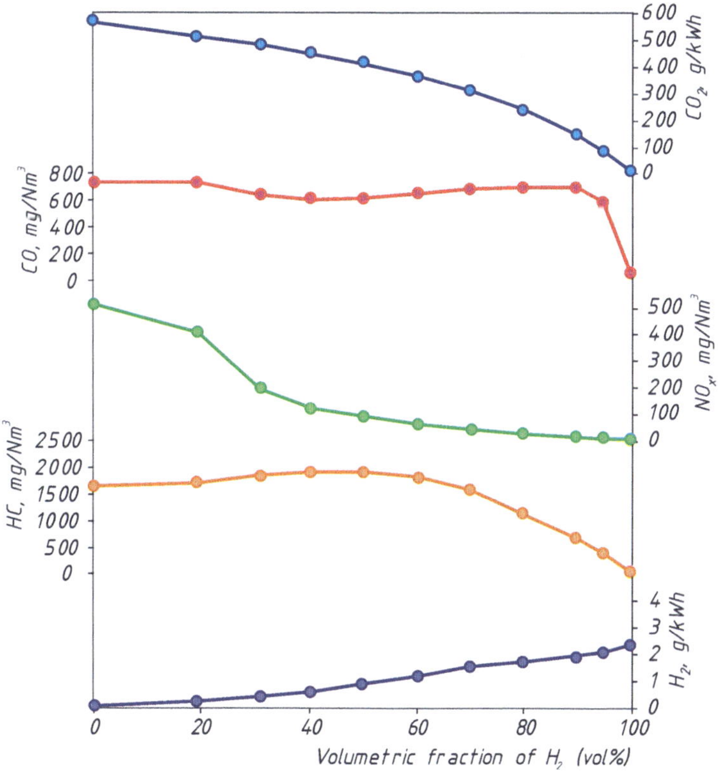

Fig. 10.6 Nature of changes in harmful emissions in a laboratory gas engine with dimensions of 30/43 cm, operating at a constant load and speed on a mixture of hydrogen and propane (adapted from [4])

References

1. Bilousov, I.V.: Internal combustion engines of modern marine vessels. Textbook for higher maritime educational institutions. Bilousov, I.V., Bulgakov, M. P.: Astroprint publishing house – Odesa, 2024. – 512 p. ISBN 978-617-7797-54-7
2. Bilousov, I.: Modern marine internal combustion engines. A technical and historical overview. Bilousov, I., Bulgakov, M., Savchuk, V.: Springer Series on Naval Architecture, Marine Engineering, Shipbuilding and Shipping (NAMESS, volume 8). Springer Nature. 2020. 385 p. ISBN 978–3–030-49751-4
3. Verification of a CFD model on hydrogen combustion in a large marine two-stroke engine. Simulation technologies, digital twins and complex system simulation. Paper No. 41. Abe, Y., Suyama, T., Ishibashi, R., Sjöholm, J., Ringsted S.: Zürich, CIMAC Congress 2025. 14 p
4. Assessment of combustion concepts and operational limits of net-zero carbon fuels. New Engine Developments – Alternative Fuels & Other New Engine Concepts. Paper No. 105. Schlick, H., Kammerdiener, T., Murakami, S., Segura Carrasco, M. I. Figer, G., Malin, M., Wermuth, N.: CIMAC Congress, Busan, 2023. 15 p

Chapter 11
Use of Biofuels for Marine Internal Combustion Engines

The term fuel of biological origin or biofuel usually refers to combustible substances of organic origin that can use as fuel for internal combustion engines [1, 2]. These substances can differ significantly in the method of production, composition, properties and aggregate state [2, 3]. Since the consideration of methods for obtaining such fuels is beyond the scope of this book, we will focus only on the key points that directly related to the use of these fuels in internal combustion engines [4].

Gaseous biofuels usually obtained by anaerobic fermentation in methane tanks or anaerobic columns in which special bacteria process organic waste (manure, food waste, wastewater, etc.), releasing biogas. The resulting biogas contains methane, carbon dioxide and other impurities that can removed during the purification process. Since biogas can contain sulfur compounds, it also subjected to desulfurization. Because of purification, a product obtained in which the methane concentration reaches 98%, which is why it called biomethane [2, 3]. In its composition and properties, biomethane is almost identical to natural gas. Therefore, its further use as a fuel for gas and gas-diesel engines is no different from the use of natural gas.

Alcohol fuels can also obtained from biological waste or specially grown crops containing sugar, starch or cellulose. The processing process includes fermentation of sugars and starches with yeast, and then distillation and purification of the resulting ethyl alcohol. Cellulose is pre-decomposed into individual sugar molecules (glucose) using acids or special enzymes. This process called hydrolysis. The sugars obtained in this way subjected to subsequent fermentation and distillation. The resulting alcohol called bioethanol, and can used in marine engines adapted for this type of fuel [3]. Dry distillation of cellulose and lignin can produce methyl alcohol (methanol), which today considered a promising fuel for marine engines, as discussed in the relevant chapters of the book.

Diesel fuels based on fatty acids of plant and animal origin. Currently, two fundamentally different methods of processing fatty acids into fuels with properties as

I. Bilousov, R. Varbanets, *Marine Internal Combustion Engines and Environmental Protection Issues*, Springer Series on Naval Architecture, Marine Engineering, Shipbuilding and Shipping 27,
https://doi.org/10.1007/978-3-032-20940-5_11

close as possible to diesel fuel have become widespread. These fuels have received the general name - biodiesel [2].

Hydrotreated Vegetable Oil (HVO) is a synthetic diesel fuel produced by hydrocracking or hydrogenating vegetable oils and animal fats with hydrogen. Hydrocracking breaks large molecules into smaller ones using hydrogen, while hydrogenation adds hydrogen atoms to the molecules. Most vegetable and animal oils are triglycerides that are suitable for processing. Raw materials for obtaining oils can be rapeseed, soybeans, sunflower, palm oil, animal fats, etc. In some sources, fuels obtained using this technology is called renewable diesel fuel.

Fatty Acid Methyl Esters (FAME) also called biodiesel. The raw material for the production of FAME is also vegetable oils, but the processing technologies and chemical composition of these two types of fuel are different. The basis for obtaining FAME is a chemical reaction known as interesterification. This reaction is a process in which fats (triglycerides) react with methanol to form fatty acid methyl esters and glycerol. Glycerol is a by-product of this process, which not formed during the production of HVO.

Regardless of the method of production, biodiesel fuels can use as independent fuels or in a mixture with traditional ones [3]. However, if their content in the mixed fuel is more than 7%, it recommended checking the compatibility of the materials of the gaskets and seals. It recommended avoiding contact of HVO and FAME with equipment and components made of materials such as bronze, brass, copper (Cu), lead (Pb), tin (Sn) and zinc (Zn). They can oxidize biofuels, which can lead to the formation of deposits.

According to ISO 8217:2017, marine fuels may contain biofuel additives that must meet the requirements of EN 14214 or ASTM D6751.

Biofuels have good autoignition characteristics [4], for the reason they increasingly being considered as pilot fuels for engines with pilot ignition of alternative fuels such as methanol and ammonia. In this case, it becomes possible to eliminate the emissions of anthropogenic carbon dioxide from the engines.

Large-scale use of biodiesel fuels as the main fuel for marine diesel engines is unlikely to expect due to limited resources for their production. However, some ships may use such fuels in the future as an alternative to petroleum fuels. The main advantage of this approach is that their use requires minimal adaptation of existing engines and marine fuel systems. It should note that biofuels usually contain a small amount of sulphur. Therefore, general lubrication recommendations for operation with maximum sulphur content of 0.1% apply to them. Careful monitoring of the cylinder condition and proper fuel cleaning are required. Ash content in the fuel and the presence of potential abrasive components that can lead to problems in the operation of the engine and its service systems should take into account. Potential problems with the use of biofuels mainly related to their poor stability during long-term storage, an increased risk of corrosion of fuel system components, as well as problems with their filtration and sealing materials. Some types of biofuels tend to absorb water, which increases the risk of microbial contamination of the fuel and increased corrosion [3]. In case of suspected microbial activity, planned measures to eliminate it should taken.

The flash point of biodiesel fuels must comply with SOLAS requirements and be at least 60 °C. The acid number should not exceed 2.5 mg KOH/g (ASTM D664 method).

References

1. Future ship powering options. Exploring alternative methods of ship propulsion. Royal Academy of Engineering, London, July 2013. 97 p. ISBN: 978-1-909327-01-6.
2. Hsieh C.C., Felby C.: Biofuels for the marine shipping sector. An overview and analysis of sector infrastructure, fuel technologies and regulations. University of Copenhagen, IEA Bioenergy Website. 2017. 84 p.
3. Marine biofuels. What to expect in the 2020s. Alfa Laval, 100003528-2-EN 2103. April 2021. 15 p.
4. Results of the Experimental Research of the Medium Speed Diesel Engine Work on Soybean Oil. Shvets, I., Hrabovenko, O., Dotsenko, S., Nesterenko, V.:Proceedings of 24th International Scientific Conference. Transport Means 2020. Pp. 671–675.

Chapter 12
Features of the Organization of Work Processes in Ship Engines Running on Alternative Fuels

12.1 Variable Compression Ratio Technology in Two-Stroke Low-Speed Gas-Diesel Engines

Using two different fuels in one engine is often associated with making compromises. Engine settings optimized for one type of fuel are usually not optimal for another type of fuel. Moreover, in most cases they are not optimal for different operating modes, even when operating on the same fuel.

The most difficult parameter to optimize for almost all engines is the geometric compression ratio, which specified by its design parameters and, in most cases, cannot change during their operation.

Since all marine engines must ensure operation at maximum operating power, the compression ratio optimized for this mode, based on the limitations of the maximum cycle pressure. At the same time, when the engine power decreases, the maximum cycle pressure decreases, and accordingly, the compression ratio can increased to achieve higher fuel efficiency. Considering that approximately 60–90% of the time, marine engines operate in modes other than MCR, the ability to change the compression ratio in accordance with the operating mode is a serious tool for optimizing the efficient and environmental performance of the engine. For dual-fuel engines, the ability to change the compression ratio is an even more pressing issue, since different fuels have different detonation resistance. For example, to operate on gas fuel, especially when feeding it to the working cylinder during the compression stroke, the compression ratio must reduce, while, for example, when operating on ammonia, it can significantly be increased.

It should note that the seemingly obvious ability to control the compression ratio by changing the closing time of the exhaust valve leads to a significant loss of charge, which is unacceptable in modes close to the nominal power.

I. Bilousov, R. Varbanets, *Marine Internal Combustion Engines and Environmental Protection Issues*, Springer Series on Naval Architecture, Marine Engineering, Shipbuilding and Shipping 27,
https://doi.org/10.1007/978-3-032-20940-5_12

Taking into account the above considerations, WinGD, together with Mitsui E&S DU Co, developed the Variable Compression Ratio (VCR) technology for X-series engines and the X-DF dual-fuel engines produced because of this series.

The essence of the VCR technology reduced to a dynamic change in the position of the piston relative to the engine cylinder cover (Fig. 12.1). This changes the volume of the combustion chamber, and accordingly, the compression ratio of the charge. By adjusting the length of the piston rod connecting the piston to the crank mechanism's crosshead, you can change the piston's position. For this purpose, the crosshead is equipped with a hydraulic mechanism installed in its crosspiece, which

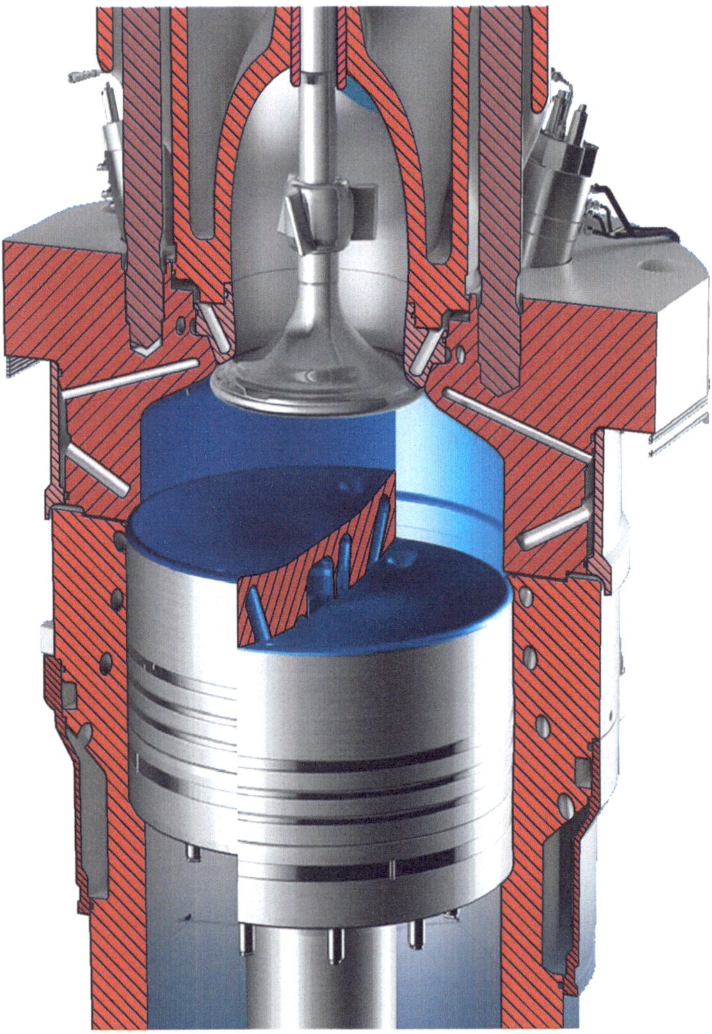

Fig. 12.1 Position of the piston relative to the cylinder cover in two-stroke low-speed engines of the X series from WinGD with variable compression ratio (adapted from [1])

controls the amount of protrusion of the piston rod from the crosspiece. A blind cylindrical hole in the crosspiece housing forms a hydraulic cylinder, into which the piston rod inserted, having several flanges. The lower flange of the rod performs the function of a hydraulic piston (Fig. 12.2), forming, together with the hole in the crosspiece, the lower hydraulic chamber. The amount of protrusion of the rod from the crosspiece depends on the amount of oil located in the lower chamber under the hydraulic piston. The amount of oil controlled by feeding oil from the engine's circulating lubrication system under pressure through the crank lever and draining it from the lower chamber through the outlet valve.

By changing the ratio between the oil supply and drain, it is possible to change the amount of rod protrusion and, accordingly, the position of the piston in the working cylinder. For example, for the X72 engine, the amount of rod protrusion can increased by 100 mm compared to the initial position, which allows increasing the geometric compression ratio from 12 to 18.

Figure 12.3 shows the nature of the change in the engine operating process for different values of the compression ratio.

A hydraulic sealing ring also installed on the piston rod to seal the hydraulic oil in the lower chamber. To prevent the possibility of the rod moving in the axial direction under the action of dynamic forces, an upper hydraulic chamber provided in the upper part of the mechanism for changing its position. It formed by the upper flange

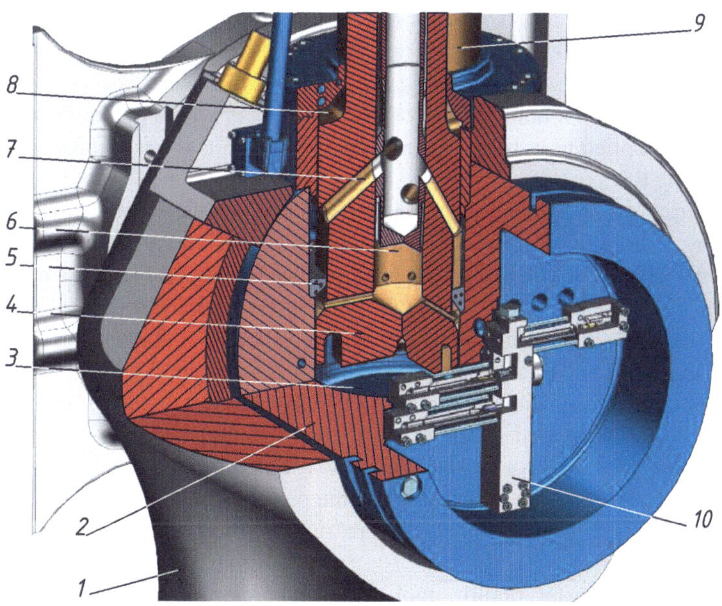

Fig. 12.2 Mechanism for changing the compression ratio of two-stroke low-speed engines of the X series by WinGD: (1) connecting rod; (2) crosshead crossmember; (3) lower hydraulic chamber; (4) hydraulic piston, (5) seal; (6) chamber for supplying oil for piston cooling; (7) channel for draining oil after piston cooling; (8) upper hydraulic chamber; (9) piston rod; (10) control valves for the hydraulic drive of the compression ratio change system (adapted from [1])

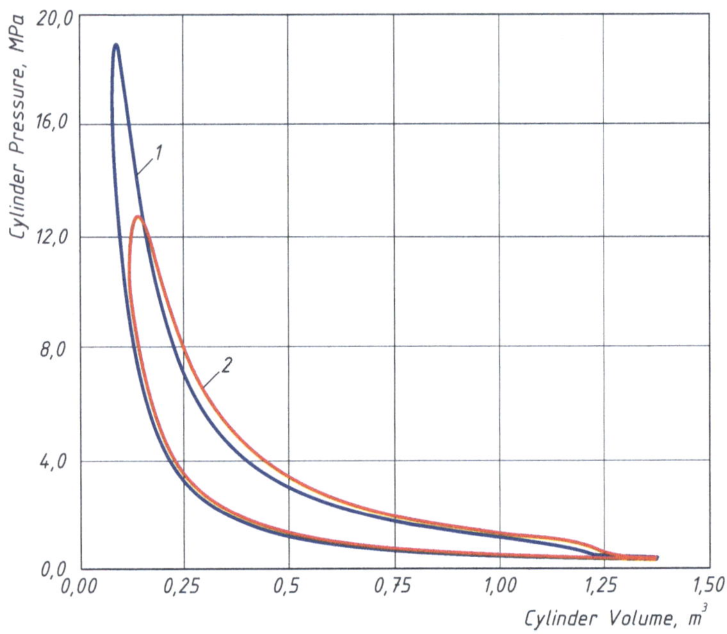

Fig. 12.3 Nature of change in the working process of the X72DF engine for different compression ratios set using the VCR system when the engine is running at full load with a rotation speed of 89 min^{-1}: (1) compression ratio 12:1; (2) compression ratio 18:1 (adapted from [1])

of the rod, a cylindrical hole in the crossbar and a cover with a hydraulic seal of the rod. The oil supplied to it, filling the annular space, limits the possibility of the rod moving along the axis.

The space between the upper and lower flanges of the piston rod divided into two cavities by a sliding seal. These cavities used to supply and remove cooling oil to the piston, through the internal drillings of the rod. Figure 12.4 shows a diagram of the variable compression ratio system of the X series engines from WinGD.

The system operates using oil taken from the general circulating lubrication system of the engine, which significantly simplifies the design. An electrically driven booster pump increases the lubricating oil pressure from 0.4–0.5 to 6.0–5.0 MPa, under which it is supplied to the system. To minimize energy consumption, the booster pump electric motor has an adjustable rotation speed.

Since the rod, extension adjusted when the piston is at BDC, high oil pressures are not required to move it, since the pressure on the piston is minimal in this position.

Oil from the booster pump supplied to the pressure accumulator, which is equipped with a valve limiting the maximum pressure in it. Then the oil supplied to the electromagnetic proportional valve, which regulates the oil supply to the lower chamber of each piston rod hydraulic cylinder. The oil supplied to the discharge valve via an additional knee-lever feed mechanism, which is similar in design to the

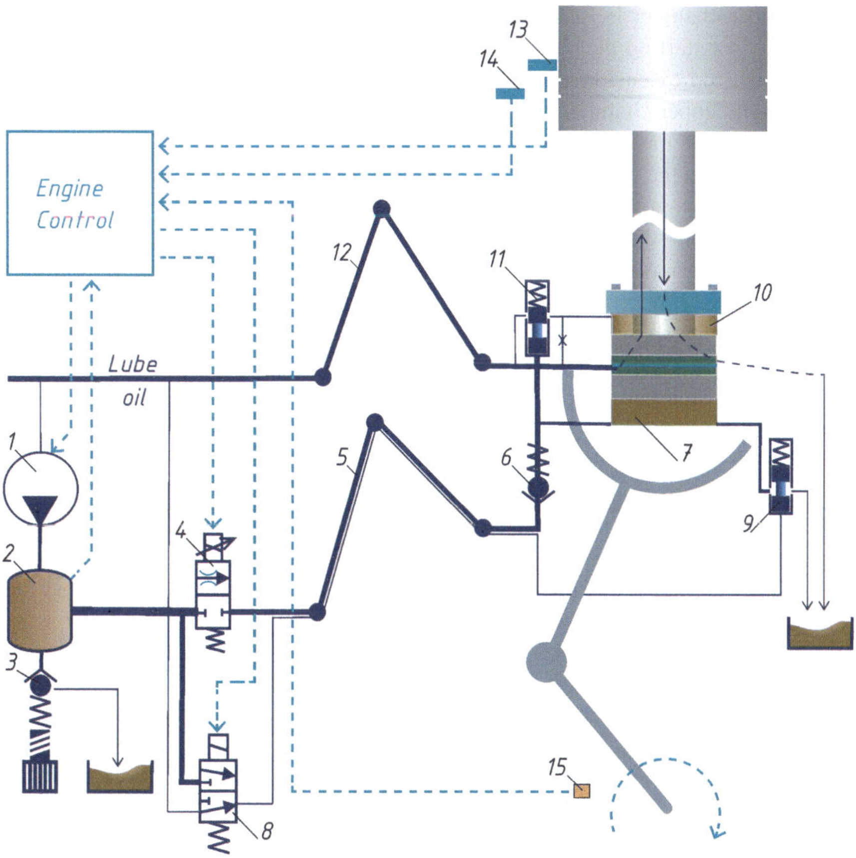

Fig. 12.4 Schematic diagram of the compression ratio control system in the two-stroke low-speed engines of the X series from WinGD: (1) booster pump; (2) pressure accumulator; (3) pressure regulating valve; (4) electromagnetic proportional valve; (5) knee lever of the VCR system; (6) discharge valve; (7) lower hydraulic chamber; (8) electromagnetic relief valve; (9) exhaust valve; (10) upper hydraulic chamber; (11) pressure relief valve; (12) knee lever of the piston cooling system; (13) piston position sensor; (14) air temperature sensor in the lower part of the piston; (15) crankshaft rotation angle sensor (adapted from [2])

mechanism for supplying oil for piston cooling (Sect. 6 of the book [3]). This valve keeps the oil in a closed volume when the piston subjected to pressure exceeding the pressure in the lower chamber. As already mentioned, the volume of oil locked in the lower chamber determines the position of the rod and, accordingly, the current value of the compression ratio.

To drain the oil from the lower chamber when the compression ratio needs to reduced, an electromagnetic relief valve used, which controls the opening and closing of the exhaust valve via a hydraulic line laid along the knee-lever feed mechanism. The exhaust valve discharges oil from the lower hydraulic chamber, lowering the piston rod.

The upper chamber holds the piston in the lower position in any situation (for example, when starting the engine or if the exhaust valve is faulty). In the event of failure of the rod extension mechanism, the pressure relief valve maintains the oil volume in the upper chamber to prevent the piston from rising. In the event of low oil pressure in the lower chamber, the rod drops to the lower position. In this case, the engine retains the ability to operate with a minimum compression ratio.

To ensure the supply of oil for piston cooling, a separate knee-lever feed mechanism for supplying oil to the crosshead installed.

To control the mechanism for changing the compression ratio and monitor its operation, each cylinder is equipped with a piston position sensor. This, in combination with the crankshaft position sensors already installed on the engine, allows the engine control unit to monitor the current value of the compression ratio and change it if necessary. To monitor external parameters and consider them when developing optimal combustion control algorithms, the engine can be equipped with additional sensors, such as the temperature at the inlet to the working cylinders.

Tests conducted on a research low-pressure gas-diesel engine showed that the use of the VCR system allows reducing gas fuel consumption by 2–6 g/(kW × h) depending on the load. At the same time, maximum savings achieved in partial load modes. In diesel mode, the ability to maintain a higher compression ratio across the entire load range makes it possible to save 8–12 g/(kW × h), depending on the engine type and its power. The compact dimensions of the developed VCR system do not require additional space for their placement, allowing all the necessary equipment to be located in the available space of a serial engine. This makes it possible not only to use this technology on newly built engines, but also to modernize engines already in operation.

12.2 Variable Valve Timing Technology in Four-Stroke Gas-Diesel Engines

We have already noted earlier what an important role for ensuring the efficient operation of gas and gas-diesel engines played by the regulation of the mixture composition in various operating modes.

In addition to the already considered methods of controlling the fuel-air ratio, the technology of controlling the mixture composition by changing the valve timing is widely introduced in modern engines. This includes ensuring the working process follows the Miller cycle.

The issues related to the organization of the operation of engines according to the Miller cycle and the methods of regulating the gas exchange phases considered in more detail in the second and fourth sections of the main book [3].

As an example of a valve timing system, Fig. 12.5 shows the design of the valve drive of the Wärtsilä 31 DF engine [4].

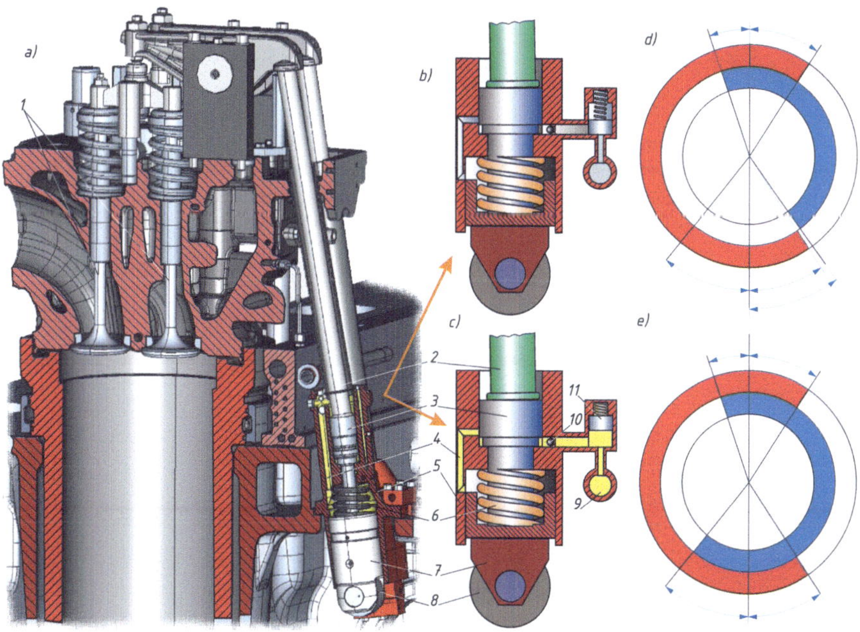

Fig. 12.5 Variable valve timing mechanism of the Wärtsilä 31 DF four-stroke medium-speed engine: (1) intake valves; (2) pushrod; (3) piston with hydraulic cavity; (4) drain channel; (5) push-rod housing; (6) return spring; (7) pushrod; (8) pushrod roller; (9) control oil supply; (10) check valve; (11) separator piston (adapted from [4])

The engine is equipped with a drive mechanism in which a roller tappet transmits force to a hydraulic piston with a flange in the lower part, forming an annular hydraulic cavity with the housing.

In the absence of oil supply to the drive mechanism, this cavity remains unfilled, and the tappet exactly follows the profile of the drive cam.

When oil supplied under pressure, the cavity under the piston filled with oil through the check valve. Thus, the oil is in a closed volume, holding the valve open for some time after the tappet roller leaves the cam lowering line.

The oil will hold in the cavity under the piston until the drain channel, blocked at the bottom of the tappet cylindrical guide, opened. The drain hole of the drain channel will begin to open only after the tappet approaches its lower position. This leads to an increase in the valve opening time, which contributes to the dynamic recharging of the working cylinder when operating at partial loads.

In combination with compressor performance control (discussed in Sect. 4 of the book [3]), this measure allows for effective control of the mixture composition over the entire range of engine operating modes.

References

1. The development of a variable compression ratio (VCR) on a large two-stroke slow-speed engine; The joint development approach of IHI, DU and WinGD. Paper No. 424. Nylund, I., Yamada, T., Moriyama, K. CIMAC Congress, Vancouver, 2019. 11 p
2. 2024. Low-speed Engines. WinGD, Winterthur, Switzerland, 2024. – 111 p
3. Bilousov I.V. Internal combustion engines of modern marine vessels. Textbook for higher maritime educational institutions. Bilousov, I.V., Bulgakov, M.P. Astroprint publishing house – Odesa, 2024. – 512 p. ISBN 978–617-7797-54-7
4. Wärtsilä 31DF Product Guide. Wärtsilä, Marine Business. Vaasa, December 2020. 192 p

The manufacturer's authorised representative in the EU is Springer Nature Customer Service Centre GmbH, Europaplatz 3, 69115 Heidelberg, Germany. If you have any concerns regarding our products, please contact ProductSafety@springernature.com

Printed and bound by CPI Group (UK) Ltd, Croydon, CR0 4YY
17/09/2026
02218430-0007